Symmetry in Classical and Fuzzy Algebraic Hypercompositional Structures

Symmetry in Classical and Fuzzy Algebraic Hypercompositional Structures

Special Issue Editor

Irina Cristea

MDPI • Basel • Beijing • Wuhan • Barcelona • Belgrade • Manchester • Tokyo • Cluj • Tianjin

Special Issue Editor
Irina Cristea
Center for Information
Technologies and Applied
Mathematics, University of
Nova Gorica
Slovenia

Editorial Office
MDPI
St. Alban-Anlage 66
4052 Basel, Switzerland

This is a reprint of articles from the Special Issue published online in the open access journal *Symmetry* (ISSN 2073-8994) (available at: https://www.mdpi.com/journal/symmetry/special_issues/Fuzzy_Algebraic_Hypercompositional_Structures).

For citation purposes, cite each article independently as indicated on the article page online and as indicated below:

LastName, A.A.; LastName, B.B.; LastName, C.C. Article Title. *Journal Name* **Year**, *Article Number*, Page Range.

ISBN 978-3-03928-708-6 (Pbk)
ISBN 978-3-03928-709-3 (PDF)

Contents

About the Special Issue Editor

Irina Cristea received her Ph.D. degree in mathematics in 2007 from University Ovidius of Constanta, Romania. After a period of post doctorate studies at University of Udine, Italy, in 2012, she became Assistant Professor at the University of Nova Gorica, Slovenia, where she currently hold the position of Associate Professor and acting head of the Center for Information Technologies and Applied Mathematics.

Her main research direction is theory of algebraic hypercompositional structures and their connections with fuzzy sets theory. In this area, she has published more than 60 reasearch articles in journals indexed by Scopus/Web of Science. She is also a co-author of the monograph "Fuzzy Algebraic Hyperstructures: An Introduction" published in 2015 by Springer.

Prof. Irina Cristea is deeply involved in the editorial activities, being member of the Editorial Board of five international journals (among them Mathematics, published by MDPI) and acting as a reviewer for numerous journals. In 2019, she was awarded by Publons the "Top 1% Reviewers in Mathematics".

Preface to "Symmetry in Classical and Fuzzy Algebraic Hypercompositional Structures"

Symmetry plays a fundamental role in our daily lives and in the study of the structure of different objects in physics, chemistry, biology, mathematics, architecture, arts, sociology, linguistics, etc. For example, the structure of molecules is well explained by their symmetry properties, described by symmetry elements and symmetry operations. A symmetry operation is a change, a transformation after which certain objects remain invariant, such as rotations, reflections, inversions, or permutation operations. Until now, the most efficient method to better describe symmetry was using mathematical tools offered by group theory.

Naturally generalizing the concept of a group, by considering the result of the interaction between two elements of a non-empty set to be a set of elements (and not only one element, as for groups), Frederic Marty, in 1934, at only 23 years old, defined the concept of a hypergroup. The law characterizing such a structure is called multi-valued operation, hyperoperation, or hypercomposition, and the theory of the algebraic structures endowed with at least one multi-valued operation is known as hyperstructure theory or hypercompositional algebra. Marty's motivation to introduce hypergroups was that the quotient of a group modulo any subgroup (not necessarily normal) is a hypergroup.

The main aim of this Special Issue was to present recent advances in hypercompositional algebra, the crisp and fuzzy algebra, where symmetry plays a crucial role. Studies are related (but not limited) to equivalence relations, orderings, permutations, symmetrical groups, graphs and hypergraphs, lattices, fuzzy sets, representations, etc. Some applications of algebraic hypercompositional structures in engineering, information technologies, computer science, where symmetry, or the lack of symmetry, is clearly specified and laid out, are also underlined.

Irina Cristea
Special Issue Editor

symmetry

MDPI

Article

Breakable Semihypergroups

Dariush Heidari [1] and Irina Cristea [2,*]

1 Faculty of Science, Mahallat Institute of Higher Education, Mahallat 37811-51958, Iran; dheidari82@gmail.com

2 Centre for Information Technologies and Applied Mathematics, University of Nova Gorica, Vipavska Cesta 13, 5000 Nova Gorica, Slovenia

* Correspondence: irina.cristea@ung.si; Tel.: +386-0533-15-395

Received: 14 December 2018; Accepted: 12 January 2019; Published: 16 January 2019

Abstract: In this paper, we introduce and characterize the breakable semihypergroups, a natural generalization of breakable semigroups, defined by a simple property: every nonempty subset of them is a subsemihypergroup. Then, we present and discuss on an extended version of Rédei's theorem for semi-symmetric breakable semihypergroups, proposing a different proof that improves also the theorem in the classical case of breakable semigroups.

Keywords: breakable semigroup; semihypergroup; hyperideal; semi-symmetry

---

1. Introduction

Breakable semigroups, introduced by Rédei [1] in 1967, have the property that every nonempty subset of them is a subsemigroup. It was proved that they are semigroups with empty Frattini-substructure [1]. For a structure S (i.e., a group, a semigroup, a module, a ring or a field), the set of those elements which may be omitted from each generating system (containing them) of S is a substructure of the same kind of S, called the Frattini-substructure of S. However, as mentioned in the book [1], there are some exceptions. The first one is when the Frattini-substructure is the empty set and this is the case of breakable semigroups, unit groups, zero modules or zero rings. The second one concerns the skew fields having the Frattini-substructure zero [1]. Based on the definition, it is easy to see that a semigroup S is breakable if and only if $xy \in \{x, y\}$ for any $x, y \in S$, i.e., the product of any two elements of the given semigroup is always one of the considered elements. Another characterization of these semigroups is given by Tamura and Shafer [2], using the associated power semigroup, i.e., a semigroup S is breakable if and only if its power semigroup $\mathcal{P}^*(S)$ is idempotent. An idempotent semigroup is a semigroup S that satisfies the identity $a^2 = a$ for any $a \in S$. A complete description of breakable semigroups was given by Rédei [1], writing them as a special decomposition of left-zero and right-zero semigroups (see Theorem 1).

The power set, i.e., the family of all subsets of the initial set, has many roles in algebra, one of them being in hyperstructures theory, where the power set $\mathcal{P}(S)$ is the codomain of any hyperoperation on a nonempty set S, i.e., a mapping $S \times S \longrightarrow \mathcal{P}(S)$. If the support set S is endowed with a binary associative operation, i.e., $(S, \cdot)$ is a semigroup, then this operation can be extended also to the set of nonempty subsets of S, denoted by $\mathcal{P}^*(S)$, in the most natural way: $A \star B = \{a \cdot b \mid a \in A, b \in B\}$. Thereby, $(\mathcal{P}^*(S), \star)$ becomes a semigroup, called the power semigroup of S. Similarly, if $(S, \circ)$ is a semihypergroup, then we can define on the power set a binary operation

$$A \star B = \bigcup_{a \in A, b \in B} a \circ b, \quad \text{for all} \quad A, B \in \mathcal{P}^*(S)$$

which is again associative (see Theorem 5). Going more in deep now, if we have a group $(G, \cdot)$ and we extend the operation to the set $\mathcal{P}^*(G)$ as before, then a new operation is defined on

$\mathcal{P}^*(G)$: $A \circ B = \{a \cdot b \mid a \in A, b \in B\}$. A nonempty subset $\mathcal{G}$ of $\mathcal{P}^*(G)$ is called an HX-group [3] on G, if $(\mathcal{G}, \circ)$ is a group. Similarly, on the group $(G, \cdot)$, one may define a hyperoperation by $a\hat{\circ}b = \{a \cdot b \mid a \in A, b \in B\}$, where $A, B \in \mathcal{P}^*(G)$, called by Corsini [4] the Chinese hyperoperation. An overview on the links between HX-groups and hypergroups has recently proposed by Cristea et al. [5].

Having in mind these connections between semigroups and semihypergroups and the importance of the power set and the decomposition of a set in the classical algebra, in this paper we would like to direct the reader's attention to a new concept, that one of breakable semihypergroup. The rest of the paper is structured as follows. In Section 2 we recall the breakable semigroups and the fundamental semigroups associated with semihypergroups. The main part of the paper is covered by Section 3, where we define the breakable semihypergroups and we present their characterizations using the power set and a generalization of Rédei's theorem for semi-symmetric semihypergroups, that permits to decompose them in a certain way. This decomposition is similar with that one proposed by Rédei's for semigroups, but slightly modified, to cover all the types of algebraic semihypergroups, by consequence all the types of algebraic semigroups. We have noticed that for some semigroups the Rédei's theorem does not work, while our proposed decomposition solves the problem. Besides we show that the set of all hyperideals of a breakable semi-symmetric semihypergroup is a chain. The semi-symmetry property plays here a fundamental role. This property holds for the classical structures, while in the hyperstructures has a significant meaning: the cardinalities of the hyperproducts of two elements $x \circ y$ and $y \circ x$ are the same for each pair of elements (x, y) in the considered hyperstructure $(H, \circ)$. Clearly this is evident for commutative hyperstructures. At the end of the paper, some conclusive ideas and new lines of research are included.

2. Preliminaries

Since we like to have the keywords of this note clearly specified and laid out, in this section we recall some definitions and properties of semigroups and semihypergroups. For more details on both arguments the reader is refereed to [1,2,6] for the classical algebraic structures and [7–10] for the algebraic hyperstructures.

A semigroup $(S, \cdot)$ is called a *left zero semigroup*, by short an l-semigroup, if each element of it is a left zero element, i.e., for any $x \in S$, we have $x \cdot y = x$ for all $y \in S$. Similarly, a *right zero semigroup*, or an *r-semigroup*, is a semigroup in which each element is a zero right element, i.e., for any $x \in S$, we have $x \cdot y = y$ for all $y \in S$.

In 1967, Rédei [1] gave the definition of *breakable semigroups*, as a subclass of the semigroups having an empty Frattini-substructure.

Definition 1. *A semigroup S is breakable if every non-empty subset of S is a subsemigroup.*

It is easy to see that a semigroup $(S, \cdot)$ is breakable if and only if $x \cdot y \in \{x, y\}$ for any $x, y \in S$.

A complete description of the structure of a breakable semigroup is given by Theorem 50 in [1].

Theorem 1. *A semigroup S is breakable if and only if, it can be partitioned into classes and the set of classes can be ordered in such a way that every class constitutes an l-semigroup or an r-semigroup, and for any two elements $x \in C$ and $y \in C'$ of two different classes C, C', with $C < C'$, we have $x \cdot y = y \cdot x = y$.*

Moreover, if $(S, \cdot)$ is a semigroup, then it is obvious that the set $\mathcal{P}^*(S)$ of all non-empty subsets of S can be endowed with a semigroup structure, too, called the *power semigroup*, where the binary operation is defined as follows: for $A, B \in \mathcal{P}^*(S)$, $A \cdot B = \{a \cdot b \mid a \in A, b \in B\}$. Then a breakable semigroup can be characterized also using properties of its power semigroup, as shown by Tamura and Shafer [2].

Theorem 2. *A semigroup S is breakable if and only if its power semigroup is idempotent, i.e.,* $X = X^2$ *for all* $X \in \mathcal{P}^*(S)$.

On the other hand, the set $\mathcal{P}^*(S)$ is the codomain of any hyperoperation defined on the support set S, i.e., a mapping $S \times S \longrightarrow \mathcal{P}^*(S)$. Now, if we start with a semihypergroup $(S, \circ)$, till now only a classical operation was defined on S, and not a hyperoperation, so the power set is again a semigroup, as we will show later on in Theorem 5.

The other natural and crucial connection between hyperstructures and classical structures is represented by the *strongly regular relations*. More exactly, on any semihypergroup $(S, \circ)$ one can define the relation β and its transitive closure β^*, and define a suitable operation on the quotient S/β^* in order to endow it with a semigroup structure, called the *fundamental semigroup* related to S. Here below we recall the construction, introduced by Koskas [11] and studied mainly by Freni [12], who proved that $\beta = \beta^*$ on hypergroups. For all natural numbers $n > 1$, define the relation β_n on a semihypergroup $(S, \circ)$, as follows: $a\beta_n b$ if and only if there exist $x_1, \ldots, x_n \in S$ such that $\{a, b\} \subseteq \prod_{i=1}^{n} x_i$. Take $\beta = \bigcup_{n \geq 1} \beta_n$, where $\beta_1 = \{(x, x) \mid x \in S\}$ is the diagonal relation on S. Denote by β^* the transitive closure of β. The relation β^* is a strongly regular relation. On the quotient S/β^* define a binary operation as follows: $\beta^*(a) \odot \beta^*(b) = \beta^*(c)$ for all $c \in \beta^*(a) \circ \beta^*(b)$. Moreover, the relation β^* is the smallest equivalence relation on a semihypergroup S, such that the quotient S/β^* is a semigroup. The quotient S/β^* is called the *fundamental semigroup*.

3. Breakable Semihypergroups

In this section, based on the notion of breakable semigroup introduced by Rédei [1], we define and characterize breakable semihypergroups. We present a generalization of Rédei's theorem for semi-symmetric semihypergroups.

In a classical structure (semigroup, monoid, group, ring, etc.) the composition of two elements is always another element of the support set. This property is not conserved in a hyperstructure, but it is extended in such a way that the result of the composition of two elements—called hypercomposition—is a subset of the support set. This means that, for two elements $x, y \in S$, the cardinalities of the compositions $x \cdot y$ and $y \cdot x$ in a classical algebraic structure are always equal (being both 1), while in a hyperstructure they could be greater than 1 and also different one from another. For this reason we introduce the next concept.

Definition 2. *A semihypergroup* $(S, \circ)$ *is called semi-symmetric if* $|x \circ y| = |y \circ x|$ *for every* $x, y \in S$.

It is clear that any commutative semihypergroup is also semi-symmetric.

Definition 3. *A semihypergroup S is called breakable if every non-empty subset of S is a subsemihypergroup.*

Obviously, every breakable semigroup can be considered as a breakable semihypergroup, by consequence l-semigroups and r-semigroups are examples of breakable semihypergroups.

A hyperoperation "$\circ$" on a nonempty set S, satisfying the property $x, y \in x \circ y$ for all elements $x, y \in S$, is called *extensive* (by J. Chvalina and his group of researchers [13–15]) or *closed* (by Ch. Massouros [16]). The most simple hyperoperation of this type was defined by the first time by Konguetsof [17] around 70's as $x \circ y = \{x, y\}$ for all $x, y \in S$. More than 20 years later, this hyperoperation was re-considered by G.G. Massouros et al. [18,19] in the framework of automata theory, proving the following result.

Theorem 3. *Let H be a non-empty set [19]. For every* $x, y \in H$ *define* $x \star_B y = \{x, y\}$. *Then* $(H, \star_B)$ *is a join hypergroup.*

G.G. Massouros called this hyperstructure a *B-hypergroup*, after the binary result that the hyperoperation gives.

Example 1. *Consider* $S = (\{1,2,3\}, \circ)$ *defined by the following Cayley table*

$\circ$	**1**	**2**	**3**
1	1	1	$\{1,3\}$
2	$\{1,2\}$	2	$\{2,3\}$
3	$\{1,3\}$	3	3

Then S is a breakable semihypergroup.

Example 2. *Consider* $S = (\{1,2,3,4,5\}, \circ)$ *defined by the following Cayley table*

$\circ$	**1**	**2**	**3**	**4**	**5**
1	1	2	3	4	5
2	2	2	$\{2,3\}$	2	$\{2,5\}$
3	3	$\{2,3\}$	3	3	$\{3,5\}$
4	4	2	3	4	5
5	5	$\{2,5\}$	$\{3,5\}$	5	5

Then S is a breakable semihypergroup.

Notice that in both examples the hyperoperation is extensive. Moreover, both are semihypergroups, but not hypergroups, since the reproduction axiom does not hold. The next theorem gives a characterization of breakable hypergroups.

Theorem 4. *A hypergroup* $(H, \circ)$ *is breakable if and only if it is a B-hypergroup.*

Proof. First, suppose that $(H, \circ)$ is a breakable hypergroup. For any two distinct elements x and y of H, by left reproducibility, there exists $z \in H$ such that $y \in x \circ z$. Since H is breakable, it follows that $\{x, z\}$ is a subsemihypergroup, so $x \circ z \subseteq \{x, z\}$. It follows that $y \in \{x, z\}$ and thus $y = z$. Therefore $y \in x \circ y$. Similarly, using the right reproducibility, one proves that $x \in x \circ y$. So we obtain $x \circ y = \{x, y\}$, i.e., $(H, \circ)$ is a B-hypergroup.

Conversely, the other implication is evident. □

Similarly to the classic case, one can characterize the breakable semihypergroups using the associated power semigroup.

Theorem 5. *Let* $(S, \circ)$ *be a semihypergroup. Then the following assertions hold:*

(I) $(\mathcal{P}^*(S), \star)$ *is a semigroup, where the binary operation* $\star$ *is defined by:*

$$A \star B = \bigcup_{a \in A, b \in B} a \circ b, \quad \text{for all} \quad A, B \in \mathcal{P}^*(S).$$

(II) $(S, \circ)$ *is breakable if and only if* $(\mathcal{P}^*(S), \star)$ *is idempotent.*

Proof. (I) The binary operation $\star$ is associative since, for every non empty subsets A, B, C of S we have

$$\begin{aligned} A \star (B \star C) &= A \star (\bigcup_{b \in B, c \in C} b \circ c) \\ &= \bigcup_{a \in A, b \in B, c \in C} a \circ (b \circ c) \\ &= \bigcup_{a \in A, b \in B, c \in C} (a \circ b) \circ c \\ &= (\bigcup_{a \in A, b \in B} a \circ b) \star C \\ &= (A \star B) \star C. \end{aligned}$$

(II) Let $(S, \circ)$ be breakable and $A \subseteq S$. Then A is a subsemihypergroup of S, that is $A \star A \subseteq A$. On the other hand, for every $a \in A$ we have $a = a \circ a \subseteq A \star A$. Thus $A \star A = A$, so $(\mathcal{P}^*(S), \star)$ is idempotent. Conversely, suppose that $(\mathcal{P}^*(S), \star)$ is idempotent. Then, for every non empty subset A of S, we have $A \star A = A$, so A is a subsemihypergroup, meaning that S is breakable. □

Proposition 1. *The fundamental semigroup of a breakable semihypergroup is breakable, too.*

Proof. Let S be a breakable semihypergroup and $(S/\beta^*, \odot)$ the associated fundamental semigroup. We know that, for $x, y \in S$, $\beta^*(x) \odot \beta^*(y) = \beta^*(z)$, whenever $z \in x \circ y \subseteq \{x, y\}$, because $(S, \circ)$ is breakable. So $\beta^*(x) \odot \beta^*(y) \in \{\beta^*(x), \beta^*(y)\}$, meaning that $(S/\beta^*, \odot)$ is breakable, too. □

Now it is the time to go back to Rédei's theorem and try to find a generalization in the broader context of semihypergroups. Notice here the significance of the notion of semi-symmetric semihypergroup.

Theorem 6. *A semi-symmetric semihypergroup $(S, \circ)$ is breakable if and only if it can be partitioned into classes, i.e., $S = \bigcup_{\gamma \in \Gamma} S_\gamma$, where Γ is a chain and all S_γ are pairwise disjoint l-semigroups, r-semigroups or B-hypergroups. Moreover, for every $x \in S_\alpha$ and $y \in S_\beta$, with $\alpha < \beta$, we have $x \circ y = y \circ x = y$.*

Proof. "$\Longrightarrow$" Suppose that $(S, \circ)$ is a breakable semi-symmetric semihypergroup. Then, for any $x, y \in S$, the sets $\{x\}$ and $\{x, y\}$ are semi-symmetric semihypergroups, so

$$x^2 = x \tag{1}$$

and

$$x \circ y = x \text{ or } x \circ y = y \text{ or } x \circ y = \{x, y\}. \tag{2}$$

We will prove the theorem in several steps.

Step 1. First we define on S three relations as follows:

$$x \sim_l y \Longleftrightarrow x \circ y = y, y \circ x = x. \tag{3}$$

$$x \sim_r y \Longleftrightarrow x \circ y = x, y \circ x = y. \tag{4}$$

$$x \sim_h y \Longleftrightarrow x \circ y = y \circ x = \{x, y\}. \tag{5}$$

In [1], it was proved that $\sim_l$ and $\sim_r$ are equivalences. We show now that also $\sim_h$ is an equivalence. The reflexivity holds because of (1) and the simmetry is evident. For proving the transitivity, take $x, y, z \in S$ such that $x \sim_h y$ and $y \sim_h z$, thus

$$x \circ y = y \circ x = \{x, y\} \quad \text{and} \quad y \circ z = z \circ y = \{y, z\}.$$

Hence it follows that

$$\begin{aligned}&\{x, y\} \cup x \circ z = x \circ y \cup x \circ z = x \circ \{y, z\} = x \circ (y \circ z) = (x \circ y) \circ z = \\ &= x \circ z \cup y \circ z = x \circ z \cup \{y, z\}.\end{aligned}$$

Thus $\{x, z\} \subseteq x \circ z \subseteq \{x, z\}$ (because S is breakable), implying that $x \circ z = \{x, z\}$, i.e., $x \sim_h z$. Therefore, $\sim_h$ is an equivalence relation on S.

Define the corresponding partitions of S related to $\sim_l, \sim_r$ and $\sim_h$ by $\mathcal{C}_l, \mathcal{C}_r$ and $\mathcal{C}_h$, respectively. Based on relations (3)–(5), we can notice that each class in $\mathcal{C}_l$, $\mathcal{C}_r$ and $\mathcal{C}_h$ is a maximal l-semigroup, r-semigroup and B-hypergroup, respectively. Indeed, for example, let H be a B-subhypergroup such that $\hat{x}_h \subseteq H \subseteq S$, where $\hat{x}_h \in \mathcal{C}_h$. Then, for every $y \in H$, we have $x \circ y = y \circ x = \{x, y\}$, meaning that $y \in \hat{x}_h$, so $\hat{x}_h = H$. Thus the class $\hat{x}_h$ represented by x is maximal.

Step 2. We show that if any two classes of $\mathcal{C}_l, \mathcal{C}_r$ or $\mathcal{C}_h$ have a common element, then one of them contains only one element. Let us assume, in contrast, that there exist two classes $\hat{x}_l \in \mathcal{C}_l$ and $\hat{z}_h \in \mathcal{C}_h$, both with more than one element, such that $\hat{x}_l \cap \hat{z}_h \neq \emptyset$. Thus there exists $y \in \hat{x}_l \cap \hat{z}_h$. It means that $\{x, y\}$ is an l-semigroup and $\{y, z\}$ is a B-hypergroup. Then

$$(x \circ z) \circ y = x \circ (z \circ y) = x \circ \{z, y\} = x \circ z \cup x \circ y = x \circ z \cup \{y\} \tag{6}$$

and

$$y \circ (x \circ z) = (y \circ x) \circ z = x \circ z. \tag{7}$$

On the other hand, because of (2), we have $x \circ z = x$ or $z \in x \circ z$. If $x \circ z = x$, using (6), we get $y = x \circ y = \{x, y\}$, which is impossible, because $x \neq y$. If $z \in x \circ z$, then by (7), we have $\{y, z\} = y \circ z \subseteq y \circ (x \circ z) = x \circ z$, so $y \in x \circ z$, which is again a contradiction, because of (2). Similarly, the other cases can be verified.

Step 3. Based on the assertion proved in the previous step, we may define on S a new partition: we take the classes, of cardinality at least 2, of $\mathcal{C}_l, \mathcal{C}_r$ and $\mathcal{C}_h$, and then the singleton classes of all the other elements of S (we read here that all the other elements are put each one in a different class). We denote the corresponding equivalence relation by $\sim$ and the class of x with respect to $\sim$ by $\overline{x}$.

Take x and y from two different classes, i.e., $x \nsim y$. Since S is a breakable semihypergroup, it follows that $\{x, y\}$ is a subsemihypergroup, so relation (2) is verified. If $x \circ y = x$, then since S is semi-symmetric, it follows that $y \circ x = x$ or $y \circ x = y$. If $y \circ x = y$, it means that $x \sim_r y$ and thus $x \sim y$, which is false. So $x \circ y = y \circ x = x$. Similarly, if $x \circ y = y$ it follows that $y \circ x = y$. Thereby, for $x \nsim y$, we get

$$x \circ y = y \circ x = x \quad \text{or} \quad x \circ y = y \circ x = y. \tag{8}$$

Step 4. We show that for any different elements x_1, x_2, y of S such that $x_1 \sim x_2 \nsim y$, we have

$$\text{either} \quad x_i \circ y = y \circ x_i = y \quad \text{or} \quad x_i \circ y = y \circ x_i = x_i \tag{9}$$

for $i = 1, 2$. Since $x_1 \sim x_2$, the set $\{x_1, x_2\}$ is an l-semigroup, an r-semigroup or a B-hypergroup. Let us assume that $\{x_1, x_2\}$ is an l-semigroup, i.e., we have $x_1 \circ x_2 = x_2$ and $x_2 \circ x_1 = x_1$. Besides, from (8), we have $x_i \circ y = y \circ x_i$, for $i = 1, 2$.

Now, by contrast, if we suppose that the assertion is false, then, because of (2) with a suitably order, we have

$$x_1 \circ y = y \circ x_1 = x_1, \;\; x_2 \circ y = y \circ x_2 = y.$$

Hence $x_1 = x_1 \circ y = x_1 \circ (y \circ x_2) = (x_1 \circ y) \circ x_2 = x_1 \circ x_2 = x_2$, which is a contradiction, so relation (9) is now proved. Similarly, relation (9) holds whenever $\{x_1, x_2\}$ is an r-semigroup or a B-hypergroup.

Step 5. On the set of all classes $\overline{x}$ define an ordering relation $<$ as follows:

$$\overline{x} < \overline{y} \Longleftrightarrow x \circ y = y \circ x = y. \tag{10}$$

First we prove that the relation is well-defined, i.e., it does not depend on the representatives x and y. Take $x \sim x'$ and $y \sim y'$. By using (9) for $x_1 = x, x_2 = x'$ and y, then for $x_1 = y, x_2 = y'$ and $y = x$, respectively, we get $x' \circ y = y \circ x' = y$ and $x \circ y' = y' \circ x = y'$.

The reflexivity and the symmetry are evident. It remains to prove the transitivity. Assume that $\overline{x} < \overline{y}$ and $\overline{y} < \overline{z}$. By definition of $<$, these two relations mean that $x \circ y = y \circ x = y$ and $y \circ z = z \circ y = z$. It follows that $x \circ z = x \circ (y \circ z) = (x \circ y) \circ z = y \circ z = z$ and similarly, $z \circ x = z$, meaning that $\overline{x} < \overline{z}$.

Besides, from (8), for $x \sim y$, it follows that either $\overline{x} < \overline{y}$ or $\overline{y} < \overline{x}$ always holds, so the order $<$ is total.

"$\Longleftarrow$" The converse implication is obvious.

Now the proof is completed. □

Remark 1. *The structure of the proof of Theorem 6 is similar to that one proposed by Rédei [1] for the decomposition of breakable semigroups, but it was obviously extended to hyperstructure environment. Moreover, in the original proof, Rédei considered in Step 3 a different partition of the initial semigroup S, i.e., he considered the classes of cardinality at least 2 of C_l and C_r, and then the class of all the other elements of S. But doing in this way, not all the breakable semigroups are decomposed as is requested by Theorem 1, as we can notice here below.*

Consider on the set $S = \{1,2\}$ the operation $x \cdot y = \max\{x, y\}$. It is clear that S is a breakable semigroup and using the above mentioned partition, we have to consider 1 and 2 in the same class (the last one, "of all the other elements," let's say), since 1 and 2 are not equivalent with respect to both relations $\sim_l$ and $\sim_r$. So the partition will be $\{\{1,2\}\}$, which is not an l-semigroup or an r-semigroup, obtaining thus a contradiction. On the other way, if we consider the partition of S as in Theorem 6 in Step 3, i.e., we take the classes, of cardinality at least 2, of C_l, C_r and C_h, and then the singleton classes of all the other elements of S, we get another partition of S as $S = S_\alpha \cup S_\beta$, where $S_\alpha = \{1\}$ and $S_\beta = \{2\}$, both being l-semigroups (or r-semigroups), so Rédei's theorem is verified also in this particular case.

In the following examples we will show the decomposition of some breakable semihypergroups obtained using Theorem 6.

Example 3. *Let $\Gamma = \{\alpha, \beta\}$, $\alpha < \beta$, $S_\alpha = \{1,2\}$ be a l-semigroup and $S_\beta = \{3,4\}$ be a B-hypergroup. Then $(\{1,2,3,4\}, \circ)$ is a breakable semihypergroup with the following Cayley table:*

$\circ$	1	2	3	4
1	1	1	3	4
2	2	2	3	4
3	3	3	3	$\{3,4\}$
4	4	4	$\{3,4\}$	4

Example 4. *Let* $\Gamma = \{\alpha, \beta, \gamma, \delta\}$, $\alpha < \beta < \gamma < \delta$, $S_\alpha = \{1,2,3\}$ *and* $S_\gamma = \{6,7\}$ *be B-hypergroups,* $S_\beta = \{4,5\}$ *be an l-semigroup and* $S_\delta = \{8,9\}$ *be an r-semigroup. Then* $(\{1,2,\ldots,9\}, \circ)$ *is a breakable semihypergroup with the following Cayley table:*

$\circ$	**1**	**2**	**3**	**4**	**5**	**6**	**7**	**8**	**9**
1	1	$\{1,2\}$	$\{1,3\}$	4	5	6	7	8	9
2	$\{1,2\}$	2	$\{2,3\}$	4	5	6	7	8	9
3	$\{1,3\}$	$\{2,3\}$	3	4	5	6	7	8	9
4	4	4	4	4	5	6	7	8	9
5	5	5	5	4	5	6	7	8	9
6	6	6	6	6	6	6	$\{6,7\}$	8	9
7	7	7	7	7	7	$\{6,7\}$	7	8	9
8	8	8	8	8	8	8	8	8	8
9	9	9	9	9	9	9	9	9	9

Example 5. *Consider the following binary hyperoperation on the set of integers:*

$$\forall m, n \in \mathbb{Z}, m \circ n = \begin{cases} n & \text{if } m, n < 0 \\ 0 & \text{if } m = n = 0 \\ \{m, n\} & \text{if } m, n > 0 \\ \max\{m, n\} & \text{otherwise.} \end{cases}$$

Then $(\mathbb{Z}, \circ)$ *is a breakable semihypergroup, since it is sufficient to take* $\Gamma = \{\alpha, \beta, \gamma\}$, *with* $\alpha < \beta < \gamma$, $S_\alpha = \mathbb{Z}^-$ *as an l-semigroup,* $S_\beta = \{0\}$ *as an r-semigroup and* $S_\gamma = \mathbb{N}$ *as a B-hypergroup.*

The notion of ideal of a semigroup was extended to the hyperstructures for the first time by Hasankhani [20], defining the concept of *left (right) ideal* in a hypergroupoid, that was after changed into *hyperideal*, in order to keep the meaning of the hyperoperation.

Definition 4. *Let* $(H, \circ)$ *be a hypergroupoid. A non empty set* A *of* H *is called a left hyperideal if, for* $x \in A$, *it follows that* $y \circ x \subseteq A$ *for any* $y \in H$. *Similarly,* A *is a right hyperideal if, for* $x \in A$, *it follows that* $x \circ y \subseteq A$ *for any* $y \in H$. *Moreover* A *is called a hyperideal of* H *if it is both a left and a right hyperideal.*

Theorem 7. *Let* S *be a breakable semi-symmetric semihypergroup. Then the set of all hyperideals of* S *together with the inclusion is a chain.*

Proof. Let $(S, \circ)$ be a breakable semi-symmetric semihypergroup. Then by Theorem 6, there exists an equivalence relation $\sim$ on S such that the set of classes $\mathcal{C}$ with respect to it can be ordered in such a way that, for every distinct classes $\overline{x}$ and $\overline{y}$, we have

$$\overline{x} < \overline{y} \Longleftrightarrow x \circ y = y \circ x = y. \tag{11}$$

Please note that the definition of $\mathcal{C}$ is equivalent with

$$x \sim y \Longrightarrow x \circ y \cup y \circ x = \{x, y\}, \quad x, y \in S. \tag{12}$$

We claim that, if I is a hyperideal of S, then

$$I = \bigcup_{x \in I} \overline{x}.$$

Indeed, for every $x, x' \in S$, where $x' \sim x \in I$, we have $\{x, x'\} = x \circ x' \cup x' \circ x \subseteq I$, so the claim is proved.

Furthermore, (11) implies that, for every hyperideal I of S, we have

$$x \in I, y \in S, \overline{x} < \overline{y} \Longrightarrow y \in I, \tag{13}$$

hence $\overline{y} \subseteq I$.

Now, let I and J be distinct hyperideals of S. We will prove that either $I \subseteq J$ or $J \subseteq I$. To do this, first we will show that either $J \setminus I \neq \emptyset$ or $I \setminus J \neq \emptyset$, i.e., just one of the assertions holds. In contrast, let us suppose that $J \setminus I \neq \emptyset$ and $I \setminus J \neq \emptyset$, therefore there exist $a_0 \in I \setminus J$ and $b_0 \in J \setminus I$. From (13) it follows that

$$\overline{a_0} < \overline{b} \text{ for any } b \in J \text{ and } \overline{b_0} < \overline{a} \text{ for any } a \in I. \tag{14}$$

Indeed, if $\overline{b} < \overline{a_0}$, with $b \in J$, then $a_0 \in J$ (since J is a hyperideal), which is false. So $\overline{a_0} < \overline{b}$, for any $b \in J$. Similarly the other relation holds. This implies that $a_0 < b_0$ and $b_0 < a_0$, hence $\{a_0, b_0\} \subseteq \overline{a_0} = \overline{b_0} \subseteq I \cap J$, which is impossible.

Thereby, for two distinct hyperideals I and J, we have either $J \setminus I \neq \emptyset$ or $I \setminus J \neq \emptyset$. Without loss of generality, let $I \setminus J \neq \emptyset$ and take $b \in J$. Then, by (14), we have $\overline{a_0} < \overline{b}$, with $a_0 \in I$; this implies that $b \in I$, thus $J \subset I$. Similarly, if $J \setminus I \neq \emptyset$, then $I \subseteq J$. We can conclude thus, that the set of the hyperideals of S is a chain with respect to the inclusion. □

Corollary 1. *The set of all ideals of a breakable semigroup is a chain.*

4. Conclusions

In this paper, we have started the study of breakable semihypergroups, based on the classical concept of breakable semigroups. In a breakable semihypergroup, each nonempty subset is a subsemihypergroup. If we search for the same property in hypergroups (so semihypergroups satisfying also the reproduction axiom), we obtain that there is only one class of breakable hypergroups and this is that of B-hypergroups. Moreover, we have proved that a breakable semi-symmetric semihypergroup can be decomposed in classes that are ordered in such a way that each class is an l-semigroup, an r-semigroup or a B-hypergroup. At the end, we have proved that the set of all hyperideals of a semi-symmetric breakable semihypergroup is a chain.

The properties of the breakable semihypergroups, in particular the proposed decomposition, suggest several new lines of research. A first one could be a generalization of the classical notion of Frattini-substructure, so the study of the Frattini-subhyperstructure. Another perspective could be related to the role of the power set, so the set of subsets of the support set. It is well known that the operation on a semigroup can be extended to the family of nonempty subsets of the semigroup, endowing it with a semigroup structure, called the associated power semigroup. Now, if the support set is a semihypergroup, then we can similarly extend the hyperoperation to an operation on the power set, which remains associative. In our future work we intend to define a hyperoperation on the power set and investigate its properties, aiming to define the power semihypergroup.

Author Contributions: Conceptualization, D.H.; Investigation, D.H.; Methodology, I.C.; Supervision, I.C.; Writing original draft, D.H.; Writing review and editing, I.C.

Funding: The second author acknowledges the financial support from the Slovenian Research Agency (research core funding No. P1-0285).

Conflicts of Interest: The authors declare no conflict of interest.

References

1. Rédei, L. *Algebra I*; Pergamon Press: Oxford, UK, 1967.
2. Tamura, T.; Shafer, J. Power semigroups. *Math. Jpn.* **1967**, *12*, 25–32.
3. Hongxing, L. *HX*-group. *Busefal* **1987**, *33*, 31–37.
4. Corsini, P. On Chinese hyperstructures. *J. Discret. Math. Sci. Cryptogr.* **2003**, *6*, 133–137. [CrossRef]
5. Cristea, I.; Novak, M.; Onasanya, B.O. An overview on the links between *HX*-groups and hypergroups. **2019**, submitted.
6. Pelikan, J. On semigroups, in which products are equal to one of the factors. *Periodica Math. Hung.* **1973**, *4*, 103–106. [CrossRef]
7. Corsini, P. *Prolegomena of Hypergroup Theory*; Aviani Editore: Tricesimo, Italy, 1993.
8. Corsini, P.; Leoreanu, V. *Applications of Hyperstructure Theory*; Kluwer Academical Publications: Dordrecht, The Netherlands, 2003.
9. Vougiouklis, T. *Hyperstructures and Their Representations*; Hadronic Press Inc.: Palm Harbor, FL, USA, 1994.
10. Davvaz, B. *Semihypergroup Theory*; Elsevier/Academic Press: London, UK, 2016.
11. Koskas, M. Groupoides, demi-hypergroupes et hypergroupes. *J. Math. Pure Appl.* **1970**, *49*, 155–192.
12. Freni, D. Une note sur le cour d'un hypergroupe et sur la clôture transitive β^* de β. [A note on the core of a hypergroup and the transitive closure β^* of β]. *Riv. Mat. Pura Appl.* **1991**, *8*, 153–156. (In French)
13. Chvalina, J.; Hoskova-Mayerova, S. Discrete transformation hypergroups and transformation hypergroups with phase tolerance space. *Discret. Math.* **2008**, *308*, 4133–4143.
14. Hoskova-Mayerova, S.; Maturo, A. Algebraic hyperstructures and social relations. *It. J. Pure Appl. Math.* **2018**, *39*, 475–484.
15. Novak, M.; Krehlik, S.; Cristea, I. Ciclicity in EL-hypergroups. *Symmetry* **2018**, *10*, 611. [CrossRef]
16. Massouros, C. On connections between vector spaces and hypercompositional structures. *It. J. Pure Appl. Math.* **2015**, *34*, 133–150.
17. Konguetsof, L. Sur les hypermonoides. *Bull. Soc. Math. Belg.* **1973**, 25, 211–224.
18. Massouros, G.G. Hypercompositional structures from the computer theory. *Ratio Math.* **1999**, *13*, 37–42.
19. Massouros, G.G.; Mittas, J. Languages-Automata and hypercompositional structures. In Proceedings of the 4th International Congress Algebraic Hyperstructures and Applications, Xanthi, Greece, 27–30 June 1990; pp. 137–147.
20. Hasankhani, A. Ideals is a semihypergroup and Green's relations. *Ratio Math.* **1999**, *13*, 29–36.

symmetry

MDPI

Article

On Hypergroups with a β-Class of Finite Height

Mario De Salvo [1], Dario Fasino [2,*], Domenico Freni [2] and Giovanni Lo Faro [1]

1 Dipartimento di Scienze Matematiche e Informatiche, Scienze Fisiche e Scienze della Terra, Università di Messina, 98166 Messina, Italy; desalvo@unime.it (M.D.S.); lofaro@unime.it (G.L.F.)
2 Dipartimento di Scienze Matematiche, Informatiche e Fisiche, Università di Udine, 33100 Udine, Italy; domenico.freni@uniud.it
* Correspondence: dario.fasino@uniud.it

Received: 13 December 2019; Accepted: 10 January 2020; Published: 15 January 2020

Abstract: In every hypergroup, the equivalence classes modulo the fundamental relation β are the union of hyperproducts of element pairs. Making use of this property, we introduce the notion of height of a β-class and we analyze properties of hypergroups where the height of a β-class coincides with its cardinality. As a consequence, we obtain a new characterization of 1-hypergroups. Moreover, we define a hierarchy of classes of hypergroups where at least one β-class has height 1 or cardinality 1, and we enumerate the elements in each class when the size of the hypergroups is $n \leq 4$, apart from isomorphisms.

Keywords: hypergroup; semihypergroup; 1-hypergroup; fundamental relation; height

1. Introduction

The term algebraic hyperstructure designates a suitable generalization of a classical algebraic structure, like a group, a semigroup, or a ring. In classical algebraic structures, the composition of two elements is an element, while in an algebraic hyperstructure the composition of two elements is a set. In the last few decades, many scholars have been working in the field of algebraic hyperstructures, also called hypercompositional algebra. In fact, algebraic hyperstructures have found applications in many fields, including geometry, fuzzy/rough sets, automata, cryptography, artificial intelligence and probability [1], relational algebras [2], and sensor networks [3].

Certain equivalence relations, called fundamental relations, introduce natural correspondences between algebraic hyperstructures and classical algebraic structures. These equivalence relations have the property of being the smallest strongly regular equivalence relations such that the corresponding quotients are classical algebraic structures [4–11]. For example, if $(H, \circ)$ is a hypergroup, then the fundamental relation β is transitive [12–14] and the quotient set H/β is a group. Moreover, if $\varphi\colon H \to H/\beta$ is the canonical projection, then the kernel $\omega_H = \varphi^{-1}(1_{H/\beta})$ is a subhypergroup, which is called the *heart* of $(H, \circ)$. The heart of a hypergroup $(H, \circ)$ plays a very important role in hypergroup theory because it gives detailed information on the partition of H determined by the relation β, since $\beta(x) = \omega_H \circ x = x \circ \omega_H$ for all $x \in H$.

In this work, we focus on the fundamental relation β in hypergroups, and we introduce a new classification of hypergroups in terms of the minimum number of hyperproducts of two elements whose union is the β-class that contains these hyperproducts. Our main aim is to deepen the understanding of the properties of the fundamental relation β in hypergroups and to enumerate the non-isomorphic hypergroups fulfilling certain conditions on the cardinality of the hearth. This task belongs to an established research field that deals with fundamental relations and enumerative problems in hypercompositional algebra [5,6,13–15]. The plan of this article is the following: After introducing some basic definitions and notations to be used throughout this article, in Section 3, we define the notion of *height* $h(\beta(x))$ of an equivalence class $\beta(x)$. We give examples of hypergroups with infinite size where the height of all β-classes is finite. Denoting cardinality by $|\cdot|$, if $(H, \circ)$ is a hypergroup with a β-class of finite size such that $|\beta(x)| = h(\beta(x))$, then $|a \circ y| = |y \circ a| = 1$, for all $a \in \omega_H$ and $y \in \beta(x)$. Moreover, when ω_H is finite, we prove that $|\omega_H| = h(\omega_H)$ if and only if $(H, \circ)$ is a 1-hypergroup. In Section 4, we use the notion of height of a β-class to introduce new classes of hypergroups. We enumerate the elements in each class when the size of the hypergroups is not larger than 4, apart from isomorphisms. In particular, we prove that there are 4023 non-isomorphic hypergroups of size $n \leq 4$ with a β-class of size 1. Moreover, excluding the hypergroups $(H, \circ)$ with $|H| = 4$ and $|H/\beta| = 1$, there exist 8154 non-isomorphic hypergroups of size $n \leq 4$ with $h(\omega_H) = 1$.

2. Basic Definitions and Results

Let H be a non-empty set and let $\mathcal{P}^*(H)$ be the set of all non-empty subsets of H. A *hyperoperation* $\circ$ on H is a map from $H \times H$ to $\mathcal{P}^*(H)$. For all $x, y \in H$, the set $x \circ y$ is called the hyperproduct of x and y. The hyperoperation $\circ$ is naturally extended to subsets as follows: If $A, B \subseteq H$, then $A \circ B = \bigcup_{x \in A, y \in B} x \circ y$.

A *semihypergroup* is a non-empty set H endowed with an associative hyperproduct $\circ$, that is, $(x \circ y) \circ z = x \circ (y \circ z)$ for all $x, y, z \in H$. A semihypergroup $(H, \circ)$ is a *hypergroup* if for all $x \in H$ we have $x \circ H = H \circ x = H$; this property is called reproducibility. A non-empty subset K of a semihypergroup $(H, \circ)$ is called a *subsemihypergroup* of $(H, \circ)$ if it is closed with respect to $\circ$ that is, $x \circ y \subseteq K$ for all $x, y \in K$. A non-empty subset K of a hypergroup $(H, \circ)$ is called a *subhypergroup* if $x \circ K = K \circ x = K$, for all $x \in K$. If a subhypergroup is isomorphic to a group, then we say that it is a subgroup of $(H, \circ)$.

Given a semihypergroup $(H, \circ)$, the relation β^* of H is the transitive closure of the relation $\beta = \cup_{n \geq 1} \beta_n$, where β_1 is the diagonal relation in H and, for every integer $n > 1$, β_n is defined as follows:

$$x \beta_n y \iff \exists (z_1, \ldots, z_n) \in H^n \, : \, \{x, y\} \subseteq z_1 \circ z_2 \circ \ldots \circ z_n.$$

The relations β and β^* are among the so-called *fundamental relations* [16]. Their relevance in semihypergroup and hypergroup theory stems from the following facts [17]: If $(H, \circ)$ is a semihypergroup (resp., a hypergroup), the quotient set H/β^* equipped with the operation $\beta^*(x) \otimes \beta^*(y) = \beta^*(z)$ for all $x, y \in H$ and $z \in x \circ y$, is a semigroup (resp., a group). The canonical projection $\varphi : H \to H/\beta^*$ is a good homomorphism, that is, $\varphi(x \circ y) = \varphi(x) \otimes \varphi(y)$ for all $x, y \in H$. If $(H, \circ)$ is a hypergroup, then H/β^* is a group and the kernel $\omega_H = \varphi^{-1}(1_{H/\beta^*})$ of φ is the *heart* of $(H, \circ)$. Moreover, if $|\omega_H| = 1$, then $(H, \circ)$ is called *1-hypergroup*.

Let A be a non-empty subset of a semihypergroup $(H, \circ)$. We say that A is a *complete part* of $(H, \circ)$ if, for every $n \in \mathbb{N} - \{0\}$ and $(x_1, x_2, \ldots, x_n) \in H^n$,

$$(x_1 \circ \cdots \circ x_n) \cap A \neq \emptyset \implies (x_1 \circ \cdots \circ x_n) \subseteq A.$$

Clearly, the set H is a complete part, and the intersection $\mathcal{C}(X)$ of all the complete parts containing a non-empty set X is called the *complete closure* of X. If X is a complete part of $(H, \circ)$ then $\mathcal{C}(X) = X$.

If $(H, \circ)$ is a semihypergroup and $\varphi : H \to H/\beta^*$ is the canonical projection, then, for every non-empty set $A \subseteq H$, we have $\mathcal{C}(A) = \varphi^{-1}(\varphi(A))$. Moreover, if $(H, \circ)$ is a hypergroup, then

$$\mathcal{C}(A) = \varphi^{-1}(\varphi(A)) = A \circ \omega_H = \omega_H \circ A.$$

A hypergroup $(H, \circ)$ is said to be *complete* if $x \circ y = \mathcal{C}(x \circ y)$, for all $(x, y) \in H^2$. If $(H, \circ)$ is a complete hypergroup, then

$$x \circ y = \mathcal{C}(a) = \beta^*(a),$$

for every $(x, y) \in H^2$ and $a \in x \circ y$.

A subhypergroup K of a hypergroup $(H, \circ)$ is said to be *conjugable* if it satisfies the following property: for all $x \in H$, there exists $x' \in H$ such that $xx' \subseteq K$. The interested reader can find all relevant definitions, many properties, and applications of fundamental relations, even in more abstract contexts, also in [18–28].

For later reference, we collect in the following theorem some classic results of hypergroup theory from [12,17,26].

Theorem 1. *Let $(H, \circ)$ be a hypergroup. Then,*

1. *The relation β is transitive, which is $\beta = \beta^*$;*
2. *$\beta(x) = x \circ \omega_H = \omega_H \circ x$, for all $x \in H$;*
3. *a subhypergroup K of $(H, \circ)$ is conjugable if and only if it is a complete part of $(H, \circ)$;*
4. *the heart of $(H, \circ)$ is the smallest conjugable subhypergroup (or complete part) of $(H, \circ)$, that is, ω_H is the intersection of all conjugable subhypergroups (or complete part) of $(H, \circ)$.*

3. Locally Finite Hypergroups

Let $(H, \circ)$ be a hypergroup and let $\sim$ be the following equivalence relation on the set $H \times H$: $(x, y) \sim (z, w) \Leftrightarrow x \circ y = z \circ w$. Let $\mathcal{T}$ be a transversal of the equivalence classes of the relation $\sim$. For every $x \in H$, there exists a non-empty set $A \subseteq \mathcal{T}$ such that $\beta(x) = \bigcup_{(a,b)\in A} a \circ b$. In fact, by reproducibility of $(H, \circ)$, if $y \in \beta(x)$, then there exist $z, w \in H$ such that $y \in z \circ w$. Clearly, we have $z \circ w \cap \beta(x) \neq \emptyset$ and $z \circ w \subseteq \beta(x)$ because $\beta(x)$ is a complete part of H. Moreover, there exists $(a, b) \in \mathcal{T}$ such that $(z, w) \sim (a, b)$ and $y \in z \circ w = a \circ b$. Hence, there exists a non-empty set $A \subseteq \mathcal{T}$ such that $\beta(x) \subseteq \bigcup_{(a,b)\in A} a \circ b$ and $a \circ b \cap \beta(x) \neq \emptyset$ for all $(a, b) \in A$. The other inclusion follows from the fact that $\beta(x)$ is a complete part of $(H, \circ)$.

In conclusion, each β-class is the union of hyperproducts of pairs of elements that can be chosen within a transversal of $\sim$. This fact suggests the following definitions.

Definition 1. *Let $(H, \circ)$ be a hypergroup and let $\mathcal{T}$ be a transversal of the equivalence classes of the relation $\sim$. For every $x \in H$, the class $\beta(x)$ is called* locally finite *if there exists a finite set $A \subseteq \mathcal{T}$ such that $\beta(x) = \bigcup_{(a,b)\in A} a \circ b$. If a class $\beta(x)$ is not locally finite, we say that it is* locally infinite.

Definition 2. *Let $\beta(x)$ be a β-class of a hypergroup $(H, \circ)$. If $\beta(x)$ is locally finite, then the minimum positive integer m such that there is a non-empty set $M \subseteq \mathcal{T}$ such that $|M| = m$ and $\beta(x) = \bigcup_{(a,b)\in M} a \circ b$ is called* height *of $\beta(x)$, and we write $h(\beta(x)) = m$. If $\beta(x)$ is locally infinite, we write $h(\beta(x)) = \infty$.*

Definition 3. *A hypergroup $(H, \circ)$ is* locally finite *if all β-classes are locally finite. In particular, $(H, \circ)$ is called* locally n-finite *if $h(\beta(x)) \leq n$ for every $x \in H$, and there is at least one element $y \in H$ such that $h(\beta(y)) = n$. Moreover, $(H, \circ)$ is* strongly locally n-finite *if $h(\beta(x)) = n$ for every $x \in H$.*

Clearly, $(H, \circ)$ is locally 1-finite if and only if $(H, \circ)$ is strongly locally 1-finite. Examples of hypergroups locally 1-finite are the complete hypergroups. Indeed, if $(H, \circ)$ is a complete hypergroup, then, for every $x \in H$, there exist $y, z \in H$ such that $x \in y \circ z$ and $\beta(x) = y \circ z$.

Example 1. *In the set $H = \{1, 2, 3, 4, 5, 6\}$, consider the hyperproducts defined by the following tables:*

$\circ_1$	1	2	3	4	5	6
1	1,2	1,3	2,3	4,5,6	4,6	5,6
2	2,3	1,2	1,3	5,6	4,5,6	4,6
3	1,3	2,3	1,2	4,6	5,6	4,5,6
4	4,5	4,6	5,6	1,2	1,3	2,3
5	5,6	4,5	4,6	2,3	1,2	1,3
6	4,6	5,6	4,5	1,3	2,3	1,2

$\circ_2$	1	2	3	4	5	6
1	1,2	1,3	2,3	4,5	4,6	5,6
2	2,3	1,2	1,3	5,6	4,5	4,6
3	1,3	2,3	1,2	4,6	5,6	4,5
4	4,5	4,6	5,6	1,2	1,3	2,3
5	5,6	4,5	4,6	2,3	1,2	1,3
6	4,6	5,6	4,5	1,3	2,3	1,2

Then, $(H, \circ_1)$ and $(H, \circ_2)$ are hypergroups such that $|H/\beta| = 2$. In particular, $(H, \circ_1)$ is a locally 2-finite hypergroup since $h(\omega_H) = 2$ and $h(\beta(4)) = 1$, while $(H, \circ_2)$ is a strongly locally 2-finite hypergroup because $h(\omega_H) = h(\beta(4)) = 2$.

Example 2. *Let $(H, \circ)$ be a hypergroup and $Aut(H)$ the automorphism group of H. If $f \in Aut(H)$, let $\langle f \rangle$ be the subgroup of $Aut(H)$, generated by f. In $H \times \langle f \rangle$, we define the following hyperproduct: For all $(a, f^m), (b, f^n) \in H \times \langle f \rangle$, let*

$$(a, f^m) \star (b, f^n) = \{(c, f^{m+n}) \mid c \in a \circ f^m(b)\} = (a \circ f^m(b)) \times \{f^{m+n}\}.$$

Firstly, we show that $(H \times \langle f \rangle, \star)$ is a hypergroup. Then, we describe its β-classes and the heart. As a consequence, we obtain that $(H, \circ)$ is a locally n-finite hypergroup (resp., strongly locally n-finite hypergroup, complete hypergroup, or 1-hypergroup) if and only if $(H \times \langle f \rangle, \star)$ is a locally n-finite hypergroup (resp., strongly locally n-finite hypergroup, complete hypergroup, or 1-hypergroup).

1. *The hyperproduct $\star$ on $H \times \langle f \rangle$ is associative. In fact, if $(a, f^m), (b, f^n), (c, f^r) \in H \times \langle f \rangle$, then we obtain:*

$$\begin{aligned}
((a, f^m) \star (b, f^n)) \star (c, f^r) &= ((a \circ f^m(b) \times \{f^{m+n}\}) \star (c, f^r) \\
&= \bigcup_{z \in a \circ f^m(b)} (z, f^{m+n}) \star (c, f^r) \\
&= \bigcup_{z \in a \circ f^m(b)} (z \circ f^{m+n}(c)) \times \{f^{m+n+r}\} \\
&= (a \circ f^m(b)) \circ f^{m+n}(c) \times \{f^{m+n+r}\} \\
&= a \circ (f^m(b) \circ f^{m+n}(c)) \times \{f^{m+n+r}\} \\
&= a \circ (f^m(b \circ f^n(c)) \times \{f^{m+n+r}\} \\
&= \bigcup_{w \in b \circ f^n(c)} (a \circ f^m(w)) \times \{f^{m+n+r}\} \\
&= \bigcup_{w \in b \circ f^n(c)} (a, f^m) \star (w, \{f^{n+r}\}) \\
&= (a, f^m) \star ((b \circ f^n(c)) \times \{f^{n+r}\}) = (a, f^m) \star ((b, f^n) \star (c, f^r)).
\end{aligned}$$

Consequently, we have that

$$(a, f^m) \star (b, f^n) \star (c, f^r) = (a \circ f^m(b) \circ f^{m+n}(c)) \times \{f^{m+n+r}\}.$$

By induction, if $(a_1, f^{n_1}), (a_2, f^{n_2}), \ldots, (a_r, f^{n_r})$ are elements in $H \times \langle f \rangle$, then the hyperproduct $(a_1, f^{n_1}) \star (a_2, f^{n_2}) \star \ldots \star (a_r, f^{n_r})$ is the set

$$(a_1 \circ f^{n_1}(a_2) \circ f^{n_1+n_2}(a_3) \circ \cdots \circ f^{n_1+n_2+\ldots+n_{r-1}}(a_r)) \times \{f^{n_1+n_2+\ldots+n_r}\}.$$

2. *The hyperproduct $\star$ is reproducible. Indeed, we have $(b, f^m) \star (H \times \langle f \rangle) \subseteq H \times \langle f \rangle$, for all elements $(b, f^m) \in H \times \langle f \rangle$. On the other hand, if $(a, f^n) \in H \times \langle f \rangle$, then, by reproducibility of $(H, \circ)$, there exists $x \in H$ such that $a \in b \circ x$. Now, if we consider $(f^{-m}(x), f^{n-m})$, then $(a, f^n) \in (b, f^m) \star (f^{-m}(x), f^{n-m})$ because $a \in b \circ x = b \circ f^m(f^{-m}(x))$ and $f^m f^{n-m} = f^n$. Hence, $H \times \langle f \rangle \subseteq (b, f^m) \star (H \times \langle f \rangle)$. Thus, we have*

$$(b, f^m) \star (H \times \langle f \rangle) = H \times \langle f \rangle, \text{ for all } (b, f^m) \in H \times \langle f \rangle.$$

In the same way, one shows that $(H \times \langle f \rangle) \star (b, f^m) = H \times \langle f \rangle$.

3. *From 1 and 2, $(H \times \langle f \rangle, \star)$ is a hypergroup. If β' and β are the fundamental relations in $(H \times \langle f \rangle, \star)$ and $(H, \circ)$ respectively, we have*
$$\beta'(x, f^n) = \beta(x) \times \{f^n\},$$
for all $(x, f^n) \in H \times \langle f \rangle$. Indeed, if $(y, f^n) \in \beta'((x, f^m))$, then there exist $r \in \mathbb{N} - \{0\}$ and $(a_1, f^{n_1}), (a_2, f^{n_2}), \ldots, (a_r, f^{n_r}) \in H \times \langle f \rangle$ such that
$$\{(y, f^n), (x, f^m)\} \subseteq (a_1, f^{n_1}) \star (a_2, f^{n_2}) \star \ldots \star (a_r, f^{n_r}).$$
By point 1, we have $\{y, x\} \subseteq a_1 \circ f^{n_1}(a_2) \circ f^{n_1+n_2}(a_3) \circ \ldots \circ f^{n_1+n_2+\ldots+n_{r-1}}(a_r)$ and $f^n = f^{n_1+n_2+\ldots+n_r} = f^m$. Hence, $y \in \beta(x)$, $f^n = f^m$ and $(y, f^n) \in \beta(x) \times \{f^n\}$. Thus, $\beta'((x, f^m)) \subseteq \beta(x) \times \{f^n\}$. On the other hand, if $(y, f^n) \in \beta(x) \times \{f^n\}$, then $y\beta x$ and there exist $r \in N - \{0\}$ and $a_1, a_2, \ldots a_r \in H$ such that $\{x, y\} \in a_1 \circ a_2 \circ \ldots \circ a_r$. Now, if in $H \times \langle f \rangle$ we consider $(a_1, f^0), (a_2, f^0), \ldots, (a_{r-1}, f^0), (a_r, f^n)$, we obtain
$$\{(x, f^n), (y, f^n)\} \subseteq (a_1, f^0) \star (a_2, f^0) \star \ldots \star (a_{r-1}, f^0) \star (a_r, f^n),$$
and we have $(y, f^n) \in \beta'((x, f^n))$. Hence, $\beta(x) \times \{f^n\} \subseteq \beta'((x, f^n))$.

4. *The set $\omega_H \times \{f^0\}$ is a subhypergroup of $(H \times \langle f \rangle, \star)$. In fact, if $(a, f^0) \in \omega_H \times \{f^0\}$, we have*
$$\begin{aligned}(a, f^0) \star (\omega_H \times \{f^0\}) &= \bigcup_{b \in \omega_H} (a, f^0) \star (b, f^0) \\ &= \bigcup_{b \in \omega_H} (a \circ f^0(b)) \times \{f^0\} \\ &= \bigcup_{b \in \omega_H} (a \circ b) \times \{f^0\} \\ &= (a \circ \omega_H) \times \{f^0\} = \omega_H \times \{f^0\}.\end{aligned}$$
Hence, $(a, f^0) \star (\omega_H \times \{f^0\}) = \omega_H \times \{f^0\}$, for all $(a, f^0) \in \omega_H \times \{f^0\}$. In the same way one proves that $(\omega_H \times \{f^0\}) \star (a, f^0) = \omega_H \times \{f^0\}$, so $\omega_H \times \{f^0\}$ is a subhypergroup.

5. *The subhypergroup $\omega_H \times \{f^0\}$ is the heart $\omega_{H \times \langle f \rangle}$ of $(H \times \langle f \rangle, \star)$.*
Indeed, for all $(a, f^n) \in H \times \langle f \rangle$, there exists $b \in H$ such that $a \circ b \subseteq \omega_H$. If we consider $(f^{-n}(b), f^{-n})$, then $(a, f^n) \star (f^{-n}(b), f^{-n}) = (a \circ f^n(f^{-n}(b)) \times \{f^0\} = (a \circ b) \times \{f^0\} \subseteq \omega_H \times \{f^0\}$, so $\omega_H \times \{f^0\}$ is conjugable. From point 4 of Theorem 1, we obtain that $\omega_{H \times \langle f \rangle} \subseteq \omega_H \times \{f^0\}$. Finally, since $\omega_{H \times \langle f \rangle}$ and $\omega_H \times \{f^0\}$ are β'-classes, we deduce that $\omega_{H \times \langle f \rangle} = \omega_H \times \{f^0\}$.

Proposition 1. *Let $(H, \circ)$ be a hypergroup. Then, $h(\beta(y)) \leq |\beta(x)|$, for all $x, y \in H$.*

Proof. Let $x, y \in H$. By reproducibility of H, there exists $b \in H$ such that $y \in x \circ b$ and so we have
$$\beta(y) = \omega_H \circ y \subseteq \omega_H \circ x \circ b = \beta(x) \circ b = \bigcup_{a \in \beta(x)} a \circ b.$$

Clearly, for every $z \in \beta(y)$, there exists $a_z \in \beta(x)$ such that $z \in a_z \circ b$. Moreover, since $\beta(y)$ is a complete part of H, we have $a_z \circ b \subseteq \beta(y)$. Hence, there exists a non-empty subset A of $\beta(x)$ such that $\beta(y) = \bigcup_{a \in A} a \circ b$ and so we obtain $h(\beta(y)) \leq |A| \leq |\beta(x)|$. □

By the previous proposition, we have the following results:

Corollary 1. *Let $(H, \circ)$ be a hypergroup. Then, $h(\beta(x)) \leq \min\{|\omega_H|, |\beta(x)|\}$, for every $x \in H$.*

Corollary 2. *Let $(H, \circ)$ be a hypergroup. If there exists a β-class of size 1, then $(H, \circ)$ is a strongly locally 1-finite hypergroup.*

Proposition 2. *Let $(H, \circ)$ be a hypergroup. If $x \in H$ is such that $\beta(x)$ is finite and $|\beta(x)| = h(\beta(x))$, then $|a \circ y| = |y \circ a| = 1$ for all $a \in \omega_H$ and $y \in \beta(x)$.*

Proof. Let $n = |\beta(x)|$. For all $a \in \omega_H$ and $y \in \beta(x)$, we have $a \circ y \subseteq \omega_H \circ y = \beta(y) = \beta(x)$. If $|a \circ y| = |\beta(x)|$, then $a \circ y = \beta(x)$ and so $n = |\beta(x)| = h(\beta(x)) = 1$. Hence, $|a \circ y| = 1$. Now, let $|a \circ y| \neq |\beta(x)|$ and by contradiction we suppose that $2 \leq |a \circ y| = k < n$. Let $\beta(x) - a \circ y = \{x_1, x_2, \dots, x_{n-k}\}$, by reproducibility of H, there exist $y_1, y_2, \dots, y_{n-k} \in H$ such that $x_i \in a \circ y_i$ for all $i \in \{1, 2, \dots, n-k\}$. Since $\beta(x)$ is a complete part of H and $a \circ y_i \cap \beta(x) \neq \emptyset$, we deduce that $a \circ y_i \subseteq \beta(x)$, for all $i \in \{1, 2, \dots, n-k\}$. Therefore, $\beta(x) = a \circ y \cup a \circ y_1 \cup \dots \cup a \circ y_{n-k}$ and so $n = h(\beta(x)) \leq n - k + 1 < n$, impossible. Thus, $|a \circ y| = 1$, for all $a \in \omega_H$ and $y \in \beta(x)$. In an analogous way, we have that $|y \circ a| = 1$, for every $y \in \beta(x)$. □

An immediate consequence of the previous proposition is the following corollary:

Corollary 3. *Let $(H, \circ)$ be a hypergroup. If ω_H is finite and $|\omega_H| = h(\omega_H)$, then ω_H is a subgroup of $(H, \circ)$.*

In the preceding corollary, the finiteness of ω_H is a critical hypothesis. Indeed, in the next example, we show a hypergroup where $|\omega_H| = h(\omega_H) = \infty$ and ω_H is not a group.

Example 3. *Let $(\mathbb{Z}, +)$ be the group of integers. In the set $H = \mathbb{Z} \times \mathbb{Z}$, we define the following hyperproduct:*

$$(a, b) \star (c, d) = \{(a, b + d), (c, b + d)\}.$$

Routine computations show that $(H, \star)$ is a hypergroup and $(\mathbb{Z} \times \{0\}, \star)$ is a subhypergroup of $(H, \star)$. Hereafter, we firstly describe the core ω_H, then we compute $h(\omega_H)$.

To prove that $\omega_H = \mathbb{Z} \times \{0\}$, we will show that $\mathbb{Z} \times \{0\}$ is the smallest conjugable subhypergroup of $(H, \star)$. For every element $(x, y) \in H$, we can consider the element $(0, -y)$. We obtain $(x, y) \star (0, -y) = \{(x, 0), (0, 0)\} \subset \mathbb{Z} \times \{0\}$, hence $\mathbb{Z} \times \{0\}$ is a conjugable subhypergroup. Now, let K be a conjugable subhypergroup of $(H, \star)$ and let $(a, 0) \in \mathbb{Z} \times \{0\}$. Since K is conjugable, there exists $(x, y) \in H$ such that $(a, 0) \star (x, y) \subseteq K$, and so $\{(a, y), (x, y)\} \subseteq K$. By reproducibility of K, there exists $(x', y') \in K$ such that $(a, y) \in (x, y) \star (x', y') = \{(x, y + y'), (x', y + y')\} \subseteq K$. Clearly, $y' = 0$ because $y = y + y'$ and $(x', 0) \in K$. Since $(x', 0), (a, y) \in K$, there exists $(z, w) \in K$ such that $(x', 0) \in (a, y) \star (z, w) = \{(a, y + w), (z, y + w)\} \subseteq K$. Consequently, we have $y + w = 0$ and $(a, 0) \in K$. Hence, $\mathbb{Z} \times \{0\} \subseteq K$ and $\omega_H = \mathbb{Z} \times \{0\}$ since $\mathbb{Z} \times \{0\}$ is conjugable. Obviously, $|\omega_H| = |\mathbb{Z}|$.

Finally, we prove that $h(\omega_H) = |\mathbb{Z}|$. By Proposition 1, we have $h(\omega_H) \leq |\omega_H| = |\mathbb{Z}|$. If $h(\omega_H) < |\mathbb{Z}|$, then there exist n hyperproducts $(a_i, b_i) \star (c_i, d_i)$ of elements in H such that

$$\mathbb{Z} \times \{0\} = \omega_H = \bigcup_{i=1}^{n} (a_i, b_i) \star (c_i, d_i) = \bigcup_{i=1}^{n} \{(a_i, b_i + d_i), (c_i, b_i + d_i)\}.$$

This result is impossible since $\bigcup_{i=1}^{n} \{a_i, c_i\} \neq \mathbb{Z}$.

Now, we give two examples of hypergroups $(H, \circ)$ whose heart is a group and $|\omega_H| \geq 2$. In particular, we have that $|H/\beta| = 2$, $h(\omega_H) = 1$ and $h(\beta(a)) = |\omega_H|$, if $a \in H - \omega_H$.

Example 4. *Consider the group $(\mathbb{Z}, +)$ and a set $A = \{a_i\}_{i \in \mathbb{Z}}$ such that $\mathbb{Z} \cap A = \emptyset$. In the set $H = \mathbb{Z} \cup A$, we define the following hyperproduct:*

- *$m \circ n = \{m + n\}$ if $m, n \in \mathbb{Z}$;*

- $m \circ a_n = a_n \circ m = \{a_{m+n}\}$ *if* $m \in \mathbb{Z}$ *and* $a_n \in A$;
- $a \circ b = \mathbb{Z}$ *if* $a, b \in A$.

Routine computations show that $(H, \circ)$ *is a hypergroup. We have* $H/\beta \cong \mathbb{Z}_2$, $\omega_H = \mathbb{Z}$ *and* $\beta(a) = A$ *if* $a \in A$. *Clearly, we have* $h(\omega_H) = 1$ *and* $h(\beta(a)) = |\beta(a)| = |\mathbb{Z}|$ *because* $m \circ a$ *is a singleton, for all* $m \in \mathbb{Z}$ *and* $a \in A$.

Example 5. *Let* $(G, \cdot)$ *be a group of size* $n \geq 2$ *and let* $G = \{g_1, g_2, \ldots, g_n\}$. *Moreover, let* σ *be a n-cycle of the symmetric group defined over* $X = \{1, 2, \ldots, n\}$. *If* $A = \{a_1, a_2, \ldots, a_n\}$ *is a set disjoint with* G, *in* $H = G \cup A$, *we can define the following hyperproduct:*

- $g_h \circ g_k = \{g_h g_k\}$,
- $g_h \circ a_k = a_k \circ g_h = \{a_{\sigma^h(k)}\}$,
- $a \circ b = G$ *if* $a, b \in A$.

Then, $(H, \circ)$ *is a hypergroup such that* $H/\beta \cong \mathbb{Z}_2$, $\omega_H = G$ *and* $\beta(a) = A$ *if* $a \in A$. *Moreover, we have* $h(\omega_H) = 1$ *and* $h(\beta(a)) = |\beta(a))| = |G|$.

In the next theorem, we characterize the locally n-finite hypergroups such that $|\omega_H| = h(\omega_H)$.

Theorem 2. *Let* $(H, \circ)$ *be a hypergroup such that* $|\omega_H|$ *is finite. The following conditions are equivalent:*

1. $(H, \circ)$ *is an* 1*-hypergroup that is* $|\omega_H| = 1$;
2. $|\omega_H| = h(\omega_H)$.

Proof. The implication 1. $\Rightarrow$ 2. is trivial, hence we prove that 2. $\Rightarrow$ 1. Let n a positive integer such that $|\omega_H| = h(\omega_H) = n$. By Corollary 3, the heart ω_H is a subgroup of $(H, \circ)$. Let e be the identity of ω_H. If there exist $a, b \in H$ such that $a \circ b = \omega_H$, then we have $|\omega_H| = h(\omega_H) = 1$ and $\omega_H = \{e\}$. Now, by contradiction, we suppose that $a \circ b \neq \omega_H$, for all $a, b \in H$, and let x be an element of H. By reproducibility of $(H, \circ)$, there exists $x' \in H$ such that $e \in x \circ x'$. Clearly, we have $x \circ x' \subset \omega_H$, since ω_H is a complete part of $(H, \circ)$. Therefore, there exists an integer k, with $1 \leq k < n$, and $n - k$ elements of ω_H such that $\emptyset \neq \omega_H - x \circ x' = \{a_1, a_2, \ldots, a_{n-k}\}$. Hence, we obtain

$$\omega_H = x \circ x' \cup \{a_1, a_2, \ldots, a_{n-k}\} = x \circ x' \cup e \circ a_1 \cup e \circ a_2 \cup \ldots \cup e \circ a_{n-k},$$

and so $k = 1$ because $n = h(\omega_H) \leq n - k + 1$. Therefore, we obtain $x \circ x' = \{e\}$ and so $\{e\}$ is a conjugable subhypergroup of $(H, \circ)$. Consequently, we have $\{e\} = \omega_H = e \circ e$, a contradiction. □

The following result is an immediate consequence of Theorem 2.

Corollary 4. *Let* $(H, \circ)$ *be a finite hypergroup, then* $(H, \circ)$ *is a group if and only if* $|\beta(x)| = h(\beta(x))$, *for all* $x \in H$.

Proof. The implication $\Rightarrow$ is obvious. On the other hand, if we suppose that $|\beta(x)| = h(\beta(x))$ for all $x \in H$, then we have $|\omega_H| = h(\omega_H)$ and so $|\omega_H| = 1$, by Theorem 2. From Corollary 1 and the hypothesis, we have $|\beta(x)| = h(\beta(x)) = 1$. Hence, $(H, \circ)$ is a group. □

4. Hypergroups with at Least One β-Class of Height Equal to 1

From Theorem 2, the 1-hypergroups are characterized by the fact that $h(\omega_H) = |\omega_H| = 1$. In this section, we use the notion of height of a β-class to introduce new classes of hypergroups. We enumerate the elements in each class when the size of hypergroups is $n \leq 4$, apart from isomorphisms. We give the following definition:

Definition 4. *Let $(H, \circ)$ be a hypergroup. We say that*

a. $(H, \circ)$ *is a* $(1, \beta)$*-hypergroup if there exists* $x \in H$ *such that* $|\beta(x)| = 1$;
b. $(H, \circ)$ *is a locally 1-finite hypergroup if* $h(\beta(x)) = 1$, *for all* $x \in H$;
c. $(H, \circ)$ *is a 1-weak hypergroup if* $h(\omega_H) = 1$;
d. $(H, \circ)$ *is a weakly locally 1-finite hypergroup if there exists* $x \in H$ *such that* $h(\beta(x)) = 1$;

In the following, we denote by $\mathfrak{U}$, $\mathfrak{A}$, $\mathfrak{B}$, $\mathfrak{C}$ and $\mathfrak{D}$ the classes of 1-hypergroups, $(1, \beta)$-hypergroups, locally 1-finite hypergroups, 1-weak hypergroups, and weakly locally 1-finite hypergroups, respectively. By Definition 4 and Corollary 2, we have the inclusions $\mathfrak{U} \subseteq \mathfrak{A} \subseteq \mathfrak{B} \subseteq \mathfrak{C} \subseteq \mathfrak{D}$. Actually, these inclusions are strict, as shown in the following example.

Example 6. *In this example, we show four hypergroups* $(A, \circ)$, $(B, \circ)$, $(C, \circ)$ *and* $(D, \circ)$ *such that*

1. $(A, \circ) \in \mathfrak{A}$ *and* $(A, \circ) \notin \mathfrak{U}$,
2. $(B, \circ) \in \mathfrak{B}$ *and* $(B, \circ) \notin \mathfrak{A}$,
3. $(C, \circ) \in \mathfrak{C}$ *and* $(C, \circ) \notin \mathfrak{B}$,
4. $(D, \circ) \in \mathfrak{D}$ *and* $(D, \circ) \notin \mathfrak{C}$.

They are the following:

1. $A = \{1, 2, 3\}$ *with the hyperproduct*

$\circ$	1	2	3
1	1	2	3
2	2	1	3
3	3	3	1,2

2. $B = \{1, 2, 3, 4\}$ *with the hyperproduct*

$\circ$	1	2	3	4
1	1	1,2	3,4	3,4
2	1,2	1,2	3,4	3,4
3	3,4	3,4	1,2	1,2
4	3,4	3,4	1,2	1,2

3. $C = \{1, 2, 3, 4\}$ *with the hyperproduct*

$\circ$	1	2	3	4
1	1	2	3	4
2	2	1	4	3
3	3	4	1,2	1,2
4	4	3	1,2	1,2

4. $D = \{1, 2, 3, 4, 5\}$ *with the hyperproduct*

$\circ$	1	2	3	4	5
1	1,2	2,3	1,3	4,5	4,5
2	2,3	1,3	1,2	4,5	4,5
3	1,3	1,2	2,3	4,5	4,5
4	4,5	4,5	4,5	1,2	2,3
5	4,5	4,5	4,5	2,3	1,2

4.1. (1, β)-Hypergroups of Size $n \leq 4$

In [26], Corsini introduced the class of 1-hypergroups and listed the 1-hypergroups of size $n \leq 4$, apart from isomorphisms. In this subsection, our interest is to study the hypergroups in class $\mathfrak{A}$ and, in particular, to determine their number, apart from isomorphisms. Since the class of 1-hypergroups is a subclass of $\mathfrak{A}$, we recall the result proved by Corsini in [26].

Theorem 3. *If $(H, \circ)$ is a 1-hypergroup with $|H| \leq 4$, then $(H, \circ)$ is a complete hypergroup. Moreover, $(H, \circ)$ is isomorphic to either a group or one of the hypergroups described in the following three hyperproduct tables:*

$\circ$	1	2	3
1	1	2,3	2,3
2	2,3	1	1
3	2,3	1	1

$\circ$	1	2	3	4
1	1	2,3,4	2,3,4	2,3,4
2	2,3,4	1	1	1
3	2,3,4	1	1	1
4	2,3,4	1	1	1

$\circ$	1	2	3	4
1	1	2,3	2,3	4
2	2,3	4	4	1
3	2,3	4	4	1
4	4	1	1	2,3

Therefore, there exist eight 1-hypergroups of size $|H| \leq 4$.

Now, we study the hypergroups $(H, \circ) \in \mathfrak{A} - \mathfrak{U}$ of size $n \leq 4$. Clearly, $|\omega_H|$ and $|H/\beta|$ can take the values 2 or 3 and so we distinguish the following cases:

1. $|H| = 3$, $|\omega_H| = 2$ and $|H/\beta| = 2$;
2. $|H| = 4$, $|\omega_H| = 2$ and $|H/\beta| = 3$;
3. $|H| = 4$, $|\omega_H| = 3$ and $|H/\beta| = 2$.

1. In this case, we can suppose $H = \{a, b, x\}$, $\omega_H = \{a, b\}$ and $\beta(x) = \{x\}$. Clearly, since $H/\beta \cong \mathbb{Z}_2$, for reproducibility of H, we have the following partial hyperproduct table of $(H, \circ)$:

$\circ$	a	b	x
a			x
b			x
x	x	x	a,b

To complete this table, the undetermined entries must correspond to the hyperproduct table of the subhypergroup ω_H. Apart from isomorphisms, there are eight hypergroups of size 2. Their hyperproduct tables were determined in [29] and are reproduced here below:

W_1:

$\circ$	a	b
a	a	b
b	b	a

W_2:

$\circ$	a	b
a	a	b
b	b	a,b

W_3:

$\circ$	a	b
a	a	a,b
b	b	a,b

W_4:

$\circ$	a	b
a	a,b	a,b
b	a	b

W_5:

$\circ$	a	b
a	a	a,b
b	a,b	b

W_6:

$\circ$	a	b
a	a	a,b
b	a,b	a,b

W_7:

$\circ$	a	b
a	a,b	a,b
b	a,b	a

W_8:

$\circ$	a	b
a	a,b	a,b
b	a,b	a,b

Hence, in this case, we have eight hypergroups.

2. Without loss of generality, we suppose that $H = \{a, b, x, y\}$, $\omega_H = \{a, b\}$, $\beta(x) = \{x\}$ and $\beta(y) = \{y\}$. Since $H/\beta \cong \mathbb{Z}_3$, we have the following partial hyperproduct table of $(H, \circ)$:

$\circ$	a	b	x	y
a			x	y
b			x	y
x	x	x	y	a,b
y	y	y	a,b	x

As in the previous case, the entries that are left empty must be determined so that ω_H is a hypergroup of order 2. Hence, also in this case, we have eight hypergroups.

3. Let $H = \{a, b, c\}$ and $\beta(x) = \{x\}$. We have the following partial hyperproduct table:

$\circ$	a	b	c	x
a				x
b				x
c				x
x	x	x	x	a, b, c

In this case, it is straightforward to see that we get as many hypergroups as there are of size three, apart from isomorphisms. In [30], this number is found to be equal to 3999.

From Theorem 3 and the preceding arguments, we summarize the number of non-isomorphic $(1, \beta)$-hypergroups with $|H| \leq 4$ in Table 1.

Table 1. The number of non-isomorphic $(1, \beta)$-hypergroups with $|H| \leq 4$.

$\|H\| = 1$	$\|H\| = 2$	$\|H\| = 3$	$\|H\| = 4$
1	1	10	4011

Result 1. *There are* 4023 *non-isomorphic* $(1, \beta)$*-hypergroups of size* $n \leq 4$.

4.2. Locally 1-Finite Hypergroups of Size $n \leq 4$

In this subsection, we focus on hypergroups $(H, \circ) \in \mathfrak{B} - \mathfrak{A}$ with $|H| \leq 4$. By Definition 4, we have that $|\beta(x)| \geq 2$, for all $x \in H$.

If $|H| = 2$, then $|H/\beta| = 1$ and $(H, \circ)$ is isomorphic to one of the hypergroups $(W_i, \circ)$ listed in the previous subsection, for $i = 2, 3, \ldots, 8$.

If $|H| = 3$, then $|H/\beta| = 1$, otherwise at least one β-class has size 1 and $(H, \circ) \in \mathfrak{A}$. Hence, to determine the hypergroups in $(H, \circ) \in \mathfrak{B} - \mathfrak{A}$ of size 3, we must assume that there exist $a, b \in H$ such that $a \circ b = H$. With the help of computer-assisted computations, we found that in this case there are exactly 3972 hypergroups, apart from isomorphisms.

If $|H| = 4$, then there are two possible cases, namely $|H/\beta| = 1$ and $|H/\beta| = 2$. In the first case, the only information we can deduce about $(H, \circ)$ is that there are at least two elements $a, b \in H$ such that $a \circ b = H$. The number of hypergroups having that property is huge, and at present we are not able to enumerate them because the computational task exceeds our available resources. A detailed analysis of this case is challenging and may be the subject of further research. In the other case, if $x \in H - \omega_H$, then we have $|\omega_H| = |\beta(x)| = 2$ with $h(\omega_H) = h(\beta(x)) = 1$. Moreover, ω_H is isomorphic to one of the hypergroups $(W_i, \circ)$ for $i = 1, 2, \ldots, 8$ listed beforehand, and there exist $a, b \in H$ such that $a \circ b = \beta(x)$. On the basis of the information gathered from the preceding arguments, we are able to perform an exhaustive search of all possible hyperproduct tables with the help of a computer algebra system. In Table 2, we report the number of the hypergroups such that $|\omega_H| = |\beta(x)| = 2$ and $h(\omega_H) = h(\beta(x)) = 1$, depending on the structure of ω_H, apart from isomorphisms.

Table 2. The number of non-isomorphic hypergroups in $\mathfrak{B} - \mathfrak{A}$ with $|H| = 4$, according to the structure of ω_H.

W_1	W_2	W_3	W_4	W_5	W_6	W_7	W_8
3	25	17	17	31	26	12	20

Since the hypergroups corresponding to the cases in which the heart is one of the $W_2, W_3, \ldots, W_8$ are quite a few, we list hereafter only those whose heart is isomorphic to W_1. Apart from isomorphisms, we have the following hyperproduct tables:

$\circ$	a	b	x	y
a	a	b	x	y
b	b	a	y	x
x	x,y	x,y	a,b	a,b
y	x,y	x,y	a,b	a,b

$\circ$	a	b	x	y
a	a	b	x,y	x,y
b	b	a	x,y	x,y
x	x	y	a,b	a,b
y	y	x	a,b	a,b

$\circ$	a	b	x	y
a	a	b	x,y	x,y
b	b	a	x,y	x,y
x	x,y	x,y	a,b	a,b
y	x,y	x,y	a,b	a,b

On the basis of the previous arguments, the number of hypergroups $(H, \circ)$ belonging to $\mathfrak{B} - \mathfrak{A}$ is summarized in Table 3, in relation to the size of H:

Table 3. The number of non-isomorphic hypergroups in $\mathfrak{B} - \mathfrak{A}$, depending on their size.

$\lvert H\rvert = 2$	$\lvert H\rvert = 3$	$\lvert H\rvert = 4$ and $\lvert H/\beta\rvert = 2$
7	3972	151

Finally, since there are 4023 hypergroups in class $\mathfrak{A}$, see Result 1, we obtain the following result:

Result 2. *Excluding the hypergroups $(H, \circ)$ such that $|H| = 4$ and $|H/\beta| = 1$, there are 8153 non-isomorphic locally 1-finite hypergroups of size $n \leq 4$.*

4.3. 1-Weak Hypergroups of Size $n \leq 4$

In this subsection, we determine the hypergroups $(H, \circ) \in \mathfrak{C}$ of size $n \leq 4$, apart from isomorphisms. We observe that, if $(H, \circ) \in \mathfrak{C} - \mathfrak{B}$, then there is at least one β-class of height different from 1. Moreover, since $\mathfrak{A} \subset \mathfrak{B}$, we have $|\beta(x)| > 1$, for all $x \in H$. Hence, if $(H, \circ) \in \mathfrak{C} - \mathfrak{B}$, then $|H| \geq 4$ and $|H/\beta| \geq 2$.

Lemma 1. *Let $(H, \circ)$ be a hypergroup in $\mathfrak{C} - \mathfrak{B}$ such that $|H| = 4$, then ω_H is isomorphic to the group $\mathbb{Z}_2$.*

Proof. By hypotheses, we have $|H/\beta| = 2$ and $|\omega_H| = 2$. By Proposition 1, if $\beta(x)$ is the class different from ω_H, then $h(\beta(x)) = 2 = |\beta(x)|$. Moreover, for Proposition 2, we have $|a \circ y| = |y \circ a| = 1$, for all $a \in \omega_H$ and $y \in \beta(x)$. Now, if by contradiction we suppose that there exist $a, b \in \omega_H$ such that $|a \circ b| \neq 1$, then $a \circ b = \omega_H$ and $|b \circ x| = |a \circ (b \circ x)| = 1$ because $b \circ x \subset \beta(x)$, $a \circ (b \circ x) \subset \beta(x)$ and $h(\beta(x)) = 2$. This fact is impossible since $a \circ (b \circ x) = (a \circ b) \circ x = \omega_H \circ x = \beta(x)$ and $|\beta(x)| = 2$. Hence, $|a \circ b| = 1$, for all $a, b \in \omega_H$, and so $\omega_H \cong \mathbb{Z}_2$. □

Theorem 4. *Let $(H, \circ)$ be a hypergroup such that the heart ω_H is isomorphic to a torsion group. If ε is the identity of ω_H, then $x \in \varepsilon \circ x \cap x \circ \varepsilon$, for all $x \in H - \omega_H$.*

Proof. Let $x \in H - \omega_H$. By reproducibility of H, there exists $e \in H$ such that $x \in x \circ e$. Clearly $e \in \omega_H$; moreover, we have $x \in x \circ e \subseteq (x \circ e) \circ e = x \circ (e \circ e) = x \circ e^2$ and so $x \in x \circ e^2$. Obviously, by induction, we obtain $x \in x \circ e^n$, for all $n \in \mathbb{N} - \{0\}$. Finally, since ω_H is isomorphic to a torsion group, there exists $m \in \mathbb{N} - \{0\}$ such that $e^m = \varepsilon$, hence $x \in x \circ \varepsilon$. In the same way, we have $x \in \varepsilon \circ x$. □

By reproducibility, Lemma 1, and Theorem 4, the hypergroups $(H, \circ)$ in $\mathfrak{C} - \mathfrak{B}$ with $|H| = 4$ have the following partial hyperproduct table, apart from isomorphisms:

$\circ$	a	b	x	y
a	a	b	x	y
b	b	a	y	x
x	x	y		
y	y	x		

Since $h(\omega_H) = 1$, we have $\omega_H \in \{x \circ x,\ x \circ y,\ y \circ x,\ y \circ y\}$. Now, we prove that $\omega_H = x \circ x = x \circ y = y \circ x = y \circ y$. In fact, if we suppose that $\omega_H = x \circ y$, then we have:

$$x \circ x = x \circ (y \circ b) = (x \circ y) \circ b = \omega_H \circ b = \omega_H;$$
$$y \circ x = (b \circ x) \circ (y \circ b) = b \circ (x \circ y) \circ b = b \circ \omega_H \circ b = \omega_H;$$
$$y \circ y = (b \circ x) \circ y = b \circ (x \circ y) = b \circ \omega_H = \omega_H.$$

We obtain the same result also if we suppose that $\omega_H = x \circ x$ or $\omega_H = y \circ x$ or $\omega_H = y \circ y$. Hence, in class $\mathfrak{C} - \mathfrak{B}$, there is only one hypergroup of size 4, apart from isomorphisms. Its hyperproduct table is the following:

$\circ$	a	b	x	y
a	a	b	x	y
b	b	a	y	x
x	x	y	a, b	a, b
y	y	x	a, b	a, b

We note that this hypergroup is a special case of the hypergroup described in Example 5. The group $(G, \cdot)$ is $\mathbb{Z}_2$ and the cycle σ is a transposition.

Result 3. *Excluding the hypergroups $(H, \circ)$ such that $|H| = 4$ and $|H/\beta| = 1$, in the class $\mathfrak{C}$ there are* 8154 *hypergroups of size $n \leq 4$, apart from isomorphisms.*

We complete this section by showing a result concerning the weakly locally 1-finite hypergroups.

Theorem 5. *If $(H, \circ) \in \mathfrak{D} - \mathfrak{C}$, then $|H| \geq 5$.*

Proof. By hypothesis, there is a class $\beta(x)$ different from ω_H such that $h(\beta(x)) = 1$ and $h(\omega_H) \geq 2$. Because of the inclusions $\mathfrak{U} \subset \mathfrak{A} \subset \mathfrak{C}$, we have $|\omega_H| \geq 2$ and $|\beta(x)| \geq 2$, otherwise $(H, \circ) \in \mathfrak{C}$. Now, if we suppose that $|\omega_H| = 2$, by Proposition 1, we obtain $2 \leq h(\omega_H) \leq |\omega_H| = 2$ and so $h(\omega_H) = |\omega_H| = 2$. Consequently, with the help of Theorem 2, we have the contradiction $(H, \circ) \in \mathfrak{U} \subset \mathfrak{C}$. Hence, $|\omega_H| \geq 3$, $|\beta(x)| \geq 2$ and $|H| \geq 5$. □

Recall that the hypergroup $(D, \circ)$ shown in Example 6 belongs to $\mathfrak{D}$ but not to $\mathfrak{C}$ due to the previous theorem that the hypergroup has the smallest cardinality, among all hypergroups sharing that property.

5. Conclusions

In hypergroup theory, the relation β is the smallest strongly regular equivalence relation whose corresponding quotient set is a group. If $(H, \circ)$ is a hypergroup and $\varphi : H \to H/\beta$ is the canonical projection, then the kernel $\omega_H = \varphi^{-1}(1_{H/\beta})$ is the hearth of $(H, \circ)$. If the hearth is a singleton, then $(H, \circ)$ is a 1-hypergroup. We remark that the hearth is a β-class and also a subhypergroup of $(H, \circ)$. In particular, if $\omega_H = \{e\}$, then we have $\omega_H = e \circ e$. More generally, every β-class is the union of hyperproducts of pairs of elements of H. In this work, we defined the height of a β-class as the minimum number of such hyperproducts. This concept yields a new characterization of 1-hypergroups, see Theorem 2, and allows us to introduce new hypergroup classes, depending on the relationship between height and cardinality of the β-classes; see Definition 4. These classes include 1-hypergroups as particular cases. Apart from isomorphisms, we were able to enumerate the elements of those classes when $|H| \leq 4$, with only one exception. In fact, the problem of enumerating the non-isomorphic hypergroups where $|H| = 4$, $|H/\beta| = 1$ and $h(\omega_H) = 1$ remains open.

In conclusion, as a direction for further research, we point out that many hypergroups that arose in the analysis of the hypergroup classes introduced in the present work are join spaces or transposition hypergroups [15,31]. For example, the 10 hypergroups of size three in Table 1 are transposition hypergroups. Transposition hyperstructures are very important in hypercompositional algebra. Hence, it would be interesting to enumerate the join spaces or the transposition hypergroups belonging to the hypergroup classes introduced in Definition 4, at least for small cardinalities. Another question that is stimulated by the concept of height concerns the height of the β-classes of the coset hypergroups, i.e., the hypergroups that are quotient of a non-commutative group with respect to a non-normal subgroup [32]. We leave these observations and suggestions as a possible subject for new works.

Author Contributions: Conceptualization, formal analysis, writing—original draft: M.D.S., D.F. (Domenico Freni), and G.L.F.; software, writing—review and editing: D.F. (Dario Fasino). All authors have read and agreed to the published version of the manuscript.

Funding: This research has been carried out in the framework of the departmental research projects "Topological, Categorical and Dynamical Methods in Algebra" and "Innovative Combinatorial Optimization in Networks", Department of Mathematics, Computer Science and Physics (PRID 2017), University of Udine, Italy. The work of Giovanni Lo Faro has been partly supported by INdAM-GNSAGA, and the work of Dario Fasino has been partly supported by INdAM-GNCS.

Conflicts of Interest: The authors declare no conflict of interest.

References

1. Corsini, P.; Leoreanu-Fotea, V. *Applications of Hyperstructures Theory*; Kluwer Academic Publisher: Dordrecht, The Netherlands, 2003.
2. Cristea, I.; Kocijan, J.; Novák, M. Introduction to Dependence Relations and Their Links to Algebraic Hyperstructures. *Mathematics* **2019**, *7*, 885. [CrossRef]
3. Novák, M.; Křehlík, Š.; Ovaliadis, K. Elements of hyperstructure theory in UWSN design and data aggregation. *Symmetry* **2019**, *11*, 734. [CrossRef]
4. Davvaz, B., Salasi, A. A realization of hyperrings. *Commun. Algebra* **2006**, *34*, 4389–4400. [CrossRef]
5. De Salvo, M.; Fasino, D.; Freni, D.; Lo Faro, G. Fully simple semihypergroups, transitive digraphs, and Sequence A000712. *J. Algebra* **2014**, *415*, 65–87. [CrossRef]
6. De Salvo, M.; Fasino, D.; Freni, D.; Lo Faro, G. A family of 0-simple semihypergroups related to sequence A00070. *J. Mult. Valued Log. Soft Comput.* **2016**, *27*, 553–572.
7. De Salvo, M.; Fasino, D.; Freni, D.; Lo Faro, G. Semihypergroups obtained by merging of 0-semigroups with groups. *Filomat* **2018**, *32*, 4177–4194.
8. De Salvo, M.; Freni, D.; Lo Faro, G. Fully simple semihypergroups. *J. Algebra* **2014**, *399*, 358–377. [CrossRef]
9. De Salvo, M.; Freni, D.; Lo Faro, G. Hypercyclic subhypergroups of finite fully simple semihypergroups. *J. Mult. Valued Log. Soft Comput.* **2017**, *29*, 595–617.
10. De Salvo, M.; Freni, D.; Lo Faro, G. On hypercyclic fully zero-simple semihypergroups. *Turk. J. Math.* **2019**, *4*, 1905–1918. [CrossRef]
11. De Salvo, M.; Lo Faro, G. On the n^*-complete hypergroups. *Discret. Math.* **1999**, *208/209*, 177–188. [CrossRef]
12. Freni, D. Une note sur le cœur d'un hypergroup et sur la clôture transitive β^* de β. *Riv. Mat. Pura Appl.* **1991**, *8*, 153–156.
13. Freni, D. Strongly transitive geometric spaces: Applications to hypergroups and semigroups theory. *Commun. Algebra* **2004**, *32*, 969–988. [CrossRef]
14. Gutan, M. On the transitivity of the relation β in semihypergroups. *Rendiconti del Circolo Matematico di Palermo* **1996**, *45*, 189–200. [CrossRef]
15. Tsitouras, C.; Massouros, C.G. On enumeration of hypergroups of order 3. *Comput. Math. Appl.* **2010**, *59*, 519–523. [CrossRef]
16. Vougiouklis, T. Fundamental relations in hyperstructures. *Bull. Greek Math. Soc.* **1999**, *42*, 113–118.

17. Koskas, H. Groupoïdes, demi-hypergroupes et hypergroupes. *J. Math. Pures Appl.* **1970**, *49*, 155–192.
18. De Salvo, M.; Lo Faro, G. A new class of hypergroupoids associated with binary relations. *J. Mult. Valued Log. Soft Comput.* **2003**, *9*, 361–375.
19. Fasino, D.; Freni, D. Fundamental relations in simple and 0-simple semi-hypergroups of small size. *Arab. J. Math.* **2012**, *1*, 175–190. [CrossRef]
20. Freni, D. Minimal order semi-hypergroups of type *U* on the right. II. *J. Algebra* **2011**, *340*, 77–89. [CrossRef]
21. Hila, K.; Davvaz, B.; Naka, K. On quasi-hyperideals in semihypergroups. *Commun. Algebra* **2011**, *39*, 4183–4194. [CrossRef]
22. Naz, S.; Shabir, M. On soft semihypergroups. *J. Intell. Fuzzy Syst.* **2014**, *26*, 2203–2213. [CrossRef]
23. Antampoufis, N.; Spartalis, S.; Vougiouklis, T. Fundamental relations in special extensions. In *Algebraic Hyperstructures and Applications*; Vougiouklis, T., Ed.; Spanidis Press: Xanthi, Greece, 2003; pp. 81–89.
24. Davvaz, B. *Semihypergroup Theory*; Academic Press: London, UK, 2016.
25. Changphas, T.; Davvaz, B. Bi-hyperideals and quasi-hyperideals in ordered semihypergroups. *Ital. J. Pure Appl. Math.* **2015**, *35*, 493–508.
26. Corsini, P. *Prolegomena of Hypergroup Theory*; Aviani Editore: Tricesimo, Italy, 1993.
27. Davvaz, B.; Leoreanu-Fotea, V. *Hyperring Theory and Applications*; International Academic Press: Palm Harbor, FL, USA, 2007.
28. De Salvo, M.; Freni, D.; Lo Faro, G. On further properties of fully zero-simple semihypergroups. *Mediterr. J. Math.* **2019**, *16*, 48. [CrossRef]
29. De Salvo, M.; Freni, D. Semi-ipergruppi e ipergruppi ciclici. *Atti Sem. Mat. Fis. Univ. Modena* **1981**, *30*, 44–59.
30. Nordo, G. An algorithm on number of isomorphism classes of hypergroups of order 3. *Ital. J. Pure Appl. Math.* **1997**, *2*, 37–42.
31. Jantosciak, J. Transposition hypergroups: Noncommutative join spaces. *J. Algebra* **1997**, *187*, 97–119. [CrossRef]
32. Eaton, J.E. Theory of cogroups. *Duke Math.* **1940**, *6*, 101–107. [CrossRef]

symmetry

MDPI

Article

The Symmetry of Lower and Upper Approximations, Determined by a Cyclic Hypergroup, Applicable in Control Theory

Štěpán Křehlík [1] and Jana Vyroubalová [2,*]

1 CDV—Transport Research Centre, Líšeňská 33a, 636 00 Brno, Czech Republic; stepan.krehlik@cdv.cz
2 Faculty of Electrical Engineering and Communication, Brno University of Technology, Technická 8, 616 00 Brno, Czech Republic
* Correspondence: xvyrou04@feec.vutbr.cz; Tel.: +420-774-177-645

Received: 9 December 2019; Accepted: 23 December 2019; Published: 26 December 2019

Abstract: In the first part of our paper, we construct a cyclic hypergroup of matrices using the Ends Lemma. Its properties are then, in the second part of the paper, used to describe the symmetry of lower and upper approximations in certain rough sets with respect to invertible subhypergroups of this cyclic hypergroup. Since our approach is widely used in autonomous robotic systems, we suggest an application of our results for the study of detection sensors, which are used especially in mobile robot mapping.

Keywords: single-power cyclic hypergroup; invertible subhypergroup; lower approximation; upper approximation; rough set

MSC: 20N20

1. Introduction

Our perception of the real world is never fully precise. Our decisions are always made with a certain level of uncertainty or lack of some pieces of information. Mathematical tools that make this decision making easier include *fuzzy sets* [1], *rough sets* [2,3] or *soft sets* [4]. In the *algebraic hyperstructure theory*, i.e., in the theory of *algebraic hypercompositional structures*, there are numerous constructions leading to such structures. One of these is the application of the *Ends lemma* [5–7] in which the hyperoperation is the principal end of a partially ordered semigroup. For some theoretical results regarding the construction, see for example Novák et al. [7–9]. In [9] the authors modify the construction in order to increase its applicability.

In our paper we develop an idea similar to [9–11]. With the help of matrix calculus, which we believe is a suitable tool, we construct cyclic hypergroups and their invertible subhypergroups. Notice that matrix calculus linked to the theory of algebraic hypercompositional structures has been used as a suitable tool in various contexts such as [12–14]. Since the notion of hypercompositional cyclicity has a rather complicated evolution, we recommend the reader to study [15], which gives a complex discussion of the topic, and [16] which is the source of our definition.

In [17] the authors consider the application of rough sets in various contexts based on establishing the set describing its upper and lower approximation. In a general case, this issue is discussed, e.g., in [11,18,19]. Suppose that we can see the Ends Lemma as a certain boundary used in various areas such as economics (the need to generate *at least* certain profit), electrical engineering (the transistor basis of a p-n junction needs *at least* certain current for the charge flow yet this must not be too great), etc. Motivated by these considerations we investigate hypercompositional structures constructed with the help of the principal end (or beginning) and their subhyperstructures. We study their cyclicity and their generators. In the context of rough sets we use cyclic hypergroups to construct the universum with the help of indiscernibility relation and its subhyperstructures describing their upper and lower approximations. In order to describe these we make use of some natural relations between matrix characteristics. In the end of our paper we demonstrate how this type of rough sets, especially the upper approximations, can be used for description of an area monitored by sensors of autonomous robotic systems.

2. Notation and Context

In our paper, we work with matrices. In a general case these are matrices over a field. However, within this field we regard a subset (infinite such as $\mathbb{N}$ or finite such as in Example 5), elements of which are actually used as entries. We denote such a set by $\mathcal{E}$, and the set of all $m \times n$ matrices with entries from a $\mathcal{E}$ by $\mathbb{M}_{m,n}(\mathcal{E})$. We also suppose that there exists a total order on $\mathcal{E}$ with the smallest element of $\mathcal{E}$ denoted (if it exists) by e. Within the set $\mathcal{E}$ we, in our hyperoperations, rely on the operation $\min\{a, b\}$, the result of which is, given our context, equal to the smaller element.

On $\mathbb{M}_{m,n}(\mathcal{E})$ we, for an arbitrary pair of matrices $\mathbf{A}, \mathbf{B} \in \mathbb{M}_{m,n}(\mathcal{E})$, define relation $\leq_M$ by

$$\mathbf{A} \leq_M \mathbf{B} \text{ if } \|\mathbf{A}\|_\infty \leq_M \|\mathbf{B}\|_\infty \tag{1}$$

where $\|\mathbf{A}\|_\infty = \max_{1 \leq i \leq m} \sum_{j=1}^{n} |a_{ij}|$, i.e., $\|\mathbf{A}\|_\infty$ is the row norm.

Example 1. *For matrices*

$$A = \begin{bmatrix} 8 & 1 & 2 \\ 4 & 0 & 1 \\ 1 & 1 & 1 \end{bmatrix}, \qquad B = \begin{bmatrix} 2 & 1 & 1 \\ 3 & 0 & 3 \\ 5 & 1 & 0 \end{bmatrix},$$

there is

$$\|\mathbf{A}\|_\infty = 11 \quad \textit{and} \quad \|\mathbf{B}\|_\infty = 6.$$

Therefore $\boldsymbol{B} \leq_M \boldsymbol{A}$.

Basic Notions of the Theory of Algebraic Hypercompositional Structures

Before we give our results, recall some basic notions of the algebraic hyperstructure theory (or theory of algebraic hypercompositional structures). For further reference see, for example, books [20,21]. A *hypergroupoid* is a pair $(H, *)$, where H is a nonempty set and the mapping $* : H \times H \longrightarrow \mathcal{P}^*(H)$ is a binary *hyperoperation* (or *hypercomposition*) on H (here $\mathcal{P}^*(H)$ denotes the system of all nonempty subsets of H). If $a * (b * c) = (a * b) * c$ holds for all $a, b, c \in H$, then $(H, *)$ is called a *semihypergroup*. If moreover the reproduction axiom, i.e., relation $a * H = H = H * a$ for all $a \in H$, is satisfied, then the semihypergroup $(H, *)$ is called *hypergroup*. Unlike in groups, in hypergroups neutral elements or inverses need not be unique. By a neutral element, or an *identity* (or *unit*) of $(H, *)$ we mean such an element $e \in H$ that $e \in x * e \cap e * x$ for all $x \in H$ while by an *inverse* of $a \in H$ we mean such an element $a' \in H$ that there exists an identity $e \in H$ such that $e \in a * a' \cap a' * a$. By *idempotence* in the sense of hypercompositional structures we mean that $a \in a * a$, i.e., that the element is included in its "second power" (which is, in general, a set).

Numerous notions of algebraic structures can be generalized for algebraic hypercompositional structures while some hypercompositional notions have no counterparts in algebraic structures. One of the key algebraic concepts, *cyclicity*, can be transferred to theory of algebraic hypercompositional structures in several ways. For a complex discussion of these approaches as well as their historical context and evolution and clarification of naming and notation, see Novák, Křehlík and Cristea [15].

In our paper we will work with the following definition introduced by Vougiouklis [16] (reworded as in [15]); for more results regarding cyclic hypergroups see Vougiouklis [22].

Definition 1 ([15,16]). *A hypergroup $(H, \circ)$ is called* cyclic *if, for some $h \in H$, there is*

$$H = h^1 \cup h^2 \cup \ldots \cup h^n \cup \ldots, \tag{2}$$

where $h^1 = \{h\}$ and $h^m = \underbrace{h \circ \ldots \circ h}_{m}$. If there exists $n \in \mathbb{N}$ such that Formula (2) is finite, we say that H is a cyclic hypergroup with finite period*; otherwise, H is a cyclic hypergroup with* infinite period. *The element $h \in H$ in Formula (2) is called* generator *of H, the smallest power n for which Formula (2) is valid is called* period *of h. If all generators of H have the same period n, then H is called* cyclic with period n. *If, for a given generator h, Formula (2) is valid but no such n exists (i.e., Formula (2) cannot be finite), then H is called* cyclic with infinite period. *If we can, for some $h \in H$, write*

$$H = h^n. \tag{3}$$

Then, the hypergroup H is called single-power cyclic *with a generator h. If Formula (2) is valid and for all $n \in \mathbb{N}$ and, for a fixed $n_0 \in \mathbb{N}$, $n \geq n_0$ there is*

$$h^1 \cup h^2 \cup \ldots \cup h^{n-1} \subsetneq h^n, \tag{4}$$

then we say that H is a single-power cyclic hypergroup with an infinite period for *h.*

Apart from the notion of cyclicity we will work with the notion of *EL–hyperstructures*, i.e., hypercompositional structures constructed from ordered (or sometimes pre-ordered) semigroups by means of what is known as the "Ends Lemma". For details and applications see, for example, [8–10,12,23].

3. Single-Power Cyclic Hypergroup of Matrices

In order to construct a single-power cyclic hypergroup of matrices, we first, for an arbitrary pair of matrices $\mathbf{A}, \mathbf{B} \in \mathbb{M}_{m,n}(\mathcal{E})$, define a hyperoperation by

$$\mathbf{A} * \mathbf{B} = \{\mathbf{A}, \mathbf{B}\} \cup [\mathbf{A} \circ_m \mathbf{B})_{\leq_M}, \tag{5}$$

where $\mathbf{A} \circ_m \mathbf{B}$ is such a matrix $\mathbf{D}$ that $\mathbf{D} = \{[d_{i,j}] \mid d_{i,j} = \min\{a_{ij}; b_{ij}\}, i \in \{1, \ldots, m\}, j \in \{1, \ldots, n\}\}$ and $[\mathbf{D})_{\leq_M}$ is the set of all matrices greater than $\mathbf{D}$, i.e., $[\mathbf{D})_{\leq_M} = \{\mathbf{X} \in \mathbb{M}_{m,n}(\mathcal{E}) \mid \mathbf{D} \leq_M \mathbf{X}\}$. Thus, using terminology of Chvalina [5], $(\mathbb{M}_{m,n}(\mathcal{E}), *)$ is an *extensive hypergroupoid*, i.e., for an arbitrary pair of matrices $\mathbf{A}, \mathbf{B} \in \mathbb{M}_{m,n}(\mathcal{E})$ there is $\{\mathbf{A}, \mathbf{B}\} \subseteq \mathbf{A} * \mathbf{B}$. Notice that some other authors, motivated by the geometrical meaning, call such hypergroupoids "closed" as contrasted to "open".

Example 2. *For matrices* **A** *and* **B** *from Example 1 we have:*

$$A * B = \left\{ \begin{bmatrix} 8 & 1 & 2 \\ 4 & 0 & 1 \\ 1 & 1 & 1 \end{bmatrix}, \begin{bmatrix} 2 & 1 & 1 \\ 3 & 0 & 3 \\ 5 & 1 & 0 \end{bmatrix} \right\} \cup \left[\begin{bmatrix} 2 & 1 & 1 \\ 3 & 0 & 1 \\ 1 & 1 & 0 \end{bmatrix} \right)_{\leq_M}$$

Remark 1. *With the above example we not only demonstrate the meaning of the hyperoperation "$*$" but also provide an example to the forthcoming Lemma 3. In this respect notice that $\|\mathbf{A}\|_\infty = 11$, $\|\mathbf{B}\|_\infty = 6$ and $\|\mathbf{A} \circ_\mathbf{m} \mathbf{B}\|_\infty = 4$, which means that $A, B \in [A \circ_m B)_{\leq_M}$. Since in our paper we regard $\mathcal{E}$ as a part of $\mathbb{N} \cup \{0\}$, writing the hyperoperation (5) explicitly in the form of union is not neccessary. However, for, as an* *example,* $\mathcal{E} = \{-6, -5, -4 \ldots\}$ *the hyperoperation would no longer be extensive (without explicitly including* $\{A, B\}$*). Indeed,*

$$\mathbf{A} = \begin{bmatrix} 3 & -1 \\ 1 & 1 \end{bmatrix}, \mathbf{B} = \begin{bmatrix} 4 & -6 \\ 2 & 1 \end{bmatrix}; \|\mathbf{A}\|_\infty = 4, \|\mathbf{B}\|_\infty = 10$$

$$\|\mathbf{A} \circ_m \mathbf{B}\|_\infty = \| \begin{bmatrix} 3 & -6 \\ 1 & 1 \end{bmatrix} \|_\infty = 9.$$

The result of the hyperopration is influenced by the absolute value in the calculations of the matrix norm. Since in our paper we restrict ourselves to positive entries of matrices, especially in Section 4 we could omit the two-element set in the definition of the hyperoperation. Yet in our paper we prefer being more general, especially as far as the construction of the hyperoperation is concerned. Since negative values violate extensivity, we prefer including the two-element set in (5). Another way of preserving extensivity would be to modify the row norm by leaving out absolute values. In this respect also notice the result proved by Massouros [24] which says that adding $\{a, b\}$ *to* $a \cdot b$*, where* $(H, \cdot)$ *is a group or a hypergroup, i.e., defining* $a \bullet b = \{a, b\} \cup a \cdot b$*, results in the fact that* $(H, \bullet)$ *is a hypergroup.*

Example 3. *Suppose we have a manufacturing company with two production lines* L_1, L_2*, such that both lines produce products A and B. Consider some specific conditions for production under which the first line* L_1 *only produces* 10 *pieces of A and* 5 *pieces of B per week, and the line* L_2 *only produces* 6 *pieces of A and* 7 *pieces of B per week. This can be denoted by:*

$$P_1 = \begin{bmatrix} 10 & 5 \\ 6 & 7 \end{bmatrix},$$

In the following week, under the same specific conditions, the production can be described by $\mathbf{P}_2 = \begin{bmatrix} 8 & 6 \\ 4 & 8 \end{bmatrix}$. *The norm of the matrix, i.e.,* 15 *in case of* $\mathbf{P}_1$ *and* 13 *in case of* $\mathbf{P}_2$*, describes the production of the better line. The result of the hyperoperation (5), i.e.,*

$$\mathbf{P}_1 * P_2 = \left[\begin{bmatrix} 8 & 5 \\ 4 & 7 \end{bmatrix} \right)_{\leq_M}$$

describes all possibilities for the minimal guaranteed production in case that in future the same conditions repeat. The associativity of the hyperoperation means that if we have more conditions of the same type, their order is not important for the value of the guaranteed minimal production.

Now we include several lemmas which will simplify proofs of our forthcoming theorems. Recall that e stands for the smallest element of $\mathcal{E}$, i.e., the set of entries of matrices in $\mathbb{M}_{m,n}(\mathcal{E})$.

Lemma 1. *The matrix* $\mathbf{E}^* = \begin{bmatrix} e & \ldots & e \\ \vdots & \ddots & \vdots \\ e & \ldots & e \end{bmatrix}$ *is a unit of* $(\mathbb{M}_{m,n}(\mathcal{E}), *)$*. Moreover, there is* $\|\mathbf{E}^*\|_\infty = n \cdot e$*.*

Proof. Obvious because for all $\mathbf{A} \in \mathbb{M}_{m,n}(\mathcal{E})$ holds $\mathbf{A} \in \mathbf{A} * \mathbf{E}^* \cap \mathbf{E}^* * \mathbf{A} = [\mathbf{E}^*)_{\leq_M}$ □

Lemma 2. *Every matrix* $\mathbf{A} \in \mathbb{M}_{m,n}(\mathcal{E})$ *is idempotent (with respect to* $*$*).*

Proof. Obvious because for all $\mathbf{A} \in \mathbb{M}_{m,n}(\mathcal{E})$ holds

$$\begin{bmatrix} a_{11} & \ldots & a_{1n} \\ \vdots & \ddots & \vdots \\ a_{m1} & \ldots & a_{mn} \end{bmatrix} * \begin{bmatrix} a_{11} & \ldots & a_{1n} \\ \vdots & \ddots & \vdots \\ a_{m1} & \ldots & a_{mn} \end{bmatrix} =$$

$$\left[\begin{bmatrix} \min\{a_{11}, a_{11}\} & \ldots & \min\{a_{1n}, a_{1n}\} \\ \vdots & \ddots & \vdots \\ \min\{a_{m1}, a_{m1}\} & \ldots & \min\{a_{mn}, a_{mn}\} \end{bmatrix}\right)_{\leq_M} = \left[\begin{bmatrix} a_{11} & \ldots & a_{1n} \\ \vdots & \ddots & \vdots \\ a_{m1} & \ldots & a_{mn} \end{bmatrix}\right)_{\leq_M}$$

which means that $\mathbf{A} \in \mathbf{A} * \mathbf{A}$. □

Lemma 3. *For an arbitrary pair of matrices $\boldsymbol{A}, \boldsymbol{B} \in \mathbb{M}_{m,n}(\mathcal{E})$, where $\mathcal{E} \subseteq \mathbb{N} \cup \{0\}$, there is*

$$\|\boldsymbol{A} \circ_m \boldsymbol{B}\|_\infty \leq_M \min\left\{\|\boldsymbol{A}\|_\infty, \|\boldsymbol{B}\|_\infty\right\}.$$

Proof. We consider that $\|\mathbf{A}\|_\infty = \max\limits_{1 \leq i \leq m} \sum\limits_{j=1}^{n} |a_{ij}|$, $\|\mathbf{B}\|_\infty = \max\limits_{1 \leq i \leq m} \sum\limits_{j=1}^{n} |b_{ij}|$ and $\|\mathbf{A} \circ_m \mathbf{B}\|_\infty = \max\limits_{1 \leq i \leq m} \sum\limits_{j=1}^{n} |\min\{a_{ij}, b_{ij}\}|$. Then we have

$$\max_{1 \leq i \leq m} \sum_{j=1}^{n} |\min\{a_{ij}, b_{ij}\}| \leq_M \min\{\max_{1 \leq i \leq m} \sum_{j=1}^{n} |a_{ij}|, \max_{1 \leq i \leq m} \sum_{j=1}^{n} |b_{ij}|\}$$

$$\max_{1 \leq i \leq m} \sum_{j=1}^{n} |\min\{a_{ij}, b_{ij}\}| \leq_M \max_{1 \leq i \leq m} \sum_{j=1}^{n} a_{ij} \wedge \max_{1 \leq i \leq m} \sum_{j=1}^{n} |\min\{a_{ij}, b_{ij}\}| \leq_M \max_{1 \leq i \leq m} \sum_{j=1}^{n} b_{ij}$$

For every row of the matrix, i.e., for every $i \in \{1, \ldots, m\}$, there is

$$\sum_{j=1}^{n} |\min\{a_{ij}, b_{ij}\}| \leq_M \sum_{j=1}^{n} |a_{ij}| \wedge \sum_{j=1}^{n} |\min\{a_{ij}, b_{ij}\}| \leq_M \sum_{j=1}^{n} |b_{ij}|.$$

Thus, it is obvious that $\|\mathbf{A} \circ_m \mathbf{B}\|_\infty \leq_M \min\left\{\|\mathbf{A}\|_\infty, \|\mathbf{B}\|_\infty\right\}$ □

Theorem 1. *The extensive hypergroupoid $(\mathbb{M}_{m,n}(\mathcal{E}), *)$ is a commutative hypergroup.*

Proof. Commutativity of the hyperoperation is obvious because the operation min is commutative. Next, we have to show that associativity axiom is satisfied, i.e., that there is $\mathbf{A} * (\mathbf{B} * \mathbf{C}) = (\mathbf{A} * \mathbf{B}) * \mathbf{C}$ for all $\mathbf{A}, \mathbf{B}, \mathbf{C} \in \mathbb{M}_{m,n}(\mathcal{E})$.

We calculate left hand side:

$$\mathbf{A} * (\mathbf{B} * \mathbf{C}) = \bigcup_{\mathbf{X} \in \{\mathbf{B},\mathbf{C}\} \cup [\mathbf{B} \circ_m \mathbf{C})_{\leq_M}} \mathbf{A} * \mathbf{X} = \bigcup_{\mathbf{X} \in \{\mathbf{B},\mathbf{C}\} \cup [\mathbf{B} \circ_m \mathbf{C})_{\leq_M}} \{\mathbf{A}, \mathbf{X}\} \cup [\mathbf{A} \circ_m \mathbf{X})_{\leq_M} =$$
$$\{\mathbf{A}, \mathbf{B}, \mathbf{C}\} \cup [\mathbf{B} \circ_m \mathbf{C})_{\leq_M} \cup [\mathbf{A} \circ_m \mathbf{B})_{\leq_M} \cup [\mathbf{A} \circ_m \mathbf{C})_{\leq_M} \cup \bigcup_{\mathbf{X} \in [\mathbf{B} \circ_m \mathbf{C})_{\leq_M}} [\mathbf{A} \circ_m \mathbf{X})_{\leq_M}.$$

For the right hand side we have:

$$(\mathbf{A}*\mathbf{B})*\mathbf{C}=\bigcup_{\mathbf{Y}\in\{\mathbf{A},\mathbf{B}\}\cup[\mathbf{A}\circ_m\mathbf{B})_{\leq_M}}\mathbf{Y}*\mathbf{C}=\bigcup_{\mathbf{Y}\in\{\mathbf{A},\mathbf{B}\}\cup[\mathbf{A}\circ_m\mathbf{B})_{\leq_M}}\{\mathbf{Y},\mathbf{C}\}\cup[\mathbf{Y}\circ_m\mathbf{C})_{\leq_M}=$$
$$\{\mathbf{A},\mathbf{B},\mathbf{C}\}\cup[\mathbf{A}\circ_m\mathbf{B})_{\leq_M}\cup[\mathbf{A}\circ_m\mathbf{C})_{\leq_M}\cup[\mathbf{B}\circ_m\mathbf{C})_{\leq_M}\cup\bigcup_{\mathbf{Y}\in[\mathbf{A}\circ_m\mathbf{B})_{\leq_M}}[\mathbf{Y}\circ_m\mathbf{C})_{\leq_M}.$$

The left hand side and the right hand side are the same except for the last part of the union. We are going to show that

$$\bigcup_{\mathbf{X}\in[\mathbf{B}\circ_m\mathbf{C})_{\leq_M}}[\mathbf{A}\circ_m\mathbf{X})_{\leq_M}=\bigcup_{\mathbf{Y}\in[\mathbf{A}\circ_m\mathbf{B})_{\leq_M}}[\mathbf{Y}\circ_m\mathbf{C})_{\leq_M}.$$

The following calculation holds for all $i=\{1,\ldots,m\}, j=\{1,\ldots,n\}$:

$$\bigcup_{\mathbf{X}\in[\mathbf{B}\circ_m\mathbf{C})_{\leq_M}}[\mathbf{A}\circ_m\mathbf{X})_{\leq_M}=\bigcup_{(x_{ij})\in\left[\min\{b_{ij};c_{ij}\}\right)_{\leq_M}}[\mathbf{A}\circ_m\mathbf{X})_{\leq_M}=$$
$$\left[min\{a_{ij};\min\{b_{ij};c_{ij}\}\}\right)_{\leq_M}=\left[min\{a_{ij};b_{ij};c_{ij}\}\right)_{\leq_M}=\left[\min\{\min\{a_{ij};b_{ij}\};c_{ij}\}\right)_{\leq_M}=$$
$$\bigcup_{(y_{ij})\in\left[\min\{a_{ij};b_{ij}\}\right)_{\leq_M}}[\mathbf{Y}\circ_m\mathbf{C})_{\leq_M}=\bigcup_{\mathbf{Y}\in[\mathbf{A}\circ_m\mathbf{B})_{\leq_M}}[\mathbf{Y}\circ_m\mathbf{C})_{\leq_M}.$$

Thus the associativity axiom holds, which means that the hypergoupoid $(\mathbb{M}_{m,n}(\mathcal{E}),*)$ is a semihypergroup. Finally, because of extensivity of the hyperoperation (5) we immediately see that reproduction axiom holds as well, i.e., the semihypergroup $(\mathbb{M}_{m,n}(\mathcal{E}),*)$ is an extensive hypergroup. □

Now we can include the result concerning cyclicity of the discussed hypergroup. Notice that since we use n to denote one of the dimensions of the matrices, we will denote period of Definition 1 by p instead of n.

Theorem 2. *If the set $\mathcal{E}$ has the smallest element e, then the hypergroup $(\mathbb{M}_{m,n}(\mathcal{E}),*)$ is single-power cyclic and all matrices containing e (other than $\mathbf{E}^*$) are generators of $\mathbb{M}_{m,n}(\mathcal{E})$ with period $p=3$.*

Proof. The proof is rather straightforward. Denote by $\mathbf{A}_e$ an arbitrary matrix from $\mathbb{M}_{m,n}(\mathcal{E})$ such that at least one of its entries (e.g., $a_{1,2}$) is e. By definition, $\mathbf{A}_e^1=\mathbf{A}_e$. By Lemma 2, we have that $\mathbf{A}_e^2=\mathbf{A}_e*\mathbf{A}_e=[\mathbf{A}_e)_{\leq_M}$. Now, consider such a matrix $\mathbf{B}$, elements of which are different from e at least at those places where $a_{ij}=e$. For example, consider matrix $\mathbf{B}=\begin{bmatrix} e & b_{12} & \ldots & e\\ \vdots & \vdots & \ddots & \vdots\\ e & e & \ldots & e\end{bmatrix}$, where $b_{12}=\max\limits_{1\leq i\leq m}\sum\limits_{j=1}^{n}|a_{ij}|$. Obviously, there is $\mathbf{B}\in[\mathbf{A}_e)_{\leq_M}$. Now, we have

$$\mathbf{A}_e^3=\mathbf{E}^**\mathbf{E}^**\mathbf{E}^*=\mathbf{E}^**[\mathbf{E}^*)_{\leq_M}=\bigcup_{\mathbf{X}\in[\mathbf{E}^*)_{\leq_M}}\mathbf{E}^**\mathbf{X}.$$

Since we know that $\mathbf{B} \in [\mathbf{E}^*)_{\leq_M}$, there is

$$\bigcup_{\mathbf{X}\in[\mathbf{E}^*)_{\leq_M}} \mathbf{E}^* * X \supseteq \mathbf{E}^* * \mathbf{B} =$$

$$\begin{bmatrix} a_{11} & e & \dots & a_{1n} \\ \vdots & \vdots & \ddots & \vdots \\ a_{n1} & a_{n2} & \dots & a_{nn} \end{bmatrix} * \begin{bmatrix} e & b_{12} & \dots & e \\ \vdots & \vdots & \ddots & \vdots \\ e & e & \dots & e \end{bmatrix} = \left[\begin{bmatrix} e & e & \dots & e \\ \vdots & \vdots & \ddots & \vdots \\ e & e & \dots & e \end{bmatrix}\right)_{\leq_M} = [\mathbf{E}^*)_{\leq_M} = \mathbb{M}_{m,n}(\mathcal{E}).$$

Thus we have that $\mathbf{A}_e^3 = \mathbb{M}_{m,n}(\mathcal{E})$, which means that $\mathbb{M}_{m,n}(\mathcal{E})$ is single-power cyclic with period $p = 3$ with generators being all matrices of the form $\mathbf{A}_e$. □

Example 4. *Suppose $\mathcal{E} = \mathbb{N}_0$, i.e., consider the semiring of natural numbers including zero. Then all matrices containing 0 are generators of $\mathbb{M}_{m,n}(\mathbb{N}_0)$. For $m = 5$ and $n = 2$ e.g., matrix $\mathbf{M} = \begin{bmatrix} 50 & 15 & 400 & 3 & 45 \\ 10 & 0 & 89 & 17 & 80 \end{bmatrix}$ generates $\mathbb{M}_{m,n}(\mathbb{N}_0)$. Indeed,*

$$\begin{bmatrix} 50 & 15 & 400 & 3 & 45 \\ 10 & 0 & 89 & 17 & 80 \end{bmatrix} * \begin{bmatrix} 50 & 15 & 400 & 3 & 45 \\ 10 & 0 & 89 & 17 & 80 \end{bmatrix} = \left[\begin{bmatrix} 50 & 15 & 400 & 3 & 45 \\ 10 & 0 & 89 & 17 & 80 \end{bmatrix}\right)_{\leq_M} \tag{6}$$

If we now denote by $\mathbf{D}$ a matrix with $d_{22} = 50 + 15 + 400 + 3 + 45 = 513$ and all other elements zero, then $\mathbf{D} \in [\mathbf{M})_{\leq_M}$. We can see that

$$\begin{bmatrix} 50 & 15 & 400 & 3 & 45 \\ 10 & 0 & 89 & 17 & 80 \end{bmatrix} * \begin{bmatrix} 0 & 0 & 0 & 0 & 0 \\ 0 & 513 & 0 & 0 & 0 \end{bmatrix} =$$

$$\left[\begin{bmatrix} min\{51,0\} & min\{15,0\} & min\{400,0\} & min\{3,0\} & min\{45,0\} \\ min\{10,0\} & min\{0,513\} & min\{89,0\} & min\{17,0\} & min\{80,0\} \end{bmatrix}\right)_{\leq_M} =$$

$$\left[\begin{matrix} 0 & 0 & 0 & 0 & 0 \\ 0 & 0 & 0 & 0 & 0 \end{matrix}\right]_{\leq_M} = \mathbb{M}_{m,n}(\mathbb{N}).$$

Theorem 3. *The unit matrix $\mathbf{E}^* \in \mathbb{M}_{m,n}(\mathcal{E})$ is a generator of $(\mathbb{M}_{m,n}(\mathcal{E}), *)$ with period 2.*

Proof. Proof is obvious to thanks relation "$\leq_M''$, the fact that we regard total order on $\mathcal{E}$ and given the proof of the Lemma 1. We have that $\mathbf{E}^{*2} = \mathbf{E}^* * \mathbf{E}^* = [\mathbf{E}^*)_{\leq_M} = \mathbb{M}_{m,n}(\mathcal{E})$. □

It will be useful to investigate cyclicity of hypergroups $(\mathbb{M}_{m,n}(\mathcal{E}), *)$, where $\mathcal{E}$ is a finite set. Notice that in the case of a finite set $\mathcal{E}$ we cannot construct an analogue of matrix $\mathbf{D}$ with entry d_{22} as we did in Example 4, simply because d_{22} need not be an element of $\mathcal{E}$. Also, since $\mathcal{E}$ is finite and we suppose that it is a chain, its smallest element e always exists.

Theorem 4. *If $\mathcal{E}$ is finite, then the hypergroup $(\mathbb{M}_{m,n}(\mathcal{E}), *)$ is single-power cyclic and all matrices $\mathbf{A}_e$ containing the smallest element of $\mathcal{E}$, denoted by e, are generators of $(\mathbb{M}_{m,n}(\mathcal{E}), *)$ with period $p = 2 + m$, where m is the number of columns of $\mathbf{A}_e$.*

Proof. The proof is analogous to the proof of Theorem 2, except for the row of $\mathbf{A}_e$ containing element e. We construct matrix $\mathbf{B}$ in the following way: the row of $\mathbf{B}$ which in $\mathbf{A}_e$ contains e, will consist of m copies of the greatest elements of $\mathcal{E}$, denoted u, while all other entries of $\mathbf{B}$ will be equal to e. In this way we have $||\mathbf{A}||_\infty \leq_M ||\mathbf{B}||_\infty$, i.e., $\mathbf{B} \in [\mathbf{A}_e)_{\leq_M}$. We calculate once again, now $\mathbf{A} * \mathbf{B} = [\mathbf{C})_{\leq_M}$:

$$\begin{bmatrix} a_{11} & e & \dots & a_{1m} \\ \vdots & \vdots & \ddots & \vdots \\ a_{n1} & a_{n2} & \dots & a_{nm} \end{bmatrix} * \begin{bmatrix} u & u & \dots & u \\ \vdots & \vdots & \ddots & \vdots \\ e & e & \dots & e \end{bmatrix} = \left[\begin{bmatrix} u & e & \dots & u \\ \vdots & \vdots & \ddots & \vdots \\ e & e & \dots & e \end{bmatrix}\right)_{\leq_M}.$$

Next, we reorganize entries in the first row, which does not affect the row norm. We get

$$\begin{bmatrix} a_{11} & e & \dots & a_{1m} \\ \vdots & \vdots & \ddots & \vdots \\ a_{n1} & a_{n2} & \dots & a_{nm} \end{bmatrix} * \begin{bmatrix} e & u & \dots & u \\ \vdots & \vdots & \ddots & \vdots \\ e & e & \dots & e \end{bmatrix} = \left[\begin{bmatrix} e & e & \dots & u \\ \vdots & \vdots & \ddots & \vdots \\ e & e & \dots & e \end{bmatrix}\right)_{\leq_M}.$$

Now it is obvious that after we do this procedure $2+m$ times, we obtain matrix $\mathbf{E}^*$, for which there is $[\mathbf{E}^*)_{\leq_M} = \mathbb{M}_{m,n}(\mathcal{E})$. □

Example 5. *Consider* $\mathcal{E} = \{1, 2, \dots, 10\}$ *and* $n = 2, m = 3$. *In this case matrix* $\mathbf{A} = \begin{bmatrix} 10 & 1 & 10 \\ 10 & 10 & 10 \end{bmatrix}$ *is a generator of* $(\mathbb{M}_{m,n}(\mathcal{E}), *)$ *with period* $p = 5$. *Indeed, we calculate:*

$$\mathbf{A}^2 = \begin{bmatrix} 10 & 1 & 10 \\ 10 & 10 & 10 \end{bmatrix} * \begin{bmatrix} 10 & 1 & 10 \\ 10 & 10 & 10 \end{bmatrix} = \left[\begin{bmatrix} 10 & 1 & 10 \\ 10 & 10 & 10 \end{bmatrix}\right)_{\leq_M} \ni \begin{bmatrix} 10 & 10 & 10 \\ 1 & 1 & 1 \end{bmatrix} = \boldsymbol{B}$$

It is obvious, following from the use of the row norm, that $\mathbf{B} \in \mathbf{A}^2$. *We get*

$$\mathbf{A}^3 = \bigcup_{\mathbf{X} \in \mathbf{A}^2} \mathbf{X} * \mathbf{A} \supseteq \begin{bmatrix} 10 & 1 & 10 \\ 10 & 10 & 10 \end{bmatrix} * \begin{bmatrix} 10 & 10 & 10 \\ 1 & 1 & 1 \end{bmatrix} = \left[\begin{bmatrix} 10 & 1 & 10 \\ 1 & 1 & 1 \end{bmatrix}\right)_{\leq_M} \ni \begin{bmatrix} 1 & 10 & 10 \\ 1 & 1 & 1 \end{bmatrix}$$

$$\mathbf{A}^4 = \bigcup_{\mathbf{Y} \in \mathbf{A}^3} \mathbf{Y} * \mathbf{A} \supseteq \begin{bmatrix} 10 & 1 & 10 \\ 10 & 10 & 10 \end{bmatrix} * \begin{bmatrix} 10 & 1 & 10 \\ 10 & 10 & 10 \end{bmatrix} = \left[\begin{bmatrix} 1 & 1 & 10 \\ 1 & 1 & 1 \end{bmatrix}\right)_{\leq_M} \ni \begin{bmatrix} 1 & 10 & 1 \\ 1 & 1 & 1 \end{bmatrix}$$

$$\mathbf{A}^5 = \bigcup_{\mathbf{Z} \in \mathbf{A}^4} \mathbf{Z} * \mathbf{A} \supseteq \begin{bmatrix} 10 & 1 & 10 \\ 10 & 10 & 10 \end{bmatrix} * \begin{bmatrix} 1 & 10 & 1 \\ 1 & 1 & 1 \end{bmatrix} = \left[\begin{bmatrix} 1 & 1 & 1 \\ 1 & 1 & 1 \end{bmatrix}\right)_{\leq_M}$$

And we see that $\mathbf{A}^5 = \mathbb{M}_{m,n}(\mathcal{E})$.

Remark 2. *The generators described by Theorem 4 are neither only ones nor with the smallest period. Indeed, if in Example 5 we consider matrix* $\mathbf{C} = \begin{bmatrix} 1 & 2 & 10 \\ 1 & 1 & 1 \end{bmatrix}$, *there is* $\mathbf{D} = \begin{bmatrix} 10 & 1 & 2 \\ 1 & 1 & 1 \end{bmatrix} \in \mathbf{C}^2$. *Then* $\mathbf{C}^3 \supseteq \mathbf{C} * \mathbf{D} = \left[\begin{bmatrix} 1 & 1 & 2 \\ 1 & 1 & 1 \end{bmatrix}\right)_{\leq_M} \ni \begin{bmatrix} 1 & 2 & 1 \\ 1 & 1 & 1 \end{bmatrix}$ *and* $\mathbf{C}^4 \supseteq \begin{bmatrix} 1 & 1 & 2 \\ 1 & 1 & 1 \end{bmatrix} * \begin{bmatrix} 1 & 2 & 1 \\ 1 & 1 & 1 \end{bmatrix} = \left[\begin{bmatrix} 1 & 1 & 1 \\ 1 & 1 & 1 \end{bmatrix}\right)_{\leq_M} = \mathbb{M}_{m,n}(\mathcal{E})$ *and we see that* $\mathbf{C}$ *is a generator of* $(\mathbb{M}_{m,n}(\mathcal{E}), *)$ *with period* $p = 4$.

4. Approximation Space Determined by the Cyclic Hypergroup

Now we rewrite some basic terminology of the rough set theory introduced by Pavlak [3] into our notation.

Let $\mathbb{M}_{m,n}(\mathcal{E})$ be a certain set called the universe, and let R_M be an equivalence relation on $\mathbb{M}_{m,n}(\mathcal{E})$. The pair $\mathfrak{A} = (\mathbb{M}_{m,n}(\mathcal{E}), R_M)$ will be called an *approximation space*. We will call R_M an *indiseernibility relation*. If $\mathbf{A}, \mathbf{B} \in \mathbb{M}_{m,n}(\mathcal{E})$ and $(\mathbf{A}, \mathbf{B}) \in R_M$, we will say that $\mathbf{A}$ and $\mathbf{B}$ are *indistinguishable* in $\mathfrak{A}$. Subsets of $\mathbb{M}_{m,n}(\mathcal{E})$ will be denoted by X, Y, Z, possibly with indices. The empty set will be denoted by 0, and the universe $\mathfrak{U}$ will also be denoted by 1. Equivalence classes of the relation R_M will be called *elementary sets (atoms)* in $\mathfrak{A}$ or, briefly, *elementary sets*. The set of all atoms in $\mathfrak{A}$ will be denoted by $(\mathbb{M}_{m,n}(\mathcal{E}), R_M)$. We assume that the empty set is also elementary in every $\mathfrak{A}$. Every finite union of elementary sets in $\mathfrak{A}$ will be called a composed set in $\mathfrak{A}$, or in short, a *composed set*. The family of all composed sets in $\mathfrak{A}$ will be denoted as $Com(\mathfrak{A})$. Obviously, $Com(\mathfrak{A})$ is a Boolean algebra, i.e., the family of all composed set is closed under intersection, union, and complement of sets.

Now, let X be a certain subset of $\mathbb{M}_{m,n}(\mathcal{E})$. The least composed set in $\mathfrak{A}$ containing X will be called the *best upper approximation* of X in $\mathfrak{A}$, in symbols $\overline{Apr}_A(X)$; the greatest composed set in $\mathfrak{A}$ contained in X will be called the *best lower approximation* of X in $\mathfrak{A}$, in symbols $\underline{Apr}_A(X)$. If $\mathfrak{A}$ is known, instead of $\overline{Apr}_A(X)$ ($\underline{Apr}_A(X)$) we will write $\overline{Apr}(X)(\underline{Apr}(X))$, respectively. The set $Bnd_A(X) = \overline{Apr}_A(X) - \underline{Apr}_A(X)$ (in short $Bnd(X)$) will be called the *boundary* of X in $\mathfrak{A}$.

Definition 2 ([20]). *Let H be a set and R be an equivalence relation on H. Let A be subset of H. A* rough set *is a pair of subsets $(\overline{R}(A), \underline{R}(A))$ of H which approximates A as closer as possible from outside and inside, respectively:*

$$\overline{R}(A) = \bigcup_{R(x)\cap A\neq\emptyset} R(x)$$
$$\underline{R}(A) = \bigcup_{R(x)\subseteq A} R(x)$$

Example 6. *Let $S = \{-12\ldots, -2, -1, 0, 1, 2, \ldots 12\}$ and R be defined on S by:*

$$aRb \text{ iff } a \equiv b \mod 6.$$

In this way we obtain the following decomposition of S:

$$R(0) = \{-12, -6, 0, 6, 12, \}$$
$$R(1) = \{-7, -1, 1, 7, \}$$
$$R(2) = \{-8, -2, 2, 8, \}$$
$$R(3) = \{-9, -3, 3, 9, \}$$
$$R(4) = \{, -10, -4, 4, 10, \}$$
$$R(5) = \{-11, -5, 5, 11, \}$$

Now, consider $A = \{-8, -3, -2, 1, 2, 8, 9\} \subset S$. Then

$$\overline{R}(A) = \bigcup_{R(x)\cap A\neq\emptyset} R(x) = R(1) \cup R(2) \cup R(3),$$
$$\underline{R}(A) = \bigcup_{R(x)\subseteq A} R(x) = R(2).$$

In what follows we consider square matrices, i.e., $\mathbb{M}_{n,n}(\mathcal{E})$ only.

In order to study links between rough sets and the above cyclic hypergroup $(\mathbb{M}_{m,n}(\mathcal{E}), *)$, or rather its special case $(\mathbb{M}_{n,n}(\mathcal{E}), *)$, we need to define a new relation R_M on $\mathbb{M}_{n,n}(\mathcal{E})$. For all $\mathbf{A}, \mathbf{B} \in \mathbb{M}_{n,n}(\mathcal{E})$ we define:

$$\mathbf{A} R_M \mathbf{B} \text{ if } \|\mathbf{A}\|_\infty = \|\mathbf{B}\|_\infty \text{ and } tr(\mathbf{A}) = tr(\mathbf{B}). \tag{7}$$

It is obvious that such a relation is reflexive, transitive and symmetric. In this way we obtain a decomposition of $\mathbb{M}_{n,n}(\mathcal{E})$ into equivalence classes by row norm of matrices and their traces.

We denote by $\mathbb{M}_{n,n}(\mathcal{SE}_{\mathbf{E}})$ an arbitrary subset of $\mathbb{M}_{n,n}(\mathcal{E})$ with entries from $\mathcal{SE}_{\mathbf{E}^*}$, where $\mathcal{SE}_{\mathbf{E}^*}$ is a set generated by the principal beginning $_{\leq}(x]$, where $x \in \mathcal{E}$ and $\leq$ is the total order defined on $\mathcal{E}$ which we had already regarded. To sum up,

$$\mathcal{SE}_{\mathbf{E}} = {}_{\leq}(x] = \{y \in \mathcal{E} : y \leq x\}. \tag{8}$$

Theorem 5. *Every hypergroup $(\mathbb{M}_{n,n}(\mathcal{SE}_{\mathbf{E}}), *)$, where $\mathcal{SE}_{\mathbf{E}^*}$ is defined by (8), is an invertible subhypergroup of the hypergroup $(\mathbb{M}_{n,n}(\mathcal{E}), *)$.*

Proof. Recall that for and invertible subhypergroup A of a hypergroup H there holds $y \in A \circ x \Rightarrow x \in A \circ y$ for every $x, y \in H$.

Consider now an arbitrary set $\mathbb{M}_{n,n}(\mathcal{SE}_{\mathbf{E}})$, where $\mathcal{SE}_{\mathbf{E}^*}$ is defined by (8). It is obvious that $\mathbf{E}^* \in \mathbb{M}_{n,n}(\mathcal{SE}_{\mathbf{E}})$. Then we have that $\mathbf{A} \in \mathbb{M}_{n,n}(\mathcal{SE}_{\mathbf{E}}) * \mathbf{B}$ for all $\mathbf{A}, \mathbf{B} \in \mathbb{M}_{n,n}(\mathcal{E})$, this is because $\mathbb{M}_{n,n}(\mathcal{SE}_{\mathbf{E}}) * \mathbf{B} = \bigcup_{\mathbf{X} \in \mathbb{M}_{n,n}(\mathcal{SE}_{\mathbf{E}})} \mathbf{X} * \mathbf{B} = [\mathbf{E}^*)_{\leq_M}$. By the proof of Theorem 3 we have that $\mathbf{A} \in [\mathbf{E}^*)_{\leq_M}$. For $\mathbf{B} \in \mathbb{M}_{m,n}(\mathcal{SE}_{\mathbf{E}}) * \mathbf{A}$ the proof is the same. Thus we obtain that $(\mathbb{M}_{n,n}(\mathcal{SE}_{\mathbf{E}}), *)$ is an invertible subhypergroup of $(\mathbb{M}_{n,n}(\mathcal{E}), *)$. □

By applying Theorems 4 and 5 we immediately obtain the following corollary.

Corollary 1. *Every $(\mathbb{M}_{n,n}(\mathcal{SE}_{\mathbf{E}}), *)$ is single power cyclic.*

In the paper we assume that $\mathcal{E}$ is a chain and an equivalence R_M. As a result we can consider the set $\mathbb{M}_{n,n}(\mathcal{SE}_{\mathbf{E}})$ as a suitable set for constructing lower and uper approximations, i.e., $\underline{R_M}(\mathbb{M}_{n,n}(\mathcal{SE}_{\mathbf{E}}))$ and $\overline{R_M}(\mathbb{M}_{n,n}(\mathcal{SE}_{\mathbf{E}}))$. When discussing our system $(\mathbb{M}_{n,n}(\mathcal{E}), R_M)$, we can see that every subset $\mathbb{M}_{n,n}(\mathcal{SE}_{\mathbf{E}})$ is in the beginning of the system, i.e., the class with the smallest trace and the smallest row norm of the matrix is included in the lower and uper approximations.

Notation 1. *Our results regarding rough sets are visualised by means of figures. In all figures, a* white *square means that there does not exist any matrix with the given properties created by the approximation set, a* coloured *square means that all matrices with the given properties belong to the approximation set, and a* partially coloured *square that there exists at least one matrix with the given properties which belong to the approximation set and at least one matrix which does not belong to it.*

Example 7. *Consider $\mathcal{E} = (\mathbb{N} \cup \{0\})$ and $\mathcal{SE}_{\mathbf{E}^*} = \{0,1\}$, i.e., $(\mathbb{M}_{2,2}(\{0,1\}))$ is the set of 2×2 Boolean matrices. Then $(\mathbb{M}_{2,2}(\{0,1\}), *)$ is invertible in $(\mathbb{M}_{2,2}(\mathcal{E}), *)$. Since $\mathbf{E}^* \in \mathbb{M}_{2,2}(\{0,1\})$, we have, for an arbitrary pair of matrices $\mathbf{A}, \mathbf{B} \in \mathbb{M}_{2,2}(\mathcal{E})$ that $\mathbf{B} * \mathbb{M}_{2,2}(\{0,1\}) = \mathbb{M}_{2,2}(\mathcal{E}) \ni \mathbf{A}$ and $\mathbf{A} * \mathbb{M}_{2,2}(\{0,1\}) = \mathbb{M}_{2,2}(\mathcal{E}) \ni \mathbf{B}$. Moreover,*

$$\overline{R_M}((\mathbb{M}_{2,2}(\{0,1\}))) = R_{[0,0]} \cup R_{[1,0]} \cup R_{[1,1]} \cup R_{[1,2]}, \cup R_{[2,1]} \cup R_{[2,2]}$$
$$\underline{R_M}((\mathbb{M}_{2,2}(\{0,1\}))) = R_{[0,0]} \cup R_{[1,0]} \cup R_{[1,1]} \cup R_{[1,2]}$$

The upper approximation, i.e., the set $\overline{R_M}((\mathbb{M}_{n,n}(\{0,1\})$, is in Figure 1 visualized as the union of all squares which include some coloured parts. The lower approximation, i.e., the set $\underline{R_M}((\mathbb{M}_{n,n}(\{0,1\}))$, is the union of all square which are fully coloured.

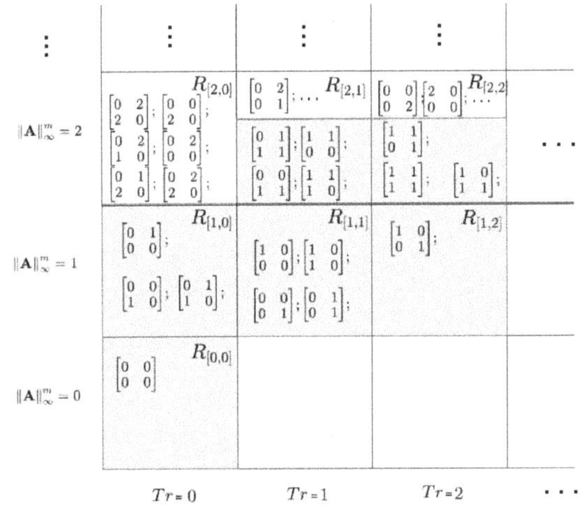

Figure 1. The lower and upper approximation with relation R_M for matrices 2×2.

Theorem 6. *Let* $\mathcal{E} = \mathbb{N}_0$ *and* $\mathbb{M}_{n,n}(\mathcal{SE}_\mathbf{E})$ *be invertible hypergroups in* $\mathbb{M}_{n,n}(\mathcal{E})$*, where* $\mathcal{SE}_{\mathbf{E}^*}$ *is defined by* $\mathcal{SE}_{\mathbf{E}^*} = (x]_\leq = \{y \in \mathcal{E} : y \leq x\}$ *for* $x = 1$, $n \in \mathbb{N}$*, with relation* R_M*. Then there is*

$$\underline{R_M}(\mathbb{M}_{n,n}(\mathcal{SE}_\mathbf{E})) = \bigcup_{i=1}^{n} R_{[1,i]} \cup R_{[0,0]}$$

$$\overline{R_M}(\mathbb{M}_{n,n}(\mathcal{SE}_\mathbf{E})) = \bigcup_{i=1}^{n} \bigcup_{j=1}^{n} R_{[i,j]} \cup \bigcup_{j=0}^{n-1} R_{[0,j]}$$

where n is the size of the matrix.

Proof. For the proof we use the idea of Example 7 and Figure 1, in which the lower approximation is the union of all squares which are fully coloured while the upper approximation is the union of all squares which contain some coloured parts. Notice that in Figure 1 rows indicate matrices with the same row norm (counted from the bottom) while columns (counted from the left) indicate matrices with the same trace. When one realizes how such a scheme in constructed, the proof becomes obvious. Indeed, take e.g., $n = 3$ and focus on Figure 2 and (9).

$$\overline{R_M}((\mathbb{M}_{3,3}(\{0,1\})) = R_{[0,0]} \cup R_{[1,0]} \cup R_{[1,1]} \cup R_{[1,2]} \cup R_{[1,3]} \cup R_{[2,0]} \cup R_{[2,1]} \cup R_{[2,2]} \cup R_{[2,3]} \cup R_{[3,1]} \cup R_{[3,2]} \cup R_{[3,3]} \quad (9)$$

$$\underline{R_M}((\mathbb{M}_{3,3}(\{0,1\})) = R_{[0,0]} \cup R_{[1,0]} \cup R_{[1,1]} \cup R_{[1,2]} \cup R_{[1,3]} \quad (10)$$

□

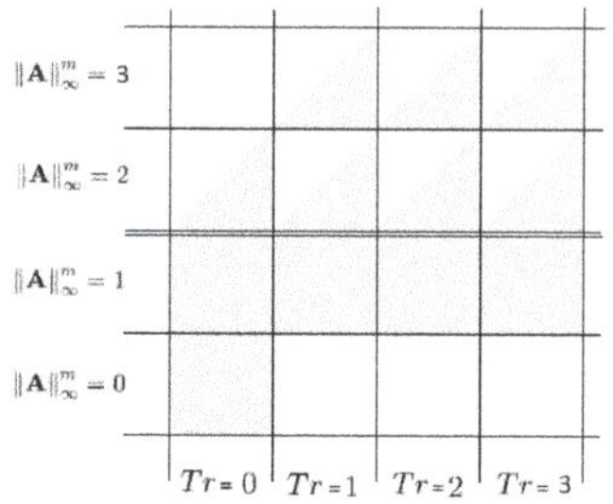

Figure 2. The lower and upper approximation with relation R_M, for matrices 3×3.

Remark 3. *Note that the boundary, defined as the difference between the upper and the lower approximation, expands across the universe so that it has only two classes for $n = 2$ and 7 classes for $n = 3$ and 14 classes for $n = 4$ and 23 classes for $n = 5$, i.e., $n^2 - 2$ classes for an $n \times n$ matrix. Also, the upper approximation has $n \cdot (n+1)$ classes and the lower approximation has $n + 2$ classes. All these formulas can be easily seen in Figures 1 and 2.*

In what follows, we will consider $\overline{R_M}(A) = \overline{R_M}(\mathbf{A} * \mathbf{B})$ and $\underline{R_M}(A) = \underline{R_M}(\mathbf{A} * \mathbf{B})$, where $\mathbf{A}, \mathbf{B} \in \mathbb{M}_{n,n}(\mathcal{E})$. In this way the construction can be considered as a dynamic system, where the set X is the hyperoperproduct defined by (5).

Corollary 2. *For $X = \boldsymbol{A} * \boldsymbol{B}$ there is $\overline{R_M}(X) = \underline{R_M}(X)$. Moreover,*

$$\overline{R_M}(X) = \bigcup_{j \geq \|\boldsymbol{A} * \boldsymbol{B}\|_\infty} \ \bigcup_{i \in \{0,1,2\dots\infty\}} R_{ij}.$$

Now we will use the relation R_M to present some basic and natural properties of the lower and upper approximations with respect to the above mentioned theorem. Recall that matrices $\mathbf{A}_e$ are generators of the cyclic hypergroup $\mathbb{M}_{m,n}(\mathcal{E})$.

Theorem 7. *The following properties hold:*

(1) $\overline{R_M}(\boldsymbol{E}^* * \boldsymbol{E}^*) = \mathbb{M}_{m,n}(\mathcal{E}) = \underline{R_M}(\boldsymbol{E}^* * \boldsymbol{E}^*)$
(2) $\underline{R_M}(\boldsymbol{A}_e) \subseteq \underline{R_M}(\boldsymbol{A}_e^2) \subseteq \underline{R_M}(\boldsymbol{A}_e^3) = \mathbb{M}_{m,n}(\mathcal{E})$
(3) $\overline{R_M}(\boldsymbol{A}_e) \subseteq \overline{R_M}(\boldsymbol{A}_e^2) \subseteq \overline{R}(\boldsymbol{A}_e^3) = \mathbb{M}_{m,n}(\mathcal{E})$

Recall that when defining the relation R_M in (7) we used the row norm and trace. Let us now modify the definition to make use of the row norm and determinant. We define:

$$\mathbf{A} R_D \mathbf{B} \text{ if } \|\mathbf{A}\|_\infty = \|\mathbf{B}\|_\infty \text{ and } |\mathbf{A}| = |\mathbf{B}|. \tag{11}$$

If we use the relation R_D instead of R_M, Figure 1 changes to Figure 3. This results in the following theorem.

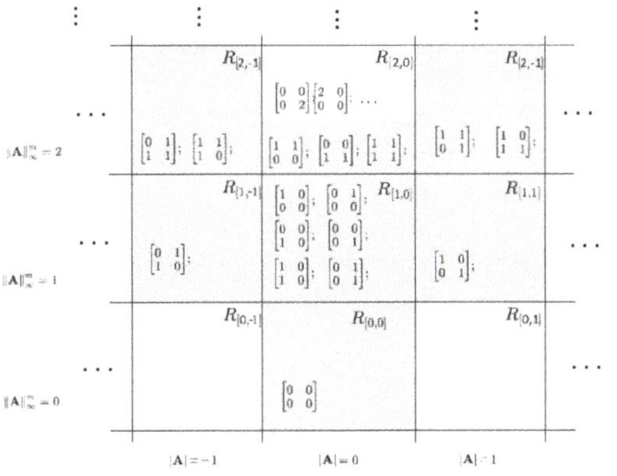

Figure 3. The lower and upper approximation with relation R_D for matrices 2×2.

Theorem 8. *Let* $\mathcal{E} = \mathbb{N}_0$ *and* $\mathbb{M}_{n,n}(\mathcal{SE}_\mathbf{E})$ *be invertible hypergroups in* $\mathbb{M}_{n,n}(\mathcal{E})$*, where* $\mathcal{SE}_{\mathbf{E}^*}$ *is defined by* $\mathcal{SE}_{\mathbf{E}^*} = (x]_\leq = \{y \in \mathcal{E} : y \leq x\}$ *for* $x = 1, n \in \mathbb{N}$*, with relation* R_D*. Than there is*

$$\underline{R_D}(\mathbb{M}_{n,n}(\mathcal{SE}_\mathbf{E})) = R_{[0,0]} \cup R_{[1,-1]} \cup R_{[1,0]} \cup R_{[1,1]} \cup R_{[-2,1]} \cup R_{[2,1]}$$

where n is the size of the matrix.

Proof. We use Figures 3 and 4 in the proof. For an arbitrary size of the matrix, n, the lower approximation always consists of 6 classes of equivalence. Obviously, for an arbitrary n and $\mathcal{E} = \{0, 1\}$ there exists only one matrix (the null one), which is in $R_{[0,0]}$, see the coloured square. For norm equal to 1, the proof is again obvious because thanks to the norm in every matrix of size n there can be maximum one 1 in every row, which means that the determinant of such a matrix can only be -1, 0 or 1. For norm 2 there exists, for an arbitrary size n a matrix with element 2 and zero determinant. Also, there exists a matrix which has one row with two 1's. If we repeat such a row, the determinant is zero, see the partially coloured square. At the same time no matrix with 2 as an entry can be in a class with norm 2 and determinant 1 oe -1 because in that case the respective row must contain only zeros as other entries. If we now expand the determinant with respect to that row, the calculation will include $2s$, where s is a subdeterminant and $s \in \mathbb{N}$, which means that the determinant can never be 1 or -1. For norm 3 we get that the numbers of matrices can include 2 and 1, which means that the calculation of the determinant will include a difference, i.e., we can get an arbitrary number. In other words, for norm greater than 2 we cannot obtain fully coloured squares, i.e., classes of equivalence which consist only of matrices with entries 0 and 1. □

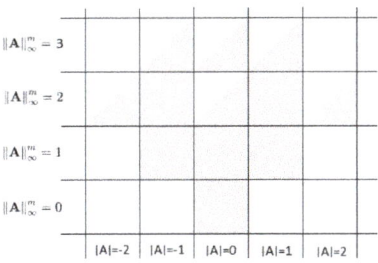

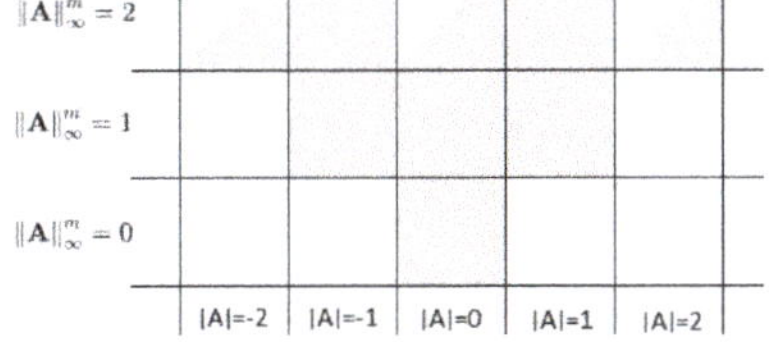

Figure 4. The lower and upper approximation with relation R_D for matrices 3×3.

Finding a general rule for the upper approximation when using the determinant is not as easy as it may seem. Even though it seems that the whole universum "behaves accordingly", it is not easy to find an algorithm which would easily define the upper approximation. Therefore we at least include a theorem which shows an important property of the upper approximation. At the same time, its proof describes the upper approximation for $n \in \{2, 3, 4, 5\}$.

Theorem 9. *For $\mathcal{E} = \mathbb{N}_0$, $\mathbb{M}_{n,n}(\mathcal{SE}_{\mathbf{E}})$ and $n \in \{2, 3, 4, 5\}$, the following holds:*

$$\overline{R_D}(\mathbb{M}_{2,2}(\mathcal{SE}_{\mathbf{E}^*})) \subset \overline{R_D}(\mathbb{M}_{3,3}(\mathcal{SE}_{\mathbf{E}^*})) \subset \overline{R_D}(\mathbb{M}_{4,4}(\mathcal{SE}_{\mathbf{E}^*})) \subset \overline{R_D}(\mathbb{M}_{5,5}(\mathcal{SE}_{\mathbf{E}^*}))$$

Proof. For the proof we use Figures 3 and 5. Results for $n = 4$ and $n = 5$, which had been computed by software means, are included below. □

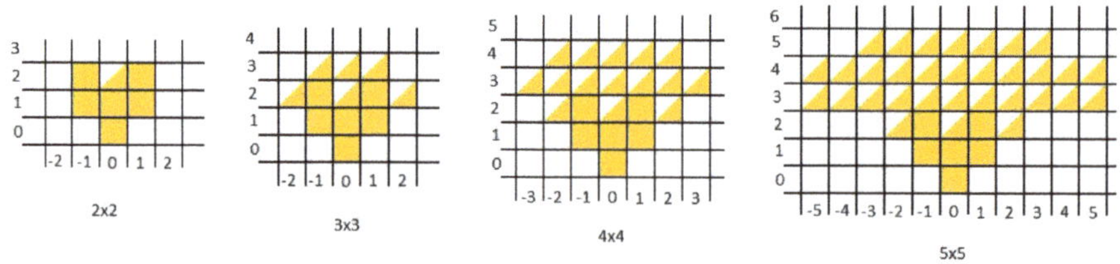

Figure 5. The lower and upper approximation with relation R_D for matrices 2×2–5×5.

Application in the Control Theory

Application of rough set theory is widely used in information technologies. This approach has fundamental importance in knowledge acquisition, cognitive science, pattern recognition, machine learning, database systems, etc. Rough sets are also used in sensors mapping in robotics.

If we examine how an autonomous mobile robot can get from point A to point B, we realize that it must have information about obstacles in front of itself to avoid collision. To find this out it uses sensors, mostly *camera and LIDAR (an abbreviation of "Light Detection And Ranging") for environment mapping* [25,26].

We can easily find surjective function from a set of $\mathbb{M}_{n,n}(\mathcal{SE}_{\mathbf{E}})$ matrices (our model) to coordinates $[x, y]$, where $x, y \in \mathbb{Z}$. These are coordinates of occupancy grid [26] cells (robotics usage), which must meet the following condition:

$$[|M|, ||M||] = [floor(x_r), floor(y_r)], \tag{12}$$

where $x_r, y_r \in \mathbb{R}$ are real coordinates that represents the interval/size of the cells in occupancy grid (see Figure 6). This process is called *quantization*. We can mark the occupancy grid as *equivalence classes*.

Both sensors can be attached to specific places on the top of robot. (Notice that all visualised information from sensors what we are working with are projected from 3D to 2D plane because of assumption that the robot can move only in 2 directions.) In our figures (for $n = 4$, for example) the *lower approximation* describes the robot's shape and pose. We can be sure for 100% occupation of these cells because of the fusion of sensor data with the known relative position between the sensors and the robot (its shape) in particular. As we can see, this approximation will never change depending on change of n (matrix size). This lower approximation is typical for an industrial warehouse robot.

Next part of LIDAR/camera scan is the *upper approximation*. This gives us information about all cells which were hit by the sensors. However, we can only say that these cells are occupied with probability between 0 and 1 (never 0 or 1) because this is how all current sensor work. The bigger the n, the bigger the range of sensors. Notice that the sensor scanning angle is *symmetric* by vertical axis of the sensor.

The above example makes use of a single sensor scan. However, we can process these scans into a whole simple mapping algorithm described in [12,26] or improved algorithm via particle filter described in [27].

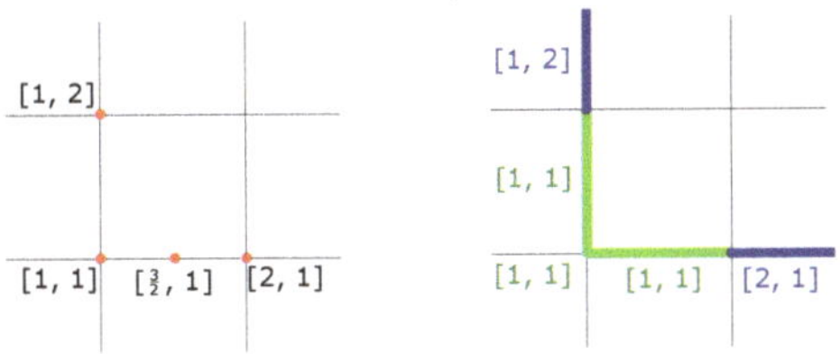

Figure 6. Quantization from real coordinates to equivalence classes.

5. Conclusions

In our paper we deal with a construction of cyclic hypergroups and present their links to rough sets. We construct subsets of such hypergroups which determine the approximation for finding the lower and upper approximation of the rough set. For this we use various types of equivalence relations. In our paper we choose the input of the matrix subhypergroups out of a two element set. Generalizing our results for arbitrary more-element sets is a topic of our further research. The constructions of the lower and upper approximation described in this paper for various n are very interesting, which can be, in case of norm and trace and $\mathcal{E} \in \{0, 1, 2\}$, visualised in a simple figure; see Figure 7. Further research can also focus on finding the approximation set. For this we can look for inspiration in the intersection of the principal end and beginning of the hyperoperation in [10]. As a result we can form an approximation set which need not include the origin of the universe, i.e., $R_{[0,0]}$.

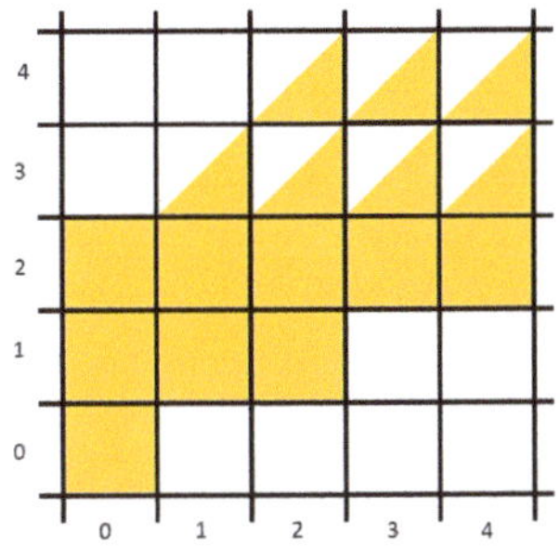

Figure 7. The lower and upper approximation with relation R_M for matrices 2×2 and $\mathcal{E} = \{0, 1, 2\}$.

Author Contributions: Investigation, Š.K. and J.V.; Methodology, Š.K. and J.V.; Software, Š.K. and J.V.; Supervision, Š.K. and J.V.; Writing—original draft, Š.K. and J.V.; Writing—review & editing,Š.K. and J.V. All authors have read and agreed to the published version of the manuscript.

Funding: The first author was supported by the Ministry of Education, Youth and Sports within the National Sustainability Program I, project of Transport R&D Centre (LO1610) and the second author was supported by the FEKT-S-17-4225 grant of Brno University of Technology.

Conflicts of Interest: The authors declare no conflict of interest.

References

1. Zadeh, L.A. Fuzzy sets. *Inf. Control* **1965**, *8*, 338–353. [CrossRef]
2. Pavlak, Z. *Rough Set: Theorerical Aspects of Reasoning Above Data;* Kluver Academic Publishers: Dordrecht, The Netherlands, 1991.

3. Pavlak, Z. Rough set. *Int. J. Comput. Inf. Sci.* **1982**, *11*, 341–356.
4. Molodtsov, D. Soft set—First results. *Comp. Math. Appl.* **1999**, *37*, 19–31. [CrossRef]
5. Chvalina, J. *Functional Graphs, Quasi-ordered Sets and Commutative Hypergroups*; Masaryk University: Brno, Czech Republic, 1995. (In Czech)
6. Hošková, Š.; Chvalina, J.; Račková, P. Transposition hypergroups of Fredholm integral operators and related hyperstructures. Part I. *J. Basic Sci.* **2008** *4*, 43–54.
7. Novák, M. On EL-semihypergroups. *Eur. J. Combin.* **2015**, *44 Pt B*, 274–286.
8. Novák, M.; Cristea, I. Composition in EL-hyperstructures. *Hacet. J. Math. Stat.* **2019**, *48*, 45–58. [CrossRef]
9. Novák, M., Křehlík, Š. EL-hyperstructures revisited. *Soft Comput.* **2018**, *22*, 7269–7280.
10. Křehlík, Š.; Novák, M. From lattices to H_v-matrices. *An. Şt. Univ. Ovidius Constanţa* **2016**, *24*, 209–222. [CrossRef]
11. Leoreanu-Foneta, V. The lower and upper approximation in hypergroup. *Inf. Sci.* **2008** *178*, 3605–3615.
12. Novák, M.; Křehlík, Š.; Staněk, D. n-ary Cartesian composition of automata. *Soft Comput.* **2019**. [CrossRef]
13. Novák, M.; Ovaliadis, K.; Křehlík, Š. A hyperstructure of Underwater Wireless Sensor Network (UWSN) design. In *AIP Conference Proceedings 1978, Proceedings of the International Conference on Numerical Analysis and Applied Mathematics (ICNAAM 2017), The MET Hotel, Thessaloniki, Greece, 25–30 September 2017*; American Institute of Physics: Thessaloniki, Greece, 2018.
14. Račková, P. Hypergroups of symmetric matrices. In Proceedings of the 10th International Congress of Algebraic Hyperstructures and Applications (AHA 2008), At Brno, Czech Republic, September 2008; University of Defence: Brno, Czech Republic, 2009; pp. 267–272.
15. Novák, M.; Křehlík, Š.; Cristea, I. Cyclicity in EL-hypergroups. *Symmetry* **2018**, *10*, 611. [CrossRef]
16. Vougiouklis, T. Cyclicity in a Special Class of Hypergroups. *Acta Universitatis Carolinae Mathematica Physica* **1981**, *22*, 3–6.
17. Skowron, A.; Dutta, S., Rough sets: Past, present, and future. *Nat. Comput.* **2018**, *17*, 855–876 [CrossRef] [PubMed]
18. Anvariyeh, S.M.; Mirvakili, S.; Davvaz, B. Pavlak's approximations in Γ-semihypergroups. *Comput. Math. Appl.* **2010**, *60*, 45–53. [CrossRef]
19. Leoreanu-Foneta, V. Approximations in hypergroups and fuzzy hypergroups. *Comput. Math. Appl.* **2011**, *61*, 2734–2741. [CrossRef]
20. Corsini, P.; Leoreanu, V. *Applications of Hyperstructure Theory*; Kluwer Academic Publishers: Dodrecht, The Netherlands; Boston, MA, USA; London, UK, 2003.
21. Davvaz, B.; Leoreanu–Fotea, V. *Applications of Hyperring Theory*; International Academic Press: Palm Harbor, FL, USA, 2007.
22. Vougiouklis, T. *Hyperstructures and Their Representations*; Hadronic Press: Palm Harbor, FL, USA, 1994.
23. Chvalina, J.; Křehlík, Š.; Novák, M. Cartesian composition and the problem of generalising the MAC condition to quasi-multiautomata. *An. Şt. Univ. Ovidius Constanţa* **2016**, *24*, 79–100.
24. Massouros, C.G. On the semi-sub-hypergroups of a hypergroup. *Int. J. Math. Math. Sci.* **1991**, *14*, 293–304. [CrossRef]
25. Kokovkina, V.A.; Antipov, V.A.; Kirnos, V.P.; Priorov, A.L. The Algorithm of EKF-SLAM Using Laser Scanning System and Fisheye Camera. In Proceedings of the 2019 Systems of Signal Synchronization, Generating and Processing in Telecommunications (SYNCHROINFO), Yaroslavl, Russia, 1–3 July 2019; pp. 1–6.
26. Vu, T.-D.; Aycard, O.; Appenrodt, N. Online Localization and Mapping with Moving Object Tracking in Dynamic Outdoor Environments. In Proceedings of the 2007 IEEE Intelligent Vehicles Symposium, Istanbul, Turkey, 13–15 June 2007; pp.190–195.
27. Zhu, D.; Sun, X.; Wang, L.; Liu, B.; Ji, K. Mobile Robot SLAM Algorithm Based on Improved Firefly Particle Filter. In Proceedings of the 2019 International Conference on Robots & Intelligent System (ICRIS), Haikou, China, 15–16 June 2019; pp. 35–38.

symmetry

MDPI

Article

Hyperhomographies on Krasner Hyperfields

Vahid Vahedi [1], Morteza Jafarpour [1] and Irina Cristea [2,*]

1 Department of Mathematics, Vali-e-Asr University of Rafsanjan, 7718897111 Rafsanjan, Iran; vvahedi@yahoo.com (V.V.); m.j@vru.ac.ir (M.J.)

2 Center for Information Technologies and Applied Mathematics, University of Nova Gorica, 5000 Nova Gorica, Slovenia

* Correspondence: irinacri@yahoo.co.uk

Received: 18 October 2019; Accepted: 19 November 2019; Published: 23 November 2019

check for updates

Abstract: In this paper, we introduce generalized homographic transformations as hyperhomographies over Krasner hyperfields.These particular algebraic hyperstructues are quotient structures of classical fields modulo normal groups. Besides, we define some hyperoperations and investigate the properties of the derived hypergroups and H_v-groups associated with the considered hyperhomographies. They are equipped hyperhomographies obtained as quotient sets of nondegenerate hyperhomographies modulo a special equivalence. Thus the symmetrical property of the equivalence relations plays a fundamental role in this constructions.

Keywords: hypergroup; hyperring; hyperfield; (hyper)homography

MSC: 20N20; 14H52; 11G05

1. Introduction

In a recently published paper [1], the authors have initiated the study of elliptic hypercurves defined on Krasner hyperfields, generalizing the elliptic curves over fields. The main idea consists in substituting the field with a hyperfield, in particular with the associated quotient Krasner hyperfield. The power of this algebraic hyperstructure has been already used in solving different problems in affine algebraic schemes [2], theory of arithmetic functions [3], tropical geometry [4], algebraic geometry [5], etc. The quotient Krasner hyperfield is practically the quotient $\bar{F} = F/G$ of a classical field F by any normal subgroup G of the multiplicative part $(F \setminus \{0\}, \cdot)$. It was introduced by Krasner in 1983 [6] and investigated from the hyperalgebraic point of view mostly by Massouros [7] around 1985. In this new environment, the definition of an elliptic curve over a field F can be naturally extended to the definition of an elliptic hypercurve over a quotient Krasner hyperfield. Besides the group operation on the set of elliptic curves is extended to a hyperoperation on a family of elliptic hypercurves. The properties of the associated hypergroup have been investigated also in relation with the Berardi's cryptographic system [8].

The study developed in this paper goes in the same direction as our recent study. This time we extend a particular quadratic equation in two variables from a field F to a Krasner hyperfield F/G. It is well known that a conic section, which is a curve obtained as the intersection between the surface of a cone and a plane, can be algebraically represented as a quadratic equation with coefficients in a filed, i.e., $g(x,y) = ax^2 + bxy + cy^2 + dx + ey + f = 0$. If $a = c = 0$ and $b \neq 0$, the equation $g(x,y) = 0$ models a *homographic transformation*. Generally, after a suitable change of variables, a homographic transformation $y = \frac{ax+b}{cx+d}$, with $ad - bc \neq 0$, can be written in the form $(X - A)(Y - B) = 1$, where $X = x$, $Y = \frac{1}{\alpha}y$, where $\alpha = \frac{bc-ad}{c^2} \neq 0$, $A = -\frac{d}{c}$, and $B = \frac{a}{\alpha c}$ are elements in the field F. Equivalently, a homography transformation is given by a function $y = f_{a,b}(x) = b + \frac{1}{x-a}$, with a, b elements in a field F. The aim of this paper is to generalize the homography transformation from the field F to the Krasner hyperfield $\bar{F}$. First, we generalize the reduced quadratic forms on Krasner hyperfields. This investigation leads us to introduce the notion of conic hypersection on a Krasner hyperfield. Secondly, using hyperconics, we define some hyperoperations and the associated hyperstructures give us the possibility of studying simultaneously some conics. As in our previous research paper [1], the results can be also applied in cryptography in relation with the Berardi's cryptographic system [8].

2. Preliminaries

We recall here some basic notions of conics and hyperstructures theory also we fix the notations used in this paper. We assign the readers to these topics in the following fundamental books [9–11].

2.1. Conic Sections

A conic is a plane affine curve of degree 2, defined by an irreducible polynomial $g(x,y) = ax^2 + bxy + cy^2 + dx + ey + f = 0$ with coefficients in a field F. Based on the number of the points at infinity (this number can be 2, 1, or 0), the irreducible conics are divided in three categories: hyperbola, parabola and ellipse. Certain sets of points on curves can form an algebraic structure, and till now it is very well known the group structure. Generally the group low on conics is defined over a field F, following the rule illustrated in Figure 1 or Figure 2. In particular, if we take $\mathcal{O}$ an arbitrary point on the conic, then for two arbitrary points p and q on the conic, their sum $p + q$ is obtained as the second point of the intersection with the conic of the parallel line through $\mathcal{O}$ to the line joining p and q. In this case $\mathcal{O}$ is the identity element of the group.

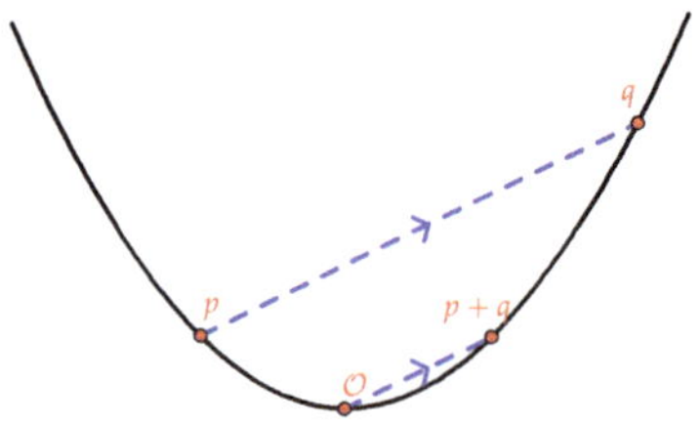

Figure 1. Addition between two points on a curve when the identity element belongs to the curve

If we consider now that the identity element $\mathcal{O}$ is at infinity, then the sum $p + q$ of two arbitrary points p and q on the conic is the image on the conic of the point obtained as intersection with the x-axis of the line passing through p and q, as shown in Figure 2.

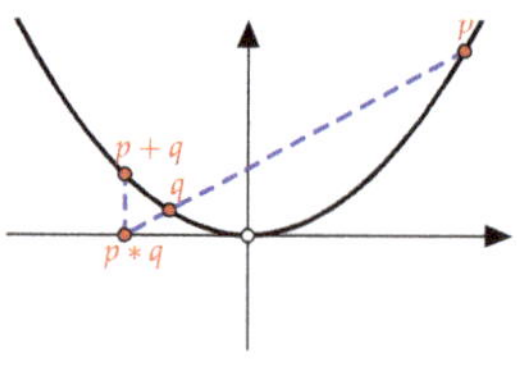

Figure 2. Addition between two points on a curve when the identity element is at infinity.

Example 1. *Consider $f(x) = x^2$ over the finite field $F = \mathbb{Z}_5$. Then we have a parabola in F and the Cayley table of its points $Q_f(F) = \{\mathcal{O}, (1,1), (2,4), (3,4), (4,1)\}$ where, $\mathcal{O} = (0,0)$ is the identity element of the group, is as follows.*

$+$	$\mathcal{O}$	$(1,1)$	$(2,4)$	$(3,4)$	$(4,1)$
$\mathcal{O}$	$\mathcal{O}$	$(1,1)$	$(2,4)$	$(3,4)$	$(4,1)$
$(1,1)$	$(1,1)$	$(2,4)$	$(3,4)$	$(4,1)$	$\mathcal{O}$
$(2,4)$	$(2,4)$	$(3,4)$	$(4,1)$	$\mathcal{O}$	$(1,1)$
$(3,4)$	$(3,4)$	$(4,1)$	$\mathcal{O}$	$(1,1)$	$(2,4)$
$(4,1)$	$(4,1)$	$\mathcal{O}$	$(1,1)$	$(2,4)$	$(3,4)$

If we take the identity element $\mathcal{O}$ at infinity, then the group operation is calculated as in the following Cayley table:

$+$	$\mathcal{O}$	$(1,1)$	$(2,4)$	$(3,4)$	$(4,1)$
$\mathcal{O}$	$\mathcal{O}$	$(1,1)$	$(2,4)$	$(3,4)$	$(4,1)$
$(1,1)$	$(1,1)$	$(3,4)$	$(4,1)$	$(2,4)$	$\mathcal{O}$
$(2,4)$	$(2,4)$	$(4,1)$	$(1,1)$	$\mathcal{O}$	$(3,4)$
$(3,4)$	$(3,4)$	$(2,4)$	$\mathcal{O}$	$(4,1)$	$(1,1)$
$(4,1)$	$(4,1)$	$\mathcal{O}$	$(3,4)$	$(1,1)$	$(2,4)$

The geometrical interpretation of the associativity of the group law is equivalent with a special case of Pascal's theorem, which is a very special case of Bezout's theorem.

Theorem 1. *For any conic and any six points $p_1, p_2, ..., p_6$ on it, the opposite sides of the resulting hexagram, extended if necessary, intersect at points lying on some straight line. More specifically, let $L(p,q)$ denote the line through the points p and q. Then the points $L(p_1,p_2) \cap L(p_4,p_5)$, $L(p_2,p_3) \cap L(p_5,p_6)$, and $L(p_3,p_4) \cap L(p_6,p_1)$ lie on a straight line, called the Pascal line of the hexagon (see Figure 3).*

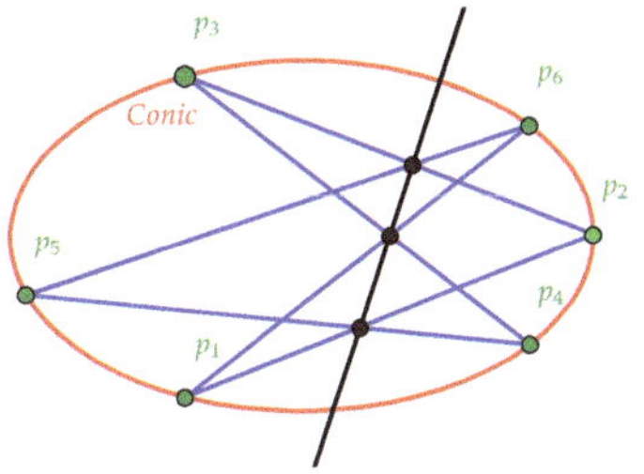

Figure 3. Pascal's theorem.

2.2. *Krasner Hyperrings and Hyperfields*

In this section we briefly recall the main definitions and properties of hyperrings and hyperfields, focussing on the concept of Krasner hyperfield.

Let H be a non-empty set and $\mathcal{P}^*(H)$ be the set of all non-empty subsets of H. Let $\circ$ be a hyperoperation (or join operation) on H, that is, a function from the cartesian product $H \times H$ into $\mathcal{P}^*(H)$. The image of the pair $(a,b) \in H \times H$ under the hyperoperation $\circ$ in $\mathcal{P}^*(H)$ is denoted by $a \circ b$. The join operation can be extended in a natural way to subsets of H as follows: for non-empty subsets A, B of H, define $A \circ B = \bigcup_{a \in A, b \in B} a \circ b$. The notation $a \circ A$ is used for $\{a\} \circ A$ and $A \circ a$ for $A \circ \{a\}$. Generally, we mean $H^k = H \times H \times \ldots \times H$ (k times), for all $k \in \mathbb{N}$ and also the singleton $\{a\}$ is identified with its element a. The hyperstructure $(H, \circ)$ is called a semihypergroup if the hyperoperation is associative, i.e., $(a \circ b) \circ c = a \circ (b \circ c)$ for all $a, b, c \in H$, which means that

$$\bigcup_{u \in a \circ b} u \circ c = \bigcup_{v \in b \circ c} a \circ v.$$

A semihypergroup $(H, \circ)$ is called a hypergroup if the reproduction law holds: $a \circ H = H \circ a = H$, for all $a \in H$.

Definition 1. *Let $(H, \circ)$ be a hypergroup and $\emptyset \neq K \subset H$. We say that $(K, \circ)$ is a subhypergroup of H, denoted by $K \leq H$, if for all $x \in K$ we have $K \circ x = K = x \circ K$.*

An element e_r (respectively e_l) of H is called a right identity (respectively left identity e_l) if for all $a \in H, a \in a \circ e_r$ (respectively $a \in e_l \circ a$). An element e is called a two side identity, or for simplicity an identity if, for all $a \in H, a \in a \circ e \cap e \circ a$. A right identity e_r (resp. left identity e_l) of H is called a scalar right identity (respectively scalar left identity) if for all $a \in H, a = a \circ e_r$ (respectively $a = e_l \circ a$). An element e is called a scalar identity if for all $a \in H, a = a \circ e = e \circ a$. An element $a' \in H$ is called a right inverse (respectively left inverse) of a in H if $e_r \in a \circ a'$, for some right identity e_r in H (respectively $e_l \in a' \circ a$, for some left identity e_l). An element $a' \in H$ is called an inverse of $a \in H$ if $e \in a' \circ a \cap a \circ a'$, for some identity e in H. We denote the set of all right inverses, left inverses and inverses of $a \in H$ by $i_r(a)$, $i_l(a)$, and $i(a)$, respectively. In addition, if H has a scalar identity, and the inverse of $a \in H$ exists, we indicate it by a^{-1}.

Definition 2. *A hypergroup H is called reversible, if the following conditions hold:*

(i) H has at least one identity e;
(ii) every element x of H has at least one inverse, that is $i(x) \neq \emptyset$;
(iii) $x \in y \circ z$ implies that $y \in x \circ z'$ and $z \in y' \circ x$, where $z' \in i(z)$ and $y' \in i(y)$.

Definition 3. *Suppose that $(H, \cdot)$ and $(K, \circ)$ are two hypergroups. A function $f : H \to K$ is called a homomorphism if $f(a \cdot b) \subseteq f(a) \circ f(b)$, for all a and b in H. We say that f is a good homomorphism if for all a and b in H, there is $f(a \cdot b) = f(a) \circ f(b)$. Moreover, $(H, \cdot)$ and $(K, \circ)$ are isomorphic, denoted by $H \cong K$, if f is a bijective good homomorphism.*

An exhaustive review for the theory of hypergroups appears in [9], while the book [12] contains a wealth of applications. The more general algebraic structure that satisfies the ring-like axioms is the hyperring. There are different kinds of hyperrings. The most general one, introduced by Vougiouklis [13], has both addition and multiplication defined as hyperoperations. If only the multiplication is a hyperoperation, then we talk about multiplicative hyperrings [14,15]. If only the addition $+$ is a hyperoperation and the multiplication $\cdot$ is a usual operation, then we say that R is an additive hyperring. A special case of this type is the hyperring introduced by Krasner [6]. An exhaustive review for the theory of hyperrings appears in [16–19].

Definition 4 ([6])**.** *A Krasner hyperring is an algebraic structure $(R, +, \cdot)$ which satisfies the following axioms:*

(1) $(R, +)$ is a canonical hypergroup, i.e.,

(i) for every $x, y, z \in R, x + (y + z) = (x + y) + z$,
(ii) for every $x, y \in R, x + y = y + x$,
(iii) there exists $0 \in R$ such that $0 + x = \{x\}$ for every $x \in R$,
(iv) for every $x \in R$ there exists a unique element $x' \in R$ such that $0 \in x + x'$; (we shall write $-x$ for x' and we call it the opposite of x.)
(v) $z \in x + y$ implies that $y \in z - x$ and $x \in z - y$.

(2) $(R, \cdot)$ is a semigroup having zero as a bilaterally absorbing element, i.e., $x \cdot 0 = 0 \cdot x = 0$.
(3) The multiplication is distributive with respect to the hyperoperation $+$.

A Krasner hyperring $(R, +, \cdot)$ is called commutative, if $(R, \cdot)$ is a commutative semigroup with unit element, i.e., a monoid. A Krasner hyperring is called a Krasner hyperfield, if the multiplicative part $(R \setminus \{0\}, \cdot)$ is a group.

In the following we recall the first construction of a Krasner hyperfield, as a quotient structure of a classical field by a normal subgroup. Let $(F, +, \cdot)$ be a field and G be a normal subgroup of $(F^*, \cdot)$, where $F^* = F \setminus \{0\}$. Take $\frac{F}{G} = \{aG \mid a \in F\}$ with the hyperoperation and the multiplication defined by:

(i) $aG \oplus bG = \{cG | c \in aG + bG\}$,
(ii) $aG \odot bG = abG$,

for all $aG, bG \in \frac{F}{G}$. Then $(\frac{F}{G}, \oplus, \odot)$ is a hyperfield. From now on, we denote $\bar{a} = aG$, for all $aG \in \frac{F}{G}$ and the constructed hyperfield $(\frac{F}{G}, \oplus, \odot)$ by $\bar{F}$, and call it the Krasner hyperfield. Moreover, we denote the inverse of $\bar{a}$ relative to $\oplus$ by $\ominus\bar{a}$ and, for $\bar{a} \neq \bar{0}$, the multiplicative inverse $\bar{a}^{-1}$ by $\dfrac{1}{\bar{a}}$. Besides, we will use the notation $\bar{S} = \{\bar{s} | s \in S\}$ and $\bar{T} = \{\bar{t} | t \in T\}$ for all $S \subseteq F, T \subseteq F^2$.

3. Hyperhomographies

In this section we define the notion of hyperhomography on a Krasner hyperfield, as a quotient structure of a classical field by a normal subgroup. Using it, we introduce some hyperoperations and investigate the properties of the associated hypergroups.

Definition 5. *Let $\bar{F}$ be the Krasner hyperfield associated with the field F and $(\bar{A}, \bar{B}) \in \bar{F}^2$. Define the generalized homography transformation on F as $\bar{1} \in (\bar{x} \ominus \bar{A}) \odot (\bar{y} \ominus \bar{B})$ on $\bar{F}$, and call it the hyperhomography relation. We call the set $H_{\bar{A},\bar{B}}(\bar{F}) = \{(\bar{x}, \bar{y}) \in \bar{F}^2 \mid \bar{1} \in (\bar{x} \ominus \bar{A}) \odot (\bar{y} \ominus \bar{B})\}$ hyperhomography, while $H_{a,b}(F) = \{(x, y) \in F^2 \mid y = f_{a,b}(x) = b + \frac{1}{x-a}\}$ is a homography, for all $a \in \bar{A}$ and $b \in \bar{B}$.*

Notice that the hyperhomography $H_{\bar{A},\bar{B}}(\bar{F})$ is a generalization of a homography $H_{a,b}(F)$, because $(x, y) \in H_{a,b}(F)$ is equivalent with $y = f_{a,b}(x) = b + \frac{1}{x-a}$, i.e., $(x - a)(y - a) = 1$. The classical operations on the field F have been extended to the hyperoperation $\oplus$ and operation $\odot$ on $\bar{F}$, where by $\bar{x} \ominus \bar{A}$ we denote the hyperaddition between $\bar{x}$ and the opposite of $\bar{A}$ with respect to the hyperoperation $\oplus$. Besides, since the result of a hyperoperation is a set, the equality relation in the definition of a homography is substitute by a "belongingness" relation in the definition of a hyperhomography.

Moreover denote $H_{\bar{A},\bar{B}}(F) = \bigcup_{a \in \bar{A}, b \in \bar{B}} H_{a,b}(F)$ and $H_{a,b}(\bar{F}) = \overline{H_{a,b}(F)} = \{(\bar{x}, \bar{y}) \mid (x, y) \in H_{a,b}(F)\}$, for all $a \in \bar{A}, b \in \bar{B}$. It follows that $\overline{H_{\bar{A},\bar{B}}(F)} = \bigcup_{a \in \bar{A}, b \in \bar{B}} H_{a,b}(\bar{F})$.

Theorem 2. *The relation between homographies and hyperhomographies is given by the following identity $H_{\bar{A},\bar{B}}(\bar{F}) = \overline{H_{\bar{A},\bar{B}}(F)}$.*

Proof. $(\Leftarrow)$. Let $\overline{(x, y)} \in \overline{H_{\bar{A},\bar{B}}(F)}$, thus there exists $(a, b) \in \bar{A} \times \bar{B}$, such that $\overline{(x, y)} = (\bar{x}, \bar{y}) \in H_{a,b}(\bar{F})$, hence $(xg_1, yg_2) \in H_{a,b}(F)$ for some $g_1, g_2 \in G$. Then $(xg_1 - a)(yg_2 - b) = 1$ and the following implications hold:

$$1 = (xg_1 - a)(yg_2 - b) \Longrightarrow \begin{cases} \bar{1} & = \overline{(xg_1 - a)(yg_2 - b)} \\ & = \overline{(xg_1 - a)} \odot \overline{(yg_2 - b)} \\ & \subseteq (\bar{x} - \bar{a}) \odot (\bar{y} - \bar{b}) \\ & = (\bar{x} \ominus \bar{a}) \odot (\bar{y} \ominus \bar{b}) \end{cases}$$
$$\Longrightarrow \bar{1} \in (\bar{x} \ominus \bar{A}) \odot (\bar{y} \ominus \bar{B})$$
$$\Longrightarrow (\bar{x}, \bar{y}) \in H_{\bar{A},\bar{B}}(\bar{F})$$
$$\Longrightarrow \overline{(x,y)} \in H_{\bar{A},\bar{B}}(\bar{F}),$$

so

$$\overline{H_{\bar{A},\bar{B}}(F)} \subseteq H_{\bar{A},\bar{B}}(\bar{F}).$$

$(\Rightarrow)$. Conversely, suppose that $(\bar{x}, \bar{y}) \in H_{\bar{A},\bar{B}}(\bar{F})$, then the following implications hold, too:

$$\begin{aligned}
(\bar{x}, \bar{y}) \in H_{\bar{A},\bar{B}}(\bar{F}) &\Longrightarrow \bar{1} \in (\bar{x} \ominus \bar{A}) \odot (\bar{y} \ominus \bar{B}) \\
&\Longrightarrow \bar{1} \in (\bar{x} \oplus \overline{(-A)}) \odot (\bar{y} \oplus \overline{(-B)}) \\
&\Longrightarrow \bar{1} \in \overline{(x - \overline{A})} \odot \overline{(y - \overline{B})} \\
&\Longrightarrow \bar{1} \in \overline{(x - \bar{A})(y - \bar{B})} \\
&\Longrightarrow \bar{1} = \overline{(x-a)(y-b)}, \qquad \textit{for some } (a,b) \in \bar{A} \times \bar{B} \\
&\Longrightarrow \bar{1} = \overline{(x-a)} \odot \overline{(y-b)} \\
&\Longrightarrow \overline{(y-b)} = \overline{(x-a)}^{-1} \\
&\Longrightarrow \overline{(y-b)} = \overline{(x-a)^{-1}} \\
&\Longrightarrow \overline{(y-b)} = \overline{(\frac{1}{x-a})} \\
&\Longrightarrow (y-b)g = \frac{1}{x-a} \quad \textit{for some } g \in G \\
&\Longrightarrow yg = b' + \frac{1}{x-a}, \; b' = bg \in \bar{B} \\
&\Longrightarrow (x, yg) \in H_{a,b'}(F) \\
&\Longrightarrow (\bar{x}, \bar{y}) = \overline{(x, gy)} \in \overline{H_{a,b'}(F)} = H_{a,b'}(\bar{F}) \\
&\Longrightarrow (\bar{x}, \bar{y}) \in \bigcup_{a \in \bar{A}, b \in \bar{B}} H_{a,b}(\bar{F}), \\
&\Longrightarrow (\bar{x}, \bar{y}) \in \overline{H_{\bar{A},\bar{B}}(F)},
\end{aligned}$$

therefore $H_{\bar{A},\bar{B}}(\bar{F}) \subseteq \overline{H_{\bar{A},\bar{B}}(F)}$ and consequently $H_{\bar{A},\bar{B}}(\bar{F}) = \overline{H_{\bar{A},\bar{B}}(F)}$. □

Thanks to Theorem 2, we call the set $H_{\bar{A},\bar{B}}(F)$ the hyperhomography on F, while the set $H_{a,b}(\bar{F})$ is a homography on $\bar{F}$.

Example 2. *Let $F = \mathbb{Z}_5$ be the field of all integers modulo 5 and $G = \{1,4\} \leqslant F^*$. Thus the quotient set F^*/G is $\bar{F} = \{\bar{0}, \bar{1}, \bar{2}\}$ and the hyperaddition $\oplus$ and the multiplication $\odot$ are defined on $\bar{F}$ as follows:*

$\oplus$	$\bar{0}$	$\bar{1}$	$\bar{2}$
$\bar{0}$	$\bar{0}$	$\bar{1}$	$\bar{2}$
$\bar{1}$	$\bar{1}$	$\bar{0}, \bar{2}$	$\bar{1}, \bar{2}$
$\bar{2}$	$\bar{2}$	$\bar{1}, \bar{2}$	$\bar{0}, \bar{1}$

$\odot$	$\bar{0}$	$\bar{1}$	$\bar{2}$
$\bar{0}$	$\bar{0}$	$\bar{0}$	$\bar{0}$
$\bar{1}$	$\bar{0}$	$\bar{1}$	$\bar{2}$
$\bar{2}$	$\bar{0}$	$\bar{2}$	$\bar{1}$

where $\bar{0} = \{0\}$, $\bar{1} = \{1,4\}$ *and* $\bar{2} = \{2,3\}$.

Consider now the hyperhomography

$$H_{\bar{0},\bar{1}}(\bar{F}) = \{(\bar{x},\bar{y}) \in \bar{F}^2 | \bar{1} \in \bar{x} \odot (\bar{y} \ominus \bar{1})\} = \{(\bar{1},\bar{0}),(\bar{1},\bar{2}),(\bar{2},\bar{1}),(\bar{2},\bar{2})\},$$

and the homographies

$$H_{0,1}(F) = \{(x,y) \in F^2 \mid y = 1 + \frac{1}{x}\} = \{(1,2),(2,4),(3,3),(4,0)\},$$

$$H_{0,4}(F) = \{(x,y) \in F^2 \mid y = 4 + \frac{1}{x}\} = \{(1,0),(2,2),(3,1),(4,3)\}.$$

Then it follows that

$$H_{0,1}(\bar{F}) = \overline{H_{0,1}(F)} = \{(\bar{1},\bar{0}),(\bar{1},\bar{2}),(\bar{2},\bar{1}),(\bar{2},\bar{2})\} = \overline{H_{0,4}(F)} = H_{0,4}(\bar{F})$$

and therefore

$$H_{\bar{0},\bar{1}}(\bar{F}) = H_{0,1}(\bar{F}) \cup H_{0,4}(\bar{F}),$$

as stated by Theorem 2.

Definition 6. *A hyperhomography* $H_{\bar{A},\bar{B}}(F)$ *in* F^2 *is called nondegenerate, if the following conditions hold, respectively:*

(i) for all $a \in \bar{A}, b \in \bar{B}$, *if* $v = a - b^{-1} \in F$, *then* $(v,0)$ *can be omitted from* $H_{a,b}(F)$,
(ii) for all $a,c \in \bar{A}$ *and* $b,d \in \bar{B}$, *there is* $H_{a,b}(F) \cap H_{c,d}(F) \neq \emptyset \Longrightarrow H_{a,b}(F) = H_{c,d}(F)$,
(iii) for all $a \in \bar{A}, b \in \bar{B}$, *the element* (a,∞) *can be added to* $H_{a,b}(F)$, *where* ∞ *is an element outside of* F.

By consequence, under the same conditions, also $H_{\bar{A},\bar{B}}(\bar{F}) = \overline{H_{\bar{A},\bar{B}}(F)}$ *is called a nondegenerate hyperhomography in* $\bar{F}^2$.

For a nondegenerate hyperhomography in F^2, we fix some new notations: $F_\infty = F \cup \{\infty\}$, $\bar{F}_\infty = \bar{F} \cup \{\infty\}$ and $\overline{(a,\infty)} = (\bar{a},\infty)$, for any $a \in F$.

Example 3. *If we go back to Example 2 and use the concepts in Definition 6, then we can omit* $(4,0)$ *from* $H_{0,1}(F)$ *and* $(1,0)$ *from* $H_{0,4}(F)$, *respectively, and add in both sets the element* $(0,\infty)$. *Then* $H_{0,1}(F) = \{(0,\infty),(1,2),(2,4),(3,3)\}$ *and* $H_{0,4}(F) = \{(0,\infty),(2,2),(3,1),(4,3)\}$ *are nondegenerate homographies, while* $H_{\bar{0},\bar{1}}(\bar{F}) = \{(0,\infty),(\bar{1},\bar{2}),(\bar{2},\bar{1}),(\bar{2},\bar{2})\}$ *is a nondegenerate hyperhomography with the property* $H_{\bar{0},\bar{1}}(\bar{F}) = H_{0,1}(\bar{F}) \cup H_{0,4}(\bar{F})$ *where,*

$$H_{0,1}(\bar{F}) = \overline{H_{0,1}(F)} = \{(\bar{0},\bar{\infty}),(\bar{1},\bar{2}),(\bar{2},\bar{4}),(\bar{3},\bar{3})\} = \{(0,\infty),(\bar{1},\bar{2}),(\bar{2},\bar{1}),(\bar{2},\bar{2})\},$$

$$H_{0,4}(\bar{F}) = \overline{H_{0,4}(F)} = \{(\bar{0},\bar{\infty}),(\bar{2},\bar{2}),(\bar{3},\bar{1}),(\bar{4},\bar{3})\} = \{(0,\infty),(\bar{2},\bar{2}),(\bar{2},\bar{1}),(\bar{1},\bar{2})\}.$$

Definition 7. *Let* $H_{\bar{A},\bar{B}}(F)$ *be a nondegenerate hyperhomography in* F^2. *Setting* $f_{a,b}(a) = \infty$, *we obtain that* $(a,\infty) = (a, f_{a,b}(a)) \in H_{a,b}(F)$. *Define*

$$\mathcal{G}^x_{a,b} = \begin{cases} \{x\}, & \text{if } G = \{1\} \\ \{x, 2a - x\}, & \text{if } G \neq \{1\}, b = 0 \\ \{y \in F | y = x \text{ or } (x-v)(y-v) = (a-v)^2\}, & \text{if } G \neq \{1\}, b \neq 0, \end{cases}$$

for all $x \in F \setminus \{v\}$ *and* $(a,b) \in \bar{A} \times \bar{B}$, *where* $v = a - b^{-1}$.

Moreover set $\widehat{X} = \{\widehat{x} \mid x \in X\}$, *where* $\widehat{x} = (x, f_{a,b}(x))$, *for all* $(a,b) \in \bar{A} \times \bar{B}$ *and* $x \in X \subseteq F$.

Corollary 1. *Let $H_{\bar{A},\bar{B}}(F)$ be a nondegenerate hyperhomography in F^2 and G be a normal subgroup of F^*. If $G \neq \{1\}$ and $0 \neq b \in \bar{B}$, then $\mathcal{G}^x_{a,b} = \{x, \frac{(ab-1)x+a(2-ab)}{bx+(1-ab)}\}$.*

Proof. According to Definition 7, if $G \neq \{1\}$ and $0 \neq b \in \bar{B}$, then $y = x$ or $(x - a + \frac{1}{b})(y - a + \frac{1}{b}) = \frac{1}{b^2}$. In the second case, solving the equation we get $y = \frac{(ab-1)x+a(2-ab)}{bx+(1-ab)}$. □

In the following, for a nondegenerate hyperhomography in F^2, we define the lines passing through two points.

Definition 8. *Let $H_{\bar{A},\bar{B}}(F)$ be a nondegenerate hyperhomography in F^2. For all $a \in \bar{A}, b \in \bar{B}$ and $\widehat{x_i}, \widehat{x_j} \in H_{a,b}(F)$, define $L_0 = \{(x,0) \mid x \in F\}$ and*

$$L_{a,b}(\widehat{x_i},\widehat{x_j}) = \begin{cases} \{(x,y) \in F^2 | y - f_{a,b}(x_i) = \frac{f_{a,b}(x_j)-f_{a,b}(x_i)}{x_j - x_i}(x - x_i)\}, & x_i \neq x_j, a \notin \{x_i, x_j\} \\ \{(x,y) \in F^2 | y - f_{a,b}(x_i) = f'_{a,b}(x_i)(x - x_i)\}, & x_i = x_j \neq a \\ \{(x,y) | y \in F_\infty, x \in \{x_i, x_j\}, x = a \Leftrightarrow y = \infty\}, & x_i \neq x_j, a \in \{x_i, x_j\} \\ \{(a,y) | y \in F_\infty\}, & x_i = x_j = a, \end{cases}$$

where $f'_{a,b}$ means the formal derivative of $f_{a,b}$. In addition we call $L_{a,b}(\widehat{x_i},\widehat{x_j})$ the line passing through the points $\widehat{x_i}$ and $\widehat{x_j}$. Intuitively, for each $a \in F$, the line passing through (a,∞) is a vertical line. In other words, (a,∞) plays an asymptotic extension role for $f_{a,b}$.

Taking two arbitrary points $\widehat{x_i}, \widehat{x_j}$ on the homography $H_{a,b}(F)$, define $x_i \bullet_{ab} x_j$ by $L_0 \cap L_{a,b}(\widehat{x_i},\widehat{x_j}) = \{(x_i \bullet_{ab} x_j, 0)\}$. Using the definition of the lines $L_{a,b}(\widehat{x_i},\widehat{x_j})$ and L_0, for $x_i \neq a \neq x_j$ we have

$$y = 0, y - f_{a,b}(x_i) = m(x - x_i) \Longrightarrow x = x_i - \frac{f_{a,b}(x_i)}{m}$$

where, $m = \begin{cases} \frac{f_{a,b}(x_j)-f_{a,b}(x_i)}{x_j - x_i}, & \text{if } x_i \neq x_j \\ f'_{a,b}(x_i), & \text{if } x_i = x_j \end{cases} = \begin{cases} \frac{-1}{(x_i-a)(x_j-a)}, & \text{if } x_i \neq x_j \\ \frac{-1}{(x_i-a)^2}, & \text{if } x_i = x_j \end{cases} = \frac{-1}{(x_i-a)(x_j-a)} \in F^*$

and hence $x = x_i \bullet_{ab} x_j \in F$. If x_i or x_j are equal to a, according to Definition 8, we have

$$L_0 \cap L_{a,b}(\widehat{x_i},\widehat{x_j}) = \begin{cases} \{(x_i,0)\}, & \text{if } x_i \neq a = x_j \\ \{(x_j,0)\}, & \text{if } x_i = a \neq x_j \\ \{(a,0)\}, & \text{if } x_i = a = x_j \end{cases} \Longrightarrow x_i \bullet_{ab} x_j = \begin{cases} x_i, & \text{if } x_i \neq a = x_j \\ x_j, & \text{if } x_i = a \neq x_j \\ a, & \text{if } x_i = a = x_j \end{cases} \Longrightarrow x_i \bullet_{ab} x_j \in F.$$

Thus $| L_0 \cap L_{a,b}(\widehat{x_i},\widehat{x_j}) | = 1$ and $x_i \bullet_{ab} x_j$ is well defined, therefore we have

$$\widehat{x_i \bullet_{ab} x_j} = (x_i \bullet_{ab} x_j, f_{a,b}(x_i \bullet_{ab} x_j)).$$

We will better illustrate the above defined notions in the following example.

Example 4. *Consider the field $F = \mathbb{R}$ and the homography transformation $f_{0,0}(x) = \frac{1}{x}$ over F, so its graph is the hyperbola $H_{0,0}(\mathbb{R})$ represented below in Figure 4. Taking on $H_{0,0}(\mathbb{R})$ two arbitrary points $\widehat{x_i} = (x_i, f_{0,0}(x_i))$ and $\widehat{x_j} = (x_j, f_{0,0}(x_j))$, we draw the line $L_{0,0}(\widehat{x_i},\widehat{x_j})$ passing through $\widehat{x_i}$ and $\widehat{x_j}$. Then $x_i \bullet_{00} x_j = L_0 \cap L_{0,0}(\widehat{x_i},\widehat{x_j})$, where L_0 is the x-axis. Then we obtain the point $\widehat{x_i \bullet_{00} x_j} = (x_i \bullet_{00} x_j, f_{0,0}(x_i \bullet_{00} x_j))$ on the hyperbola $H_{0,0}(\mathbb{R})$.*

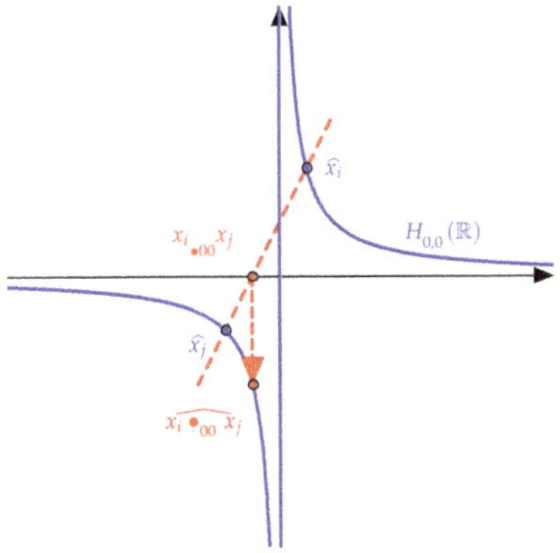

Figure 4. Hyperbola $H_{0,0}(\mathbb{R})$

Proposition 1. *Let $H_{\bar{A},\bar{B}}(F)$ be a nondegenerate hyperhomography in F^2, and $\widehat{x_i}, \widehat{x_j} \in H^2_{a,b}(F)$. Then, it follows that*

$$x_i \bullet_{ab} x_j = bx_ix_j - (ab-1)(x_i + x_j - a).$$

Proof. Based on Definition 8 and on the fact that $f_{a,b}(x) = b + \frac{1}{x-a}$ and $f'_{a,b}(x) = -\frac{1}{(x-a)^2}$, by simple computations, we obtain

$$x_i \bullet_{ab} x_j = \begin{cases} \dfrac{x_i f_{a,b}(x_j) - x_j f_{a,b}(x_i)}{f_{a,b}(x_j) - f_{a,b}(x_i)} & x_i \neq x_j, a \notin \{x_i, x_j\}, \\ x_i - \dfrac{f_{a,b}(x_i)}{f'_{a,b}(x_i)} & x_i = x_j \neq a, \\ x_i & x_i \neq a = x_j, \\ x_j & x_j \neq a = x_i, \\ a & x_i = a = x_j \end{cases}$$

$$= \begin{cases} x_i + (x_i - a)(x_j - a)(b + \frac{1}{x_i - a}) & x_i \neq x_j, a \notin \{x_i, x_j\} \\ x_i + (x_i - a)(x_i - a)(b + \frac{1}{x_i - a}) & x_i = x_j \neq a \\ x_i & x_i \neq a = x_j \\ x_j & x_j \neq a = x_i \\ a & x_i = a = x_j \end{cases}$$

$$= \begin{cases} bx_ix_j - (ab-1)(x_i + x_j - a) & x_i \neq x_j, a \notin \{x_i, x_j\}, \\ bx_i^2 - (ab-1)(2x_i - a) & x_i = x_j \neq a, \\ x_i & x_i \neq a = x_j, \\ x_j & x_j \neq a = x_i, \\ a & x_i = a = x_j. \end{cases}$$

$$= bx_ix_j - (ab-1)(x_i + x_j - a).$$

□

Remark 1. $(H_{a,b}(F), \bullet_{ab})$ *is a homography group, for all $(a,b) \in \bar{A} \times \bar{B}$. Moreover, notice that "$\bullet_{ab}$" is the group operation on the homography $H_{a,b}(F)$.*

On a nondegenerate hyperhomography $H_{\bar{A},\bar{B}}(F)$ in F^2 we introduce the equivalence relation "$\sim$" by considering

$$(x,y) \sim (x',y') \Longleftrightarrow \begin{cases} x = x', \\ y = y' \notin \{\infty\}, \end{cases} \quad \text{or} \quad y, y' \in \{\infty\}$$

and denote the set of the equivalence classes of $H_{\bar{A},\bar{B}}(F)$ and $H_{a,b}(F)$ by $\mathcal{H}_{\bar{A},\bar{B}}(F)$ and $\mathcal{H}_{a,b}(F)$, respectively. It follows that $\frac{(x,y)}{\sim} = \begin{cases} (x,y), & \text{if } x \notin \bar{A} \\ (\bar{A},\infty), & \text{if } x \in \bar{A}. \end{cases}$

Furthermore, if we introduce the notation $\mathcal{O} = \bar{A}$ and $\hat{\mathcal{O}} = (\mathcal{O},\infty)$. We will have $\bar{\mathcal{O}} = \mathcal{O}, \overline{\hat{\mathcal{O}}} = \hat{\mathcal{O}}$ and $\overline{\mathcal{H}_{\bar{A},\bar{B}}(F)} = \mathcal{H}_{\bar{A},\bar{B}}(\bar{F})$.

Thus $\mathcal{H}_{\bar{A},\bar{B}}(F)$ and $\mathcal{H}_{\bar{A},\bar{B}}(\bar{F})$ are called the equipped hyperhomographies in F^2 and $\bar{F}^2$, respectively. Besides, if we admit that $\bar{A} \bullet_{ab} x = x = x \bullet_{ab} \bar{A}$ for all $a \in \bar{A}, b \in \bar{B}$ and $(x,y) \in H_{a,b}(F)$, then the bijective map $\Pi : H_{a,b}(F) \longrightarrow \mathcal{H}_{a,b}(F)$ defined by $\Pi(x,y) = \frac{(x,y)}{\sim}$, where $\frac{(x,y)}{\sim} = \begin{cases} (x,y), & \text{if } x \neq a \\ \mathcal{O}, & \text{if } x = a \end{cases}$, equip the quotient $\mathcal{H}_{a,b}(F)$ with a group structure and gives us a group isomorphism $(H_{a,b}(F), \bullet_{ab}) \overset{\Pi}{\cong} (\mathcal{H}_{a,b}(F), \bullet_{ab})$.

In addition, the concepts in Definition (8) can be similarly defined on $\mathcal{H}_{a,b}(F)$, only by substituting a with $\mathcal{O}$.

Definition 9. *Let $\mathcal{H}_{\bar{A},\bar{B}}(F)$ be an equipped hyperhomography. We define the hyperoperation "$\circ$" on $\mathcal{H}_{\bar{A},\bar{B}}(F)$ as follows.*
Let $(x,y),(x',y') \in \mathcal{H}_{\bar{A},\bar{B}}(F)$. If $(x,y) \in \mathcal{H}_{a,b}(F)$ and $(x',y') \in \mathcal{H}_{a',b'}(F)$ for some $a, a' \in \bar{A}$ and $b, b' \in \bar{B}$, then

$$(x,y) \circ (x',y') = \begin{cases} \{\widehat{x_i \bullet_{ab} x_j} | (x_i, x_j) \in \mathcal{G}^x_{a,b} \times \mathcal{G}^{x'}_{a',b'}\}, & \text{if } \mathcal{H}_{a,b}(F) = \mathcal{H}_{a',b'}(F) \\ \\ (\mathcal{H}_{a,b}(F) \cup \mathcal{H}_{a',b'}(F)) \setminus \{\hat{\mathcal{O}}\}, & \text{otherwise.} \end{cases}$$

Theorem 3. *If $\mathcal{H}_{\bar{A},\bar{B}}(\bar{F})$ is an equipped hyperhomography, then $(\mathcal{H}_{\bar{A},\bar{B}}(F), \circ)$ has a hypergroup structure.*

Proof. Suppose that $\{X,Y,Z\} \subseteq \mathcal{H}_{\bar{A},\bar{B}}(F)$ such that $X = (x,y) \in \mathcal{H}_{a,b}(F), Y = (x',y') \in \mathcal{H}_{a',b'}(F)$ and $Z = (x'',y'') \in \mathcal{H}_{a'',b''}(F)$ where, $J = \{(a,b),(a',b'),(a'',b'')\} \subseteq \bar{A} \times \bar{B}$. First we notice that $(x,y) \circ (x',y') \subseteq \mathcal{P}^*(\mathcal{H}_{\bar{A},\bar{B}}(F))$, because $(x,x') \in \mathcal{G}^x_{a,b} \times \mathcal{G}^{x'}_{a',b'}$ implies that $\widehat{x \bullet_{ab} x'} \in (x,y) \circ (x',y')$, i.e $(x,y) \circ (x',y')$ is a non-empty set and belongs to $\mathcal{P}^*(\mathcal{H}_{\bar{A},\bar{B}}(F))$. Besides, if $(x,y) = (x_1,y_1)$ and $(x',y') = (x'_1,y'_1)$, then $x = x_1$ and $x' = x'_1$, meaning that $\mathcal{G}^x_{a,b} = \mathcal{G}^{x_1}_{a,b}$ and $\mathcal{G}^{x'}_{a,b} = \mathcal{G}^{x'_1}_{a,b}$. Hence we have $\mathcal{G}^x_{a,b} \times \mathcal{G}^{x'}_{a,b} = \mathcal{G}^{x_1}_{a,b} \times \mathcal{G}^{x'_1}_{a,b}$ and therefore $\{\widehat{z \bullet_{ab} w} | (z,w) \in \mathcal{G}^x_{a,b} \times \mathcal{G}^{x'}_{a,b}\} = \{\widehat{z \bullet_{ab} w} | (z,w) \in \mathcal{G}^{x_1}_{a,b}(f_{a,b}) \times \mathcal{G}^{x'_1}_{a,b}\}$, equivalently with $(x,y) \circ (x',y') = (x_1,y_1) \circ (x'_1,y'_1)$. By consequence, the hyperoperation "$\circ$" is well defined.

If $X = (a,\infty)$ or $Y = (a,\infty)$ or $Z = (a,\infty)$, then the associativity is obvious. If not, we have the following cases.

Case 1: $|J| = 1$.
This means that $\mathcal{H}_{a,b}(F) = \mathcal{H}_{a',b'}(F) = \mathcal{H}_{a'',b''}(F)$ and we have

$$\begin{aligned} [(x,y) \circ (x',y')] \circ (x'',y'') &= \left\{ \widehat{(x_i \bullet_{ab} x'_j)} | (x_i, x'_j) \in \mathcal{G}^x_{a,b} \times \mathcal{G}^{x'}_{a,b} \right\} \circ (x'',y'') \\ &= \left\{ \widehat{(x_i \bullet_{ab} x'_j) \bullet_{ab} x''_k} | (x_i, x'_j, x''_k) \in \mathcal{G}^x_{a,b} \times \mathcal{G}^{x'}_{a,b} \times \mathcal{G}^{x''}_{a,b} \right\}. \end{aligned}$$

Similarly, it holds that

$$(x,y)\circ[(x',y')\circ(x'',y'')]=\left\{x_i\bullet_{ab}\widehat{(x'_j\bullet_{ab}x''_k)}|(x_i,x'_j,x''_k)\in\mathcal{G}^x_{a,b}\times\mathcal{G}^{x'}_{a,b}\times\mathcal{G}^{x''}_{a,b}\right\}.$$

On the other hand we have

$$L_{a,b}(\widehat{x_i},\widehat{x'_j})\cap L_{a,b}(\widehat{x_i\bullet_{ab}x'_j},\widehat{\mathcal{O}})=\{(x_i\bullet_{ab}x'_j,0)\}\subseteq L_0,$$
$$L_{a,b}(\widehat{x'_j},\widehat{x''_k})\cap L_{a,b}(\widehat{\mathcal{O}},\widehat{x'_j\bullet_{ab}x''_k})=\{(x'_j\bullet_{ab}x''_k,0)\}\subseteq L_0.$$

In other words, for the six points $p_1=\widehat{x_i}$, $p_2=\widehat{x'_j}$, $p_3=\widehat{x''_k}$, $p_4=\widehat{x_i\bullet_{ab}x'_j}$, $p_5=\widehat{\mathcal{O}}$ and $p_6=\widehat{x'_j\bullet_{ab}x''_k}$ on the curve we have $L_{a,b}(p_1,p_2)\cap L_{a,b}(p_4,p_5)\subseteq L_0$ and $L_{a,b}(p_2,p_3)\cap L_{a,b}(p_5,p_6)\subseteq L_0$ and therefore, by Pascal's theorem (see Theorem 1), it follows also that $L_{a,b}(p_3,p_4)\cap L_{a,b}(p_6,p_1)\subseteq L_0$, equivalently with

$$L_{a,b}(\widehat{x''_k},\widehat{x_i\bullet_{ab}x'_j})\cap L_{a,b}(\widehat{x'_j\bullet_{ab}x''_k},\widehat{x_i})\subseteq L_0.$$

By Definition 8 we know that

$$\{(((x_i\bullet_{ab}x'_j)\bullet_{ab}x''_k,0)\}=L_0\cap L_{a,b}(\widehat{x_i\bullet_{ab}x'_j},\widehat{x''_k}),$$

$$\{(x_i\bullet_{ab}(x'_j\bullet_{ab}x''_k),0)\}=L_0\cap L_{a,b}(\widehat{x_i},\widehat{x'_j\bullet_{ab}x''_k})$$

where, by the associativity of the group operation "$\bullet_{ab}$", it holds $(x_i\bullet_{ab}x'_j)\bullet_{ab}x''_k=x_i\bullet_{ab}(x'_j\bullet_{ab}x''_k)$. This leads to the equality

$$L_0\cap L_{a,b}(\widehat{x_i\bullet_{ab}x'_j},\widehat{x''_k})=L_{a,b}(\widehat{x''_k},\widehat{x_i\bullet_{ab}x'_j})\cap L_{a,b}(\widehat{x'_j\bullet_{ab}x''_k},\widehat{x_i})=L_0\cap L_{a,b}(\widehat{x_i},\widehat{x'_j\bullet_{ab}x''_k}),$$

implying that

$$\left(\widehat{(x_i\bullet_{ab}x'_j)\bullet_{ab}x''_k}\right)=\left(x_i\bullet_{ab}\widehat{(x'_j\bullet_{ab}x''_k)}\right)\text{ for all }(x_i,x'_j,x''_k)\in\mathcal{G}^x_{a,b}\times\mathcal{G}^{x'}_{a,b}\times\mathcal{G}^{x''}_{a,b}.$$

Case 2: $|J|=2$.
(i) If $\mathcal{H}_{a,b}(F)=\mathcal{H}_{a',b'}(F)\neq\mathcal{H}_{a'',b''}(F)$, then we have

$$\begin{aligned}[(x,y)\circ(x',y')]\circ(x'',y'')&=\left\{\widehat{z\bullet_{ab}w}|(z,w)\in\mathcal{G}^x_{a,b}\times\mathcal{G}^{x'}_{a,b}\right\}\circ(x'',y'')\\&=\bigcup_{(u,v)\in(x,y)\circ(x',y')}(u,v)\circ(x'',y'')\\&=\mathcal{H}_{a,b}(F)\cup\mathcal{H}_{a'',b''}(F).\end{aligned}$$

On the other hand

$$\begin{aligned}(x,y)\circ[(x',y')\circ(x'',y'')]&=(x,y)\circ(\mathcal{H}_{a',b'}(F)\cup\mathcal{H}_{a'',b''}(F))\\&=(x,y)\circ\mathcal{H}_{a',b'}(F)\cup(x,y)\circ\mathcal{H}_{a'',b''}(F)\\&=\mathcal{H}_{a,b}(F)\cup\mathcal{H}_{a'',b''}(F).\end{aligned}$$

(ii) If $\mathcal{H}_{a,b}(F) \neq \mathcal{H}_{a',b'}(F) = \mathcal{H}_{a'',b''}(F)$, then the associativity holds, similarly as in the case (i).
(iii) If $\mathcal{H}_{a,b}(F) = \mathcal{H}_{a'',b''}(F) \neq \mathcal{H}_{a',b'}(F)$, then we have

$$\begin{aligned}[(x,y) \circ (x',y')] \circ (x'',y'') &= (\mathcal{H}_{a,b}(F) \cup \mathcal{H}_{a',b'}(F)) \circ (x'',y'') \\ &= \mathcal{H}_{a,b}(F) \cup \mathcal{H}_{a',b'}(F) \cup \mathcal{H}_{a'',b''}(F) \\ &= \mathcal{H}_{a,b}(F) \cup \mathcal{H}_{a',b'}(F)\end{aligned}$$

and similarly

$$\begin{aligned}(x,y) \circ [(x',y') \circ (x'',y'')] &= (x,y) \circ (\mathcal{H}_{a',b'}(F) \cup \mathcal{H}_{a'',b''}(F)) \\ &= \mathcal{H}_{a,b}(F) \cup \mathcal{H}_{a',b'}(F) \cup \mathcal{H}_{a'',b''}(F) \\ &= \mathcal{H}_{a,b}(F) \cup \mathcal{H}_{a',b'}(F).\end{aligned}$$

Case 3: $|J| = 3$.
In this case we have

$$\begin{aligned}[(x,y) \circ (x',y')] \circ (x'',y'') &= (\mathcal{H}_{a,b}(F) \cup \mathcal{H}_{a',b'}(F)) \circ (x'',y'') \\ &= \mathcal{H}_{a,b}(F) \cup \mathcal{H}_{a',b'}(F) \cup \mathcal{H}_{a'',b''}(F).\end{aligned}$$

On the other hand

$$\begin{aligned}(x,y) \circ [(x',y') \circ (x'',y'')] &= (x,y) \circ (\mathcal{H}_{a',b'}(F) \cup \mathcal{H}_{a'',b''}(F)) \\ &= \mathcal{H}_{a,b}(F) \cup \mathcal{H}_{a',b'}(F) \cup H_{a'',b''}(F).\end{aligned}$$

Therefore the hyperoperation "$\circ$" is associative.

In order to prove the reproduction axiom, we consider two cases as below:

Case 1. If $|\bar{A} \times \bar{B}| = 1$, then $\bar{F} = F$ and $\mathcal{H}_{\bar{A},\bar{B}}(F) = \mathcal{H}_{a,b}(F)$, where $a \in \bar{A}, b \in \bar{B}$. It follows that $(\mathcal{H}_{a,b}(F), \circ)$ is a homography group, so the reproduction axiom holds.

Case 2. If $|\bar{A} \times \bar{B}| > 1$, consider an arbitrary element $\widehat{x} \in \mathcal{H}_{a,b}(F) \subseteq \mathcal{H}_{\bar{A},\bar{B}}(F)$. Then

$$\begin{aligned}\widehat{x} \circ \mathcal{H}_{\bar{A},\bar{B}}(F) &= (\widehat{x} \circ \bigcup_{a\neq i\in\bar{A}, b\neq j\in\bar{B}} \mathcal{H}_{i,j}(F)) \cup (\widehat{x} \circ \mathcal{H}_{a,b}(F)), \\ &= (\bigcup_{a\neq i\in\bar{A}, b\neq j\in\bar{B}} \widehat{x} \circ \mathcal{H}_{i,j}(F)) \cup \mathcal{H}_{a,b}(F), \\ &= (\bigcup_{i\in\bar{A}, j\in\bar{B}} \mathcal{H}_{i,j}(F)) \cup \mathcal{H}_{a,b}(F), \\ &= \mathcal{H}_{\bar{A},\bar{B}}(F).\end{aligned}$$

Similarly, $(\mathcal{H}_{\bar{A},\bar{B}}(F)) \circ \widehat{x} = \mathcal{H}_{\bar{A},\bar{B}}(F)$ and thus the reproduction axiom is proved. Therefore, $(\mathcal{H}_{\bar{A},\bar{B}}(F), \circ)$ is a hypergroup. □

Remark 2. *If $G = \{1\}$, then the hyperhomography and the associated hypergroup are the classical homography and the homography group, respectively.*

Example 5. *Let us consider again Example 3, where we deal with the nondegenerate hyperhomography $H_{\bar{0},\bar{1}}(F)$ as a subset of F^2, having the form $H_{\bar{0},\bar{1}}(F) = H_{0,1}(F) \cup H_{0,4}(F)$ where, $H_{0,1}(F) = \{(0,\infty),(1,2),(2,4),(3,3)\}$ and $H_{0,4}(F) = \{(0,\infty),(2,2),(3,1),(4,3)\}$, while the associated equipped hyperhomography is $\mathcal{H}_{\bar{0},\bar{1}}(F) = \mathcal{H}_{0,1}(F) \cup \mathcal{H}_{0,4}(F)$ where,*

$$\mathcal{H}_{0,1}(F) = \{\widehat{\mathcal{O}}, (1,2), (2,4), (3,3)\}, \ \mathcal{H}_{0,4}(F) = \{\widehat{\mathcal{O}}, (2,2), (3,1), (4,3)\},$$

for $\widehat{\mathcal{O}} = (\mathcal{O}, f_{a,b}(\mathcal{O})) = (\bar{0}, f_{a,b}(\bar{0})) = (\bar{0}, \infty) = (0, \infty)$.

Now let $T = \mathcal{H}_{0,1}(F)$ *and* $K = \mathcal{H}_{0,4}(F)$. *Then* $(T, \circ)$ *and* $(K, \circ)$ *are reversible subhypergroups of* $(\mathcal{H}_{\bar{0},\bar{1}}(F), \circ)$, *which are defined by the following Cayley tables, respectively*

$(T, \circ)$	$(0,\infty)$	$(1,2)$	$(2,4)$	$(3,3)$
$(0,\infty)$	$(0,\infty)$	$(1,2),(2,4)$	$(1,2),(2,4)$	$(3,3)$
$(1,2)$	$(1,2),(2,4)$	$(0,\infty),(3,3)$	$(0,\infty),(3,3)$	$(1,2),(2,4)$
$(2,4)$	$(1,2),(2,4)$	$(0,\infty),(3,3)$	$(0,\infty),(3,3)$	$(1,2),(2,4)$
$(3,3)$	$(3,3)$	$(1,2),(2,4)$	$(1,2),(2,4)$	$(0,\infty)$

$(K, \circ)$	$(0,\infty)$	$(2,2)$	$(3,1)$	$(4,3)$
$(0,\infty)$	$(0,\infty)$	$(2,2)$	$(3,1),(4,3)$	$(3,1),(4,3)$
$(2,2)$	$(2,2)$	$(0,\infty)$	$(3,1),(4,3)$	$(3,1),(4,3)$
$(3,1)$	$(3,1),(4,3)$	$(3,1),(4,3)$	$(0,\infty),(2,2)$	$(0,\infty),(2,2)$
$(4,3)$	$(3,1),(4,3)$	$(3,1),(4,3)$	$(0,\infty),(2,2)$	$(0,\infty),(2,2)$

For a better understanding, we will explain all details in computing, for example, in the table of T *the hyperproduct* $(1,2) \circ (2,4)$. *For doing this, since* $T = \mathcal{H}_{0,1}(F)$, *we use the function* $f_{0,1}(x) = 1 + \frac{1}{x}$ *and the field* $F = \mathbb{Z}_5$. *Based on Corollary 1, we obtain*

$$\mathcal{G}^1_{\bar{0},1} = \{1, \frac{-1}{1+1}\} = \{1, -2^{-1}\} = \{1, -3\} = \{1, 2\},$$

$$\mathcal{G}^2_{\bar{0},1} = \{2, \frac{-2}{2+1}\} = \{2, -2 \cdot 3^{-1}\} = \{2, -4\} = \{2, 1\},$$

and therefore,

$$(1,2) \circ (2,4) = \{\widehat{x_i \bullet_{01} x_j} \mid x_i \in \mathcal{G}^1_{0,1}, x_j \in \mathcal{G}^2_{0,1}\} = \{\widehat{1 \bullet_{01} 1}, \widehat{1 \bullet_{01} 2}, \widehat{2 \bullet_{01} 1}, \widehat{2 \bullet_{01} 2}\}.$$

Based on Proposition 1, we have

$$\begin{aligned} 1 \bullet_{01} 1 &= 1 \cdot 1 \cdot 1 - (-1) \cdot (1+1-0) = 3 \\ 1 \bullet_{01} 2 &= 1 \cdot 1 \cdot 2 - (-1) \cdot (1+2-0) = 5 = 0 \\ 2 \bullet_{01} 1 &= 1 \cdot 2 \cdot 1 - (-1) \cdot (2+1-0) = 5 = 0 \\ 2 \bullet_{01} 2 &= 1 \cdot 2 \cdot 2 - (-1) \cdot (2+2-0) = 8 = 3 \end{aligned}$$

which imply that

$$(1,2) \circ (2,4) = \{(3, f_{0,1}(3)), (0, f_{0,1}(0)\} = \{(3, 1 + \frac{1}{3}), (0, \infty)\} = \{(3,3), (0, \infty)\}.$$

Similarly, all the other hyperproducts in both tables can be obtained.

The next result gives a characterization of the subhypergroups of the equipped hyperhomographies in F^2.

Theorem 4. *Let H be a non-empty subset of the hypergroup $\mathcal{H}_{\bar{A},\bar{B}}(F)$. Then H is a subhypergroup of the equipped hyperhomography $\mathcal{H}_{\bar{A},\bar{B}}(F)$ if and only if it can be written as $H = \bigcup_{(i,j)\in I\subseteq \bar{A}\times\bar{B}} \mathcal{H}_{i,j}(F)$, where $I = \{(i,j) \in \bar{A}\times\bar{B} \mid H\cap\mathcal{H}_{i,j}(F) \neq \emptyset\}$, or H is a subhypergroup of $\mathcal{H}_{i,j}(F)$, for some $(i,j) \in \bar{A}\times\bar{B}$.*

Proof. $(\Rightarrow)$. Suppose that H is a subhypergroup of $\mathcal{H}_{\bar{A},\bar{B}}(F)$ and $H \nleqslant \mathcal{H}_{i,j}(F)$, for every (i,j) in $\bar{A}\times\bar{B}$. There exist $(i',j') \neq (s',t')$ in $\bar{A}\times\bar{B}$ such that $H\cap\mathcal{H}_{i',j'}(F) \neq \emptyset \neq H\cap\mathcal{H}_{s',t'}(F)$. Now let I=$\{(i,j) \in \bar{A}\times\bar{B} \mid H\cap\mathcal{H}_{i,j}(F) \neq \emptyset\}$. Thus we have $H \subseteq \bigcup_{(i,j)\in I} \mathcal{H}_{i,j}(F) \subseteq \bigcup_{(i',j'),(s',t')\in I} (\mathcal{H}_{i',j'}(F)\cap H)\circ (\mathcal{H}_{s',t'}(F)\cap H) \subseteq H$. Hence $H = \bigcup_{(i,j)\in I} \mathcal{H}_{i,j}(F)$.

$(\Leftarrow)$. It is obvious. □

Theorem 5. *Let H be a subhypergroup of the hypergroup $\mathcal{H}_{\bar{A},\bar{B}}(F)$. Then H is reversible if and only if H is a subhypergroup of $\mathcal{H}_{a,b}(F)$, for some $(a,b) \in \bar{A}\times\bar{B}$.*

Proof. $(\Leftarrow)$. First we prove that any subhypergroup H of $\mathcal{H}_{a,b}(F)$ is a regular reversible hypergroup, for any $(a,b) \in \bar{A}\times\bar{B}$. The regularity is clear, because $\widehat{O}$ is an identity and each element is an inverse for itself. In order to prove the reversibility, let $\hat{x} = (x,y)$ and $\hat{x'} = (x',y')$ be arbitrary elements in $\mathcal{H}_{a,b}(F)$. We distinguish three different situations.

Case 1. If $x' \notin \mathcal{G}^x_{a,b} = \{x,\alpha\}$, where $x \bullet_{ab} \alpha = a$, then

$$\begin{aligned}
\widehat{x''} = (x'',y'') \in (x,y)\circ(x',y') &\Longrightarrow (x'',y'') = \widehat{z\bullet_{ab} w}, \text{ with } (z,w) \in \mathcal{G}^x_{a,b}\times\mathcal{G}^{x'}_{a,b}\\
&\Longrightarrow x'' = z\bullet_{ab} w,\\
&\Longrightarrow z = x''\bullet_{ab} h, \text{ where } w\bullet_{ab} h = a,\\
&\Longrightarrow (z, f_{a,b}(z)) = \widehat{x''\bullet_{ij} h} \text{ and } h \in \mathcal{G}^w_{a,b} = \mathcal{G}^{x'}_{a,b}, z \in \mathcal{G}^x_{a,b}\\
&\Longrightarrow (z, f_{a,b}(z)) \in (x'', f_{a,b}(x''))\circ(h, f_{a,b}(h))\\
&\Longrightarrow (x,y) \in (x'', f_{a,b}(x''))\circ(h, f_{a,b}(h))
\end{aligned}$$

Case 2. If $x' \in \mathcal{G}^x_{a,b} = \{x,\alpha\}$, then $\widehat{x''} = (x'',y'') \in (x,y)\circ(x',y') \Longrightarrow (x'',y'') = \widehat{z\bullet_{ab} w}$, with $z,w \in \mathcal{G}^x_{a,b}$. Thus $(x'',y'') \in \{\widehat{x\bullet_{ab} x}, \widehat{\alpha\bullet_{ab}\alpha}, \widehat{O}\}$. It follows that $\widehat{x} \in \widehat{x''}\circ\widehat{\alpha}$.

Case 3. If $(x,y) = \mathcal{O}$, then $Y \in \widehat{O}\circ X = X\circ\widehat{O}$, implying that $\widehat{O} \in Y\circ X$ and $X \in \widehat{O}\circ Y$. Notice that $\widehat{O} \in X\circ X$, for all $X \in \mathcal{H}_{i,j}(F)$ (i.e. every element is one of its inverses).

$(\Rightarrow)$. Suppose that H is a reversible subhypergroup of $\mathcal{H}_{\bar{A},\bar{B}}(F)$ such that it is not a subhypergroup of any $\mathcal{H}_{a,b}(F)$, with $(a,b) \in \bar{A}\times\bar{B}$. Based on Theorem 4, we have $H = \bigcup_{(i,j)\in I\subseteq\bar{A}\times\bar{B}} \mathcal{H}_{i,j}(F)$, where $I = \{(i,j) \in \bar{A}\times\bar{B} \mid H\cap\mathcal{H}_{i,j}(F) \neq \emptyset\}$. Let $(x,y),(x',y')$ be arbitrary elements in $H\cap\mathcal{H}_{i,j}(F)$ and $H\cap\mathcal{H}_{s,t}(F)$, respectively, that are not equal to $\widehat{O}$, with $(i,j) \neq (s,t)$. If $(x'',y'') \in ((x,y)\circ(x',y'))\cap\mathcal{H}_{i,j}(F)$, then, based on the reversibility, we have $(x',y') \in (z,w)\circ(x'',y'') \subseteq \mathcal{H}_{i,j}(F)$, where $z \in \mathcal{G}^x_{a,b}$, hence $(x',y') \in \mathcal{H}_{i,j}(F)\cap\mathcal{H}_{s,t}(F) = \{\widehat{O}\}$. Thus $(x',y') = \widehat{O}$, which is in contradiction with the supposition that $(x',y') \neq \widehat{O}$. Therefore $H \leqslant \mathcal{H}_{i,j}(F)$, for some $(i,j) \in \bar{A}\times\bar{B}$. □

In the following we will present two new hypergroup structures isomorphic with the equipped homography $\mathcal{H}_{a,b}(F)$ in the case when $b \neq 0$ and $b = 0$, respectively.

Theorem 6. *Consider the field F and define on* $F^* = F \setminus \{0\}$ *the hyperoperation*

$$\forall\, x, x' \in F^*, x \circledcirc x' = \{xx', \frac{x}{x'}, \frac{x'}{x}, \frac{1}{xx'}\}.$$

Then, for every $b \neq 0$*, there is the homomorphism* $(\mathcal{H}_{a,b}(F), \circ) \cong (F^*, \circledcirc)$.

Proof. It is easy to see that $(F^*, \circledcirc)$ is a hypergroup. Now, taking $v = a - b^{-1}$, consider the bijective function $\varphi : F \setminus \{v\} \longrightarrow F^*$ defined by $\varphi(x) = bx + 1 - ab$ and the function $\xi : \mathcal{H}_{a,b}(F) \longrightarrow \Gamma(\varphi)$ defined by $\xi((x,y)) = (x, \varphi(x))$, where $\Gamma(\varphi) = \{(x, \varphi(x)) \mid x \in F \setminus \{v\}\}$ and $\xi((a,\infty)) = (a,1) = \xi((\bar{A},\infty)) = \xi(\widehat{O})$. Geometrically, $\Gamma(\varphi)$ is the graph of the function φ, thus it is the line passing through the points of $(v,0)$ and $(a,1)$, while ξ is the map that projects the points of the hyperhomography $\mathcal{H}_{a,b}(F)$ on the above mentioned line.

Thus, using Proposition 1, for all $x_i, x_j \in F \setminus \{v\}$, we have

$$\begin{aligned}
\varphi(x_i \underset{a,b}{\bullet} x_j) &= b(x_i \underset{a,b}{\bullet} x_j) + 1 - ab, \\
&= b(bx_ix_j - (ab-1)(x_i + x_j - a)) + 1 - ab \\
&= (bx_i + 1 - ab)(bx_j + 1 - ab) \\
&= \varphi(x_i)\varphi(x_j).
\end{aligned}$$

Now suppose that $(x,y), (x',y')$ are arbitrary elements in $\mathcal{H}_{a,b}(F)$. It follows that

$$\begin{aligned}
\xi((x,y) \circ (x',y')) &= \{\xi(\widehat{x_i \underset{a,b}{\bullet} x_j}) \mid x_i \in \mathcal{G}^x_{a,b}, x_j \in \mathcal{G}^{x'}_{a,b}\} \\
&= \{(x_i \underset{a,b}{\bullet} x_j, \varphi(x_i \underset{a,b}{\bullet} x_j)) \mid x_i \in \mathcal{G}^x_{a,b}, x_j \in \mathcal{G}^{x'}_{a,b}\} \\
&= \{(x_i \underset{a,b}{\bullet} x_j, \varphi(x_i)\varphi(x_j)) \mid x_i \in \mathcal{G}^x_{a,b}, x_j \in \mathcal{G}^{x'}_{a,b}\}.
\end{aligned}$$

Take now $\Pi : F \times F^* \longrightarrow F^*$ with $\Pi((x,y)) = y$ as the projection map on the second component and define $\psi : \mathcal{H}_{a,b}(F) \longrightarrow F^*$ by $\psi = \Pi \circ \xi$.

We have $\psi((x,y)) = \varphi(x)$, for all $(x,y) \in \mathcal{H}_{a,b}(F)$, thus ψ is a bijective map and also a homomorphism because

$$\begin{aligned}
\psi((x,y) \circ (x',y')) &= \Pi(\xi((x,y) \circ (x',y'))) \\
&= \{\varphi(x_i)\varphi(x_j) \mid x_i \in \mathcal{G}^x_{a,b}, x_j \in \mathcal{G}^{x'}_{a,b}\} \\
&= \{\varphi(x_i)\varphi(x_j) \mid \varphi(x_i) \in \varphi(\mathcal{G}^x_{a,b}), \varphi(x_j) \in \varphi(\mathcal{G}^{x'}_{a,b})\} (\varphi \text{ is bijective map}) \\
&= \{\varphi(x_i)\varphi(x_j) \mid \varphi(x_i) \in \{\varphi(x), \frac{1}{\varphi(x)}\}, \varphi(x_j) \in \{\varphi(x'), \frac{1}{\varphi(x')}\}\} \\
&= \varphi(x) \circledcirc \varphi(x') \\
&= \psi((x,y)) \circ \psi((x',y')).
\end{aligned}$$

Therefore $(\mathcal{H}_{a,b}(F), \circ)$ is isomorphic to $(F^*, \circledcirc)$. □

Theorem 7. *Consider the field F and define on* $F^* = F \setminus \{0\}$ *the hyperoperation*

$$\forall\, x, x' \in F,\ x \circledcirc x' = \{x + x', x - x', -x + x', -x - x'\}.$$

Then, if $b = 0$*, there is the homomorphism* $(\mathcal{H}_{a,b}(F), \circ) \cong (F, \circledast)$.

Proof. Clearly, $(F, \circledast)$ is a hypergroup. Consider the bijective function $\varphi : F \longrightarrow F$ defined by $\varphi(x) = x - a$ and be $\Gamma(\varphi) = \{(x, \varphi(x)) \mid x \in F\}$ its graph. Besides define $\xi : \mathcal{H}_{a,b}(F) \longrightarrow \Gamma(\varphi)$ by $\xi((x,y)) = (x, \varphi(x))$, where $\xi((a,\infty)) = (a,0) = \xi((\bar{A},\infty))$. Therefore, for all $x_i, x_j \in F$, we have

$$\begin{aligned}\varphi(x_i \underset{a,b}{\bullet} x_j) &= (x_i \underset{a,b}{\bullet} x_j) - a,\\ &= (x_i + x_j - a) - a\\ &= (x_i - a) + (x_j - a)\\ &= \varphi(x_i) + \varphi(x_j)\end{aligned}$$

and for all $(x,y), (x',y') \in \mathcal{H}_{a,b}(F)$

$$\begin{aligned}\xi((x,y) \circ (x',y')) &= \{\xi(\widehat{x_i \underset{a,b}{\bullet} x_j}) \mid x_i \in \mathcal{G}^x_{a,b}, x_j \in \mathcal{G}^{x'}_{a,b}\},\\ &= \{(x_i \underset{a,b}{\bullet} x_j, \phi(x_i \underset{a,b}{\bullet} x_j)) \mid x_i \in \mathcal{G}^x_{a,b}, x_j \in \mathcal{G}^{x'}_{a,b}\},\\ &= \{(x_i \underset{a,b}{\bullet} x_j, \varphi(x_i) + \varphi(x_j)) \mid x_i \in \mathcal{G}^x_{a,b}, x_j \in \mathcal{G}^{x'}_{a,b}\}.\end{aligned}$$

As in the previous theorem, let $\Pi : F \times F \longrightarrow F$, $\Pi((x,y)) = y$ be the projection map on second component and define $\psi : \mathcal{H}_{a,b}(F) \longrightarrow F$ by $\psi = \Pi \circ \xi$.

Therefore, for all $(x,y) \in \mathcal{H}_{a,b}(F)$, $\psi((x,y)) = \varphi(x)$ and thus ψ is a bijective map. We claim that ψ is a homomorphism, too, because

$$\begin{aligned}\psi((x,y) \circ (x',y')) &= \Pi(\xi((x,y) \circ (x',y')))\\ &= \{\varphi(x_i) + \varphi(x_j) \mid x_i \in \mathcal{G}^x_{a,b}, x_j \in \mathcal{G}^{x'}_{a,b}\}\\ &= \{\varphi(x_i) + \varphi(x_j) \mid \varphi(x_i) \in \varphi(\mathcal{G}^x_{a,b}), \varphi(x_j) \in \varphi(\mathcal{G}^{x'}_{a,b})\} (\varphi \text{ is bijective map})\\ &= \{\varphi(x_i) + \varphi(x_j) | \varphi(x_i) \in \{\varphi(x), -\varphi(x)\}, \varphi(x_j) \in \{\varphi(x'), -\varphi(x')\}\}\\ &= \varphi(x) \circledcirc \varphi(x')\\ &= \psi((x,y)) \circ \psi((x',y')).\end{aligned}$$

Therefore $(\mathcal{H}_{a,b}(F), \circ)$ is isomorphic with $(F^*, \circledast)$. □

4. Associated H_v-Groups

Vougiouklis [13] introduced the notion of H_v-group as a generalization of the notion of hypergroup, substituting the associativity of the hyperoperation with the weak associativity, i.e., $a \circ (b \circ c) \cap (a \circ b) \circ c \neq \emptyset$ for all $a, b, c \in H$. The motivation of introducing this hyperstructure is the following one. We know that the quotient of a group with respect to a normal subgroup is a group, while the quotient of a group with respect to any subgroup is a hypergroup. Vougiouklis stated that the quotient of a group with respect to any partition of the group is an H_v-group.

In the following we equip the hyperhomography $\mathcal{H}_{\bar{A},\bar{B}}(\bar{F}) = \bigcup\limits_{(a,b) \in \bar{A} \times \bar{B}} \mathcal{H}_{a,b}(\bar{F})$ as a subset of $\bar{F}^2 \cup \{\widehat{\mathcal{O}}\}$ with an H_v-group structure, by defining the following hyperoperation

$$(\bar{x}, \bar{y}) \bar{\circ} (\bar{x'}, \bar{y'}) = \{(\bar{u}, \bar{v}) \mid (u,v) \in (\bar{x} \times \bar{y}) \circ (\bar{x'} \times \bar{y'})\}.$$

Notice that

$$(x,y) = \widehat{\mathcal{O}} \Longleftrightarrow (\bar{x}, \bar{y}) = \widehat{\mathcal{O}} \Longleftrightarrow \bar{x} \times \bar{y} = \widehat{\mathcal{O}},$$

and
$$(c,d) \notin (\bar{x} \times \bar{y}) \cap \mathcal{H}_{a,b}(F) \text{ or } (c',d') \notin (\bar{x}' \times \bar{y}') \cap \mathcal{H}_{a',b'}(F) \Longrightarrow (c,d) \circ (c',d') = \emptyset.$$
Moreover according with Thorem 2, the hyperoperation $\bar{\circ}$ is well defined on $\mathcal{H}_{\bar{A},\bar{B}}(\bar{F})$ and in addition we have
$$\overline{(x,y) \circ (x',y')} \subseteq (\bar{x},\bar{y})\bar{\circ}(\bar{x'},\bar{y'}).$$

Proposition 2. $(\mathcal{H}_{\bar{A},\bar{B}}(\bar{F}),\bar{\circ})$ *is an* H_v*-group.*

Proof. Let (x,y), (x',y') and (x'',y'') be elements in $\mathcal{H}_{\bar{A},\bar{B}}(\bar{F})$. Then we have
$$\overline{(x,y) \circ (x',y') \circ (x'',y'')} \subseteq [(\bar{x},\bar{y})\bar{\circ}(\bar{x'},\bar{y'})]\bar{\circ}(\bar{x''},\bar{y''}) \cap (\bar{x},\bar{y})\bar{\circ}[(\bar{x'},\bar{y'})\bar{\circ}(\bar{x''},\bar{y''})].$$

□

Proposition 3. *Let* $\psi_{\bar{A},\bar{B}} : \mathcal{H}_{\bar{A},\bar{B}}(F) \longrightarrow \mathcal{H}_{\bar{A},\bar{B}}(\bar{F})$, $\psi_{\bar{A},\bar{B}}(x,y) = (\bar{x},\bar{y})$. *Then* $\psi_{\bar{A},\bar{B}}$ *is an epimorphism of* H_v*-groups.*

Proof. Suppose that (x,y) and (x',y') belong to $\mathcal{H}_{\bar{A},\bar{B}}(F)$. We have
$$\begin{aligned}\psi_{\bar{A},\bar{B}}((x,y) \circ (x',y')) &= \{(\bar{u},\bar{v}) | (u,v) \in (x,y) \circ (x',y')\}\\ &\subseteq (\bar{x},\bar{y})\bar{\circ}(\bar{x}',\bar{y}')\\ &= \psi_{\bar{A},\bar{B}}(x,y)\bar{\circ}\psi_{\bar{A},\bar{B}}(x',y').\end{aligned}$$

□

Example 6. *If we consider the hyperhomography* $\mathcal{H}_{\bar{0},\bar{1}}(\bar{F}) = \{\widehat{\mathcal{O}},(\bar{1},\bar{2}),(\bar{2},\bar{1}),(\bar{2},\bar{2})\}$, *then after long calculations similarly s those in Example 5, we get the following* H_v*-group table.*

$\bar{\circ}$	$\widehat{\mathcal{O}}$	$(\bar{1},\bar{2})$	$(\bar{2},\bar{1})$	$(\bar{2},\bar{2})$
$\widehat{\mathcal{O}}$	$\widehat{\mathcal{O}}$	$(\bar{1},\bar{2}),(\bar{2},\bar{1})$	$(\bar{1},\bar{2}),(\bar{2},\bar{1})$	$(\bar{2},\bar{2})$
$(\bar{1},\bar{2})$	$(\bar{1},\bar{2}),(\bar{2},\bar{1})$	$\mathcal{H}_{\bar{0},\bar{1}}(\bar{F})$	$\mathcal{H}_{\bar{0},\bar{1}}(\bar{F})$	$(\bar{1},\bar{2}),(\bar{2},\bar{1}),(\bar{2},\bar{2})$
$(\bar{2},\bar{1})$	$(\bar{1},\bar{2}),(\bar{2},\bar{1})$	$\mathcal{H}_{\bar{0},\bar{1}}(\bar{F})$	$\mathcal{H}_{\bar{0},\bar{1}}(\bar{F})$	$(\bar{1},\bar{2}),(\bar{2},\bar{1}),(\bar{2},\bar{2})$
$(\bar{2},\bar{2})$	$(\bar{2},\bar{2})$	$(\bar{1},\bar{2}),(\bar{2},\bar{1}),(\bar{2},\bar{2})$	$(\bar{1},\bar{2}),(\bar{2},\bar{1}),(\bar{2},\bar{2})$	$\mathcal{H}_{\bar{0},\bar{1}}(\bar{F})$

Proposition 4. *On* $\mathcal{H}_{a,a}(F)$, *as a subset of* $\mathcal{H}_{\bar{A},\bar{A}}(F)$, *define the hyperoperation*
$$(x,y) \centerdot (x',y') = \begin{cases} \{\widehat{x \bullet_{aa} x'}, \widehat{y \bullet_{aa} y'}\}, & \text{if } \mathcal{O} \notin \{(x,y),(x',y')\}\\ (x,y), & \text{if } (x',y') = \mathcal{O}\\ (x',y'), & \text{if } (x,y) = \mathcal{O}.\end{cases}$$

Then $(\mathcal{H}_{a,a}(F),\centerdot)$ *is an* $H_v - group$.

Proof. First we prove that, if $\mathcal{O} \neq \widehat{x}$ and $\widehat{x} \in \mathcal{H}_{a,a}(F)$, then $\widehat{f_{a,a}(x)} \in \mathcal{H}_{a,a}(F)$. To this aim, consider an arbitrary element $\widehat{x} \in \mathcal{H}_{a,a}(F)$ not equal to $\mathcal{O}$ and notice that $f_{a,a}^2 = id_F$. Then

$$\begin{aligned}\widehat{x} \in \mathcal{H}_{a,a}(F) &\Longrightarrow (x,y) \in \mathcal{H}_{a,a}(F), y = f_{a,a}(x)\\ &\Longrightarrow f_{a,a}(y) = f_{a,a}(f_{a,a}(x))\\ &\Longrightarrow f_{a,a}(y) = (f_{a,a} \circ f_{a,a})(x)\\ &\Longrightarrow f_{a,a}(y) = x\\ &\Longrightarrow (y,x) \in \mathcal{H}_{a,a}(F),\ \ x = f_{a,a}(y)\\ &\Longrightarrow \widehat{y} \in \mathcal{H}_{a,a}(F)\\ &\Longrightarrow \widehat{f_{a,a}(x)} \in \mathcal{H}_{a,a}(F).\end{aligned}$$

Consequently, $(x,y) \centerdot (x',y') \subseteq \mathcal{H}_{a,a}(F)$, for all $(x,y),(x',y') \in \mathcal{H}_{a,a}(F)$ and "$\centerdot$" is well defined. Now, let $(x,y),(x',y'),(x'',y'')$ belong to $\mathcal{H}_{a,a}(F)$. We get

$$\{x \widehat{\bullet_{aa} x' \bullet_{aa}} x'', y \widehat{\bullet_{aa} y' \bullet_{aa}} y''\} \subseteq ([(x,y) \centerdot (x',y')] \centerdot (x'',y'')) \cap ((x,y) \centerdot [(x',y') \centerdot (x'',y'')]) \neq \emptyset,$$

thus the weak associativity condition holds. It can easily be seen that the reproduction axiom is valid, too. □

Example 7. *The Cayley table of the H_v−group $(\mathcal{H}_{0,0}(F), \centerdot)$ where, $F = \mathbb{Z}_5$ is again the field of order 5, is as follows:*

$\centerdot$	$\widehat{O}$	$(1,1)$	$(2,3)$	$(3,2)$	$(4,4)$
$\widehat{O}$	$\widehat{O}$	$(1,1)$	$(2,3)$	$(3,2)$	$(4,4)$
$(1,1)$	$(1,1)$	$(2,3)$	$(3,2),(4,4)$	$(3,2),(4,4)$	$\widehat{O}$
$(2,3)$	$(2,3)$	$(3,2),(4,4)$	$(1,1),(4,4)$	$\widehat{O}$	$(1,1),(2,3)$
$(3,2)$	$(3,2)$	$(3,2),(4,4)$	$\widehat{O}$	$(1,1),(4,4)$	$(2,3),(1,1)$
$(4,4)$	$(4,4)$	$\widehat{O}$	$(1,1),(2,3)$	$(1,1),(2,3)$	$(3,2)$

In this table, we have $\widehat{O} \centerdot (x,y) = (x,y) = (x,y) \centerdot \widehat{O}$, for all $(x,y) \in \mathcal{H}_{0,0}(F)$ and

$$(x,y) \centerdot (x',y') = \{(x+x',(x+x')^{-1}),(y+y',(y+y')^{-1})\},$$

for all $(x,y),(x',y') \in \mathcal{H}_{0,0}(F) \setminus \{\widehat{O}\}$. The hyperoperation is not associative, but only weak associative, as we can notice here below:

$$\widehat{O} = [[(1,1) \centerdot (1,1)] \centerdot (3,2)] \neq [(1,1) \centerdot [(1,1) \centerdot (3,2)]] = \{\widehat{O},(3,2),(4,4)\}.$$

5. Conclusions

In the last few years, researchers in the hypercompositional structure theory have investigated, principally from a theoretical point of view, all types of hyperrings: general hyperrings [20], multiplicative hyperrings [15], additive hyperrings [21], superrings [22], but till now, only the Krasner hyperrings have found interesting and useful applications in number theory, algebraic geometry, scheme theory, as mentioned in the introductory part of this article. Here the authors continue the study on the research topic started in [1] about elliptic hypercurves defined on quotient Krasner hyperfield, with applications in cryptography [8]. In a similar way, the notion of a homography on a field is extended to hyperhomography over Krasner hyperfields. More exactly, considering an arbitrary field F and a normal subgroup G of its multiplicative group, we get a Krasner hyperfield $\bar{F} = F/G$. Then the homography $H_{a,b}(F) = \{(x,y) \in F^2 \mid y = f_{a,b}(x) = b + \frac{1}{x-a}\}$, where $a, b \in F$ is naturally extended to the hyperhomography $H_{\bar{A},\bar{B}}(\bar{F}) = \{(\bar{x},\bar{y}) \in \bar{F}^2 \mid \bar{1} \in (\bar{x} \ominus \bar{A}) \odot (\bar{y} \ominus \bar{B})\}$ over the hyperfield $(\bar{F}, \oplus, \odot)$. Besides, the group operation on a homography leads to a hyperoperation on the associated equipped hyperhomography $\mathcal{H}_{\bar{A},\bar{B}}(\bar{F})$, that becomes a hypergroup. Then, all reversible subhypergroups of an equipped hyperhomography are characterized. In the last part of the paper, other hyperoperations are defined on hyperhomographies and their properties are investigated in connection with weak associativity.

Author Contributions: Conceptualization, V.V., M.J. and I.C.; Funding acquisition, I.C.; Investigation, V.V. and M.J.; Methodology, I.C.; Writing —original draft, V.V. and M.J.; Writing—review & editing, I.C.

Funding: The third author acknowledges the financial support from the Slovenian Research Agency (research core funding No. P1 - 0285).

Conflicts of Interest: The authors declare no conflict of interest.

References

1. Vahedi, V.; Jafarpour, M.; Aghabozorgi, H.; Cristea, I. Extension of elliptic curves on Krasner hyperfields. *Commun. Alg.* **2019**, *47*, 4806–4823. [CrossRef]
2. Connes, A.; Consani, C. The hyperring of adele classes. *J. Number Theory* **2011**, *131*, 159–194. [CrossRef]
3. Al Tahan, M.; Davvaz, B. On the existence of hyperrings associated with arithmetic functions. *J. Number Theory* **2017**, *174*, 136–149. [CrossRef]
4. Viro, O. On basic concepts of tropical geometry. *Proc. Steklov Inst. Math.* **2011**, *273*, 252–282. [CrossRef]
5. Jun, J. Algebraic geometry over hyperrings. *Adv. Math.* **2018**, *323*, 142–192. [CrossRef]
6. Krasner, M. A class of hyperrings and hyperfields. *Int. J. Math. Math. Sci.* **1983**, *6*, 307–311. [CrossRef]
7. Massouros, C.G. On the theory of hyperrings and hyperfields. *Algebra i Logika* **1985**, *24*, 728–742. [CrossRef]
8. Berardi, L.; Eugeni, F.; Innamorati, S. Remarks on Hypergroupoids and Criptography. *J. Combin. Inf. Syst. Sci.* **1992**, *17*, 217–231.
9. Corsini, P. *Prolegomena of Hypergroup Theory*; Aviani Editore: Tricesimo, Italy, 1993.
10. Hankerson, D.; Menezes, A.; Vanstone, S.A. *Guide to Elliptic Curve Cryptography*, Springer: Berlin, Germany, 2004.
11. Koblitz, N. *Introduction to Elliptic Curves and Modular Forms*; Volume 97 of Graduate Texts in Mathematics; Springer: New York, NY, USA, 1984.
12. Corsini, P.; Leoreanu, V. *Applications of Hyperstructure Theory*; Kluwer Academic Publications: Dordrecht, The Netherlands, 2003.
13. Vougiouklis, T. The fundamental relation in hyperrings. The general hyperfield. In Proceedings of the 4th International Congress on Algebraic Hypergroups and Applications, Xanthi, Greece, 27–30 June 1990; World Scientific: Singapore, 1991; pp. 209–217.
14. Rota, R. Sugli iperanelli moltiplicativi. *Rend. Mate* **1982**, *7*, 711–724.
15. Ameri, R.; Kordi, A.; Hoskova-Mayerova, S. Multiplicative hyperring of fractions and coprime hyperideals. *An. Ştiinţ. Univ. Ovidius Constanţa Ser. Mater.* **2017**, *25*, 5–23. [CrossRef]
16. Davvaz, B.; Leoreanu-Fotea, V. *Hyperring Theory and Applications*; International Academic Press: Cambridge, MA, USA, 2007.
17. Nakassis, A. Recent results in hyperring and hyperfield theory. *Int. J. Math. Math. Sci.* **1988**, *11*, 209–220. [CrossRef]
18. Cristea, I., Jančić-Rašović, S., Composition hyperrings. *An. Şt. Univ. Ovidius Constanţa* **2013**, *21*, 81–94. [CrossRef]
19. Norouzi, M.; Cristea, I. Hyperrings with n -ary composition hyperoperation. *J. Alg. Appl.* **2018**, *17*, 1850022. [CrossRef]
20. Norouzi, M.; Cristea, I. Fundamental relation on m-idempotent hyperrings. *Open Math.* **2017**, *15*, 1558–1567. [CrossRef]

21. Bordbar, H.; Cristea, I.; Novak, M. Height of hyperideals in Noetherian Krasner hyperrings. *Politehn. Univ. Bucharest Sci. Bull. Ser. A Appl. Math. Phys.* **2017**, *79*, 31–42.
22. Ameri, R.; Eyvazi, M.; Hoskova-Mayerova, S. Superring of polynomials over a hyperring. *Mathematics* **2019**, *7*, 902. [CrossRef]

symmetry

MDPI

Article

Some Results on (Generalized) Fuzzy Multi-H_v-Ideals of H_v-Rings

Madeline Al Tahan [1,†], **Šarka Hošková-Mayerova** [2,*,†] **and Bijan Davvaz** [3,†]

[1] Department of Mathematics, Lebanese International University, 1803 Beirut, Lebanon; madeline.tahan@liu.edu.lb
[2] Department of Mathematics and Physics, University of Defence in Brno, Kounicova 65, 66210 Brno, Czech Republic
[3] Department of Mathematics, Yazd University, 89136 Yazd, Iran; davvaz@yazd.ac.ir
* Correspondence: sarka.myerova@unob.cz; Tel.: +42-0973-442-225
† These authors contributed equally to this work.

Received: 11 October 2019; Accepted: 1 November 2019; Published: 6 November 2019

check for updates

Abstract: The concept of fuzzy multiset is well established in dealing with many real life problems. It is possible to find various applications of algebraic hypercompositional structures in natural, technical and social sciences, where symmetry, or the lack of symmetry, is clearly specified and laid out. In this paper, we use fuzzy multisets to introduce the concept of fuzzy multi-H_v-ideals as a generalization of fuzzy H_v-ideals. Moreover, we introduce the concept of generalized fuzzy multi-H_v-ideals as a generalization of generalized fuzzy H_v-ideals. Finally, we investigate the properties of these new concepts and present different examples.

Keywords: H_v-structures; H_v-ring; fundamental equivalence relation; H_v-ideal; multiset; fuzzy multiset; fuzzy multi-H_v-ideal.

MSC: 20N25, 20N20

---

1. Introduction

Symmetry is one of the central concepts of science, especially theoretical physics, mathematics and geometry of the 20th century. A given phenomenon or object is symmetrical if it is possible to introduce or consider a certain symmetry operation by which the phenomenon or object becomes in a certain sense identical to itself. The notion of symmetry has fascinated thinkers since antiquity (e.g., Pythagoreans). Later, in the so-called Erlangen program, Felix Klein tied a group of symmetry to each geometry. Mathematically, these symmetry operations are most often described by the term "group". We distinguish continuous symmetry, which are described mathematically mainly by the term "Lie groups", and discrete symmetry, which are described mainly by the term "discrete group". In mathematics, a symmetric relation is one in which variables can be exchanged or index permutations can be made without changing the relation (understood as a geometric object). The natural generalization of classical group theory is the approach of algebraic hyperstructures, introduced by F. Marty [1] during the eighth Congress of Scandinavian Mathematicians that was held in 1934. Marty generalized the notion of a group (which is a non-empty set with a binary operation satisfying some axioms and the operation of two elements is an element) to that of a hypergroup. A hypergroup is a non-empty set equipped with an associative and reproductive hyperoperation, where the composition of any two elements in it is a non-empty set. Since then, researchers started studying different kinds of hyperstructures such as: hyperrings, hypermodules, hypervector spaces, and many others by considering both parts: theoretical part as well as their applications to different subjects of science. Later in 1990, Th. Vougiouklis introduced weak hyperstructures (or H_v-structures) as a generalization of the concept of algebraic hyperstructures (hypergroups, hyperrings, hypermodules). The name "weak hyperstructures" is due to having some axioms of classical algebraic hyperstructures are replaced by their corresponding weak axioms in weak hyperstructures. Many researchers such as Corsini [2], Corsini and Leoreanu [3], Davvaz [4,5], Davvaz and Leoreanu-Fotea [6], Davvaz and Cristea [7] and Vougiouklis [8] wrote books related to (weak) hyperstructure theory and their applications. An overview about hyperstructure theory was published by Hoskova and Chvalina in [9].

On the other hand, fuzzy mathematics is an almost new branch in mathematics which was introduced in 1965 by Zadeh (see [10]). It is an extension of the classical notion of set and it is related to fuzzy set theory and fuzzy logic. Fuzzy sets are sets whose elements have degrees of membership that vary between 0 and 1 both inclusive. In classical set theory, the elements' membership in a certain set is usually identified by the condition that an element either belongs to the set or does not belong to it. By contrast, fuzzy set theory enables the gradual evaluation of the membership of elements in a set with values ranging between 0 and 1. If the membership function of a fuzzy set takes only the values 0, 1 then we go back to the classical notion of a set. As a generalization of fuzzy sets, Yager [11] introduced the concept of Fuzzy Multiset and investigated a calculus for them. Fuzzy Multiset permits the occurrence of an element more than once and each occurrence may have the same or different membership values.

In [12], Onasanya and Hoskova-Mayerova introduced multi-fuzzy groups induced by multisets. In [13,14], the authors studied fuzzy multi-polygroups and fuzzy multi-hypergroups. Moreover, Davvaz [15] and Davvaz et al. [16] discussed fuzzy H_v-ideals and generalized fuzzy H_v-ideals and investigated their properties. Our paper generalizes the work in [12,13,15,17] to combine H_v-rings and fuzzy multisets. More specifically, it is concerned about fuzzy multi-H_v-ideals and generalized fuzzy multi-H_v-ideals and it is constructed subsequently: Our motivation is described in Introduction, Section 2 presents basic notions with respect to (weak) hyperstructures and fuzzy multisets that are used throughout the paper. Section 3 defines and studies the properties of fuzzy multi-H_v-ideals and their relation to H_v-ideals. Finally, Section 4 defines generalized fuzzy multi-H_v-ideals and studies their properties.

2. Basic Definitions

In this section, we present some preliminary definitions and results related to hyperstructure theory [3,4,6] and fuzzy multisets [18] that are used throughout the paper.

2.1. (Weak) Hyperstructure Theory

Let H be a non-empty set and $\mathcal{P}^*(H)$ be the set of all non-empty subsets of H. Then, a mapping $\circ : H \times H \to \mathcal{P}^*(H)$ is called a *binary hyperoperation* on H. The couple $(H, \circ)$ is called a *hypergroupoid*. In this definition, if X and Y are two non-empty subsets of H and $h \in H$, then we define:

$$X \circ Y = \bigcup_{\substack{x \in X \\ y \in Y}} x \circ y,\ h \circ X = \{h\} \circ X \text{ and } X \circ h = X \circ \{h\}.$$

H_v-structures were introduced by T. Vougiouklis, and studied in detail in [8,19,20], as a generalization of the ordinary algebraic hyperstructures. The equalities presented in some axioms of classical algebraic hyperstructures are substituted by non-empty intersection in H_v-structures. A hypergroupoid $(H, \circ)$ is called a *quasi-hypergroup* if $a \circ H = H \circ a = H$ for all $a \in H$. And it is called an *H_v-semigroup* if $(x \circ (y \circ z)) \cap ((x \circ y) \circ z) \neq \emptyset$ for all $x, y, z \in H$. A hypergroupoid $(H, \circ)$ is called an *H_v-group* if it is a quasi-hypergroup and an H_v-semigroup. A multivalued system $(R, +, \cdot)$ is an *H_v-ring* if (1) $(R, +)$ is an H_v-group; (2) $(R, \cdot)$ is is an H_v-semigroup; (3) "$\cdot$" is weak distributive with respect to $+$.

Let $\{R_\alpha : \alpha \in \Gamma\}$ be a collection of H_v-rings (See [7]) and $\prod_{\alpha \in \Gamma} R_\alpha = \{< x_\alpha >: x_\alpha \in R_\alpha\}$. Then $(\prod_{\alpha \in \Gamma} R_\alpha, \oplus, \otimes)$ is an H_v-ring, where

$$< x_\alpha > \oplus < y_\alpha >= \{< z_\alpha >: z_\alpha \in x_\alpha + y_\alpha, \alpha \in \Gamma\},$$

$$< x_\alpha > \otimes < y_\alpha >= \{< z_\alpha >: z_\alpha \in x_\alpha \cdot y_\alpha, \alpha \in \Gamma\}.$$

A subset S of an H_v-ring $(R,+,\cdot)$ is called an *H_v-subring* if $(S,+,\cdot)$ is an H_v-ring. To prove that $(S,+,\cdot)$ is an H_v-subring of $(R,+,\cdot)$, it suffices to show that $x+S = S+x = S$ and $x \cdot y \subseteq S$ for all $x,y \in R$. An H_v-subring S of $(R,+,\cdot)$ is called an *H_v-ideal* of R if $R \cdot S \subseteq S$ and $S \cdot R \subseteq S$.

Let $(R,+,\star)$ and $(S,+',\star')$ be two H_v-rings. Then $f : R \to S$ is said to be *strong homomorphism* if $f(x+y) = f(x) +_1 f(y)$ and $f(x \star y) = f(x) \star' f(y)$ for all $x,y \in R$. $(R,+,\star)$ and $(S,+',\star')$ are called *isomorphic H_v-rings*, and written as $R \cong S$, if there exists a bijective function $f : R \to S$ that is also a strong homomorphism.

Fundamental relations are used as a tool to connect and relate the classes of hyperstuctures and algebraic structures together. In [8], Vougiouklis defined the notion of fundamental relation on H_v-rings. Koskas [21] introduced the fundamental relation $\beta^\star$ on hypergroups and later in 1990, Vougiouklis [8] introduced the fundamental relation $\gamma^\star$ on hyperrings. These fundamental relations $\beta^\star$ (for hypergroups (H_v-groups)) and $\gamma^\star$ (for hyperrings (H_v-rings)) are defined as the smallest strongly regular equivalence relations so that the quotient would be group and ring respectively. Many authors studied fundamental relations such as: Antampoufis and Hoskova-Mayerova [22], Corsini [2], Cristea and Norouzi [23–26], Davvaz [16], Freni [27], etc..

For all $n > 1$, we define the relation γ on an H_v-ring $(R,+,\cdot)$ as follows:

$$a\gamma b \iff \{a,b\} \subseteq u, u \text{ is any finite sum of finite products of elements in } R.$$

Clearly, the relation γ is reflexive and symmetric. The $\gamma^\star$, the transitive closure of γ, is called the *fundamental equivalence relation* on R and $(R/\gamma^\star, \oplus, \odot)$ is its *fundamental ring*, where for all $a,b \in R$,

$$\gamma^\star(a) \oplus \gamma^\star(b) = \gamma^\star(c) \text{ for all } c \in \gamma^\star(a) + \gamma^\star(b),$$

$$\gamma^\star(a) \odot \gamma^\star(b) = \gamma^\star(c) \text{ for all } c \in \gamma^\star(a) \cdot \gamma^\star(b).$$

2.2. Fuzzy multisets

A multiset (or bag) is a set containing repeated elements. [28,29] A fuzzy multiset is a generalization of fuzzy set and it was introduced by Yager in [11] under the name *fuzzy bag*. In these fuzzy bags the count of the number of elements itself becomes a crisp bag.

Definition 1 ([10])**.** *Let U be any non-empty set. A fuzzy set on U is characterized by a membership function $\mu_A(x)$ that assigns any element in U a grade of membership in A. The fuzzy set may be represented by the set of ordered pairs $A = \{(x, \mu_A(x)) : x \in U\}$, where $\mu_A(x) \in [0,1]$.*

Definition 2 ([30])**.** *Let X be a non-empty set and Q be the set of all crisp multisets drawn from the interval $[0,1]$. A fuzzy multiset A drawn from X is represented by a function $CM_A : X \to Q$.*

In the above definition, the value $CM_A(x)$ is a crisp multiset drawn from $[0,1]$. For each $x \in X$, $CM_A(x)$ is defined as the decreasingly ordered sequence of elements and it is denoted by:

$$\{\mu_A^1(x), \mu_A^2(x), \ldots, \mu_A^p(x)\} : \mu_A^1(x) \geq \mu_A^2(x) \geq \ldots \geq \mu_A^p(x).$$

A fuzzy set on a set X can be considered as a special case of fuzzy multiset where $CM_A(x) = \{\mu_A^1(x)\}$ for all $x \in X$.

Example 1. *Let $X = \{a,b,c,d\}$. Then $A = \{(0.7,0.5)/b, (0.7,0.2,0.1,0.1)/c, (0.3,0.1)/d\}$ and $B = \{(1,1)/a, (0.7,0.6,0.5,0.1)/b, (0.3,0.1)/c, (0.5,0.4,0.1)/d\}$ are fuzzy multisets of X.*

In Example 1, by $(0.7, 0.5)/b$ we mean that $CM_A(b) = (0.7, 0.5)$.

Definition 3 ([31]). *Let X, Y be non-empty sets, $f : X \to Y$ be a mapping, and A a fuzzy multiset of X and B a fuzzy multiset of Y. Then*

1. *The image of A under f is denoted by $f(A)$ or*

$$CM_{f(A)}(y) = \begin{cases} \bigvee_{f(x)=y} CM_A(x) & \text{if } f^{-1}(y) \neq \emptyset \\ 0 & \text{otherwise.} \end{cases}$$

2. *The inverse image of B under f is denoted by $f^{-1}(B)$ where $CM_{f^{-1}(B)}(x) = CM_B(f(x))$.*

Example 2. *Let X be a non-empty set, S be a non-empty subset of X, and A be a fuzzy multiset of S. By considering the inclusion map $f : S \to X$, $f(x) = x$ for all $x \in S$, we get that*

$$CM_{f(A)}(x) = \begin{cases} CM_A(x) & \text{if } x \in S \\ 0 & \text{otherwise.} \end{cases}$$

is a fuzzy multiset of X.

3. Fuzzy Multi-H_v-Ideal

In this section, we introduce for the first time the notion of fuzzy multi-H_v-ideal as a generalization of fuzzy H_v-ideal, present several examples and results related to this new concept. The results in [15] related to fuzzy H_v-ideals can be considered as a special case of the results of this section.

Definition 4. *Let $(R, +, \cdot)$ be an H_v-ring. A fuzzy multiset A (with fuzzy count function CM_A) over R is a fuzzy multi-H_v-ideal of R if for all $x, y \in R$, the following conditions hold.*

1. $CM_A(x) \wedge CM_A(y) \leq \inf\{CM_A(z) : z \in x + y\}$;
2. *for every $x, a \in R$ there exists $y \in R$ such that $x \in a + y$ and $CM_A(x) \wedge CM_A(a) \leq CM_A(y)$;*
3. *for every $x, a \in R$ there exists $z \in H$ such that $x \in z + a$ and $CM_A(x) \wedge CM_A(a) \leq CM_A(z)$;*
4. *$CM_A(x) \vee CM_A(y) \leq CM_A(z)$ for all $z \in x \cdot y$.*

Remark 1. *Let $(R, +, \cdot)$ be an H_v-ring with "$+$" a commutative hyperoperation and A be a fuzzy multiset over R. To prove that A is a fuzzy multi-H_v-ideal of R, it suffices to prove Conditions 1, 2, and 4 or Conditions 1, 3, and 4 of Definition 4. This is clear as in the case of commutative H_v-group, Conditions 2 and 3 are equivalent to each other.*

Example 3. *Let $(R, +, \cdot)$ be an H_v-ring with a fixed element $a \in R$ and A be a fuzzy multiset of R defined as $CM_A(x) = CM_A(a)$ for all $x \in R$. Then A is a fuzzy multi-H_v-ideal of R (the constant fuzzy multi-H_v-ideal.).*

Remark 2. *Let $(R, +, \cdot)$ be an H_v-ring. Then we can define at least one fuzzy multi-H_v-ideal of R which is mainly the one that is defined in Example 3.*

We present some examples on non-constant fuzzy multi-H_v-ideals.

Example 4. *Let $(R_1, +_1, \cdot_1)$ be the H_v-ring defined as follows:*

$+_1$	0	1
0	0	R_1
1	R_1	1

$\cdot_1$	0	1
0	0	0
1	0	R_1

It is clear that $A = \{(0.8, 0.6, 0.6, 0.1)/0, (0.5, 0.4, 0.4)/1\}$ *is a fuzzy multi-*H_v*-ideal of* R_1.

Example 5. *Let* $(R_2, +_2, \cdot_2)$ *be the* H_v*-ring defined by the following tables:*

$+_2$	a	b	c
a	a	b	c
b	b	b	R_2
c	c	R_2	c

$\cdot_2$	a	b	c
a	a	a	a
b	a	b	c
c	a	b	c

It is clear that $A = \{(0.9, 0.7, 0.6, 0.6, 0.1)/a, (0.8, 0.4, 0.2)/b, (0.8, 0.4, 0.2)/c\}$ *is a fuzzy multi-*H_v*-ideal of* R_2.

Example 6. *Let* $(R_3, +_3, \cdot_3)$ *be the* H_v*-ring defined by the following tables:*

$+_3$	d	e	f
d	d	e	f
e	e	$\{e, f\}$	d
f	f	e	d

$\cdot_3$	d	e	f
d	d	d	d
e	d	e	f
f	d	f	d

It is clear that both: $A = \{(0.9, 0.7, 0.6, 0.6, 0.1)/d, (0.9, 0.7, 0.6, 0.6, 0.1)/f\}$ *and* $B = \{(0.9, 0.8, 0.8, 0.1)/d\}$ *are fuzzy multi-*H_v*-ideals of* R_3.

Proposition 1. *Let* $(R, +)$ *be an* H_v*-group and "*$\cdot$*" be any hyperoperation on* R *with* $\{x, y\} \subseteq x \cdot y$ *for all* $x, y \in R$. *Then* A *is a fuzzy multi–*H_v*-ideal of the* H_v*-ring* $(R, +, \cdot)$ *if and only if* A *is the constant fuzzy multi-*H_v*-ideal of* R.

Proof. It is clear that if A is the fuzzy multiset described in Example 3 then A is a fuzzy multi-H_v-ideal of R. Let A be a fuzzy multi-H_v-ideal of R and $a \in R$. Having $x, a \in x \cdot a$ for all $x \in R$ and Condition 4 of Definition 4 implies that both $CM(x)$ and $CM(a)$ are greater than or equal $CM(x) \vee CM(a)$. Thus, $CM_A(x) = CM_A(a)$ for all $x \in R$. □

Example 7. *Let $(R, +, \cdot)$ be the H_v-ring defined by the following tables:*

$+$	0	1	2
0	0	1	2
1	1	2	0
2	2	0	1

$\cdot$	0	1	2
0	$\{0,1\}$	$\{0,1\}$	$\{0,2\}$
1	$\{0,1\}$	1	$\{1,2\}$
2	R	$\{1,2\}$	$\{1,2\}$

Using Proposition 1, we get that the constant fuzzy multi-H_v-ideal of R is the only fuzzy multi-H_v-ideal of R.

Notation 1. *Let $(R, +, \cdot)$ be an H_v-ring, A be a fuzzy multiset of R and $CM_A(x) = (\mu_A^1(x), \mu_A^2(x), \ldots, \mu_A^p(x))$. Then*

- $CM_A(x) = 0$ *if* $\mu_A^1(x) = 0$,
- $CM_A(x) > 0$ *if* $\mu_A^1(x) > 0$,
- $CM_A(x) = \underline{1}$ *if* $CM_A(x) = (\underbrace{1, \ldots, 1}_{s\ times})$ *where*

$$s = \max\{k \in \mathbb{N} : CM_A(y) = (\mu_A^1(y), \mu_A^2(y), \ldots, \mu_A^k(y)), \mu_A^k(y) \neq 0, y \in R\}.$$

Definition 5. *Let $(R, +, \cdot)$ be an H_v-ring and A be a fuzzy multiset of R. Then $A_\star = \{x \in R : CM_A(x) > 0\}$ and $A^\star = \{x \in R : CM_A(x) = \underline{1}\}$.*

Proposition 2. *Let $(R, +, \cdot)$ be an H_v-ring and A be a fuzzy multi-H_v-ideal of R. Then $A_\star$ is either the empty set or an H_v-ideal of R.*

Proof. Let $a \in A_\star \neq \emptyset$. First, we show that $a + A_\star = A_\star + a = A_\star$. We prove $a + A_\star = A_\star$ and $A_\star + a = A_\star$ is done similarly. For all $x \in A_\star$ and $z \in a + x$, we have $CM_A(z) \geq CM_A(a) \wedge CM_A(x) > 0$. The latter implies that $z \in A_\star$ and hence, $A_\star + a \subseteq A_\star$. Moreover, for all $x \in A_\star$, Condition 2 of Definition 4 implies that there exist $y \in R$ such that $x \in a + y$ and $CM_A(y) \geq CM_A(x) \wedge CM_A(a) > 0$. The latter implies that $y \in A_\star$ and $x \in a + A_\star$. Thus, $A_\star \subseteq a + A_\star$. Now, we prove that $R \cdot A_\star \subseteq A_\star$ and $A_\star \cdot R \subseteq A_\star$. We prove that $R \cdot A_\star \subseteq A_\star$ and $A_\star \cdot R \subseteq A_\star$ is done similarly. Let $r \in R$ and $x \in A_\star$. Then for all $z \in r \cdot x$, Condition 4 of Definition 4 implies that $CM(z) \geq CM(r) \vee CM(x) > 0$. Thus, $z \in A_\star$. □

Proposition 3. *Let $(R, +, \cdot)$ be an H_v-ring and A be a fuzzy multi-H_v-ideal of R. Then $A^\star$ is either the empty set or an H_v-ideal of R.*

Proof. Let $a \in A^\star \neq \emptyset$. First, we show that $a + A^\star = A^\star + a = A^\star$. We prove $a + A^\star = A^\star$ and $A^\star + a = A^\star$ is done similarly. For all $x \in A^\star$ and $z \in a + x$, we have $CM_A(z) \geq CM_A(a) \wedge CM_A(x) = \underline{1}$. The latter implies that $z \in A^\star$ and hence, $A^\star + a \subseteq A^\star$. Moreover, for all $x \in A^\star$, Condition 2 of Definition 4 implies that there exist $y \in R$ such that $x \in a + y$ and $CM_A(y) \geq CM_A(x) \wedge CM_A(a) = \underline{1}$. The latter implies that $y \in A^\star$ and $x \in a + A^\star$. Thus, $A^\star \subseteq a + A^\star$. Now, we prove that $R \cdot A^\star \subseteq A^\star$ and $A^\star \cdot R \subseteq A^\star$. We prove that $R \cdot A^\star \subseteq A^\star$ and $A^\star \cdot R \subseteq A^\star$ is done similarly. Let $r \in R$ and $x \in A^\star$. Then for all $z \in r \cdot x$, Condition 4 of Definition 4 implies that $CM(z) \geq CM(r) \vee CM(x) = \underline{1}$. Thus, $z \in A^\star$. □

Example 8. *Let $(R_3, +_3, \cdot_3)$ be the H_v-ring presented in Example 6. Having $A = \{(0.9, 0.7, 0.6, 0.6, 0.1)/d, (0.9, 0.7, 0.6, 0.6, 0.1)/f\}$, $B = \{(0.9, 0.8, 0.8, 0.1)/d\}$ fuzzy multi-H_v-ideals of R_3, we get that $A_\star = \{d, f\}$ and $B_\star = \{d\}$ are H_v-ideals of R_3. Also, $A^\star = B^\star = \emptyset$.*

Notation 2. *Let $(R, +, \cdot)$ be an H_v-ring, A be a fuzzy multiset of R and $CM_A(x) = (\mu_A^1(x), \mu_A^2(x), \ldots, \mu_A^p(x))$. We say that $CM_A(x) \geq (t_1, \ldots, t_k)$ if $p \geq k$ and $\mu_A^i(x) \geq t_i$ for all $i = 1, \ldots, k$. If $CM_A(x) \ngeq (t_1, \ldots, t_k)$ and $(t_1, \ldots, t_k) \ngeq CM_A(x)$ then we say that $CM_A(x)$ and $(t_1, \ldots, t_k)$ are not comparable.*

Theorem 1. *Let $(R, +, \cdot)$ be an H_v-ring, A a fuzzy multiset of R with fuzzy count function CM and $t = (t_1, \ldots, t_k)$ where $t_i \in [0, 1]$ for $i = 1, \ldots, k$ and $t_1 \geq t_2 \geq \ldots \geq t_k$. Then A is a fuzzy multi-H_v-ideal of R if and only if CM_t is either the empty set or an H_v-ideal of R.*

Proof. Let CM_t be an H_v-ideal of R and $x, y \in R$. By setting $t_0 = CM(x) \wedge CM(y)$, we get that $x, y \in CM_{t_0}$. Having CM_{t_0} an H_v-ideal of R implies that for all $z \in x + y$, $CM(z) \geq t_0 = CM(x) \wedge CM(y)$. We prove Condition 2 of Definition 4 and Condition 3 is done similarly. Let $a, x \in R$ and $t_0 = CM(x) \wedge CM(a)$. Then $a, x \in CM_{t_0}$. Having CM_{t_0} an H_v-ideal of R implies that $a + CM_{t_0} = CM_{t_0}$. The latter implies that there exist $y \in CM_{t_0}$ such that $x \in a + y$. Thus, $CM(y) \geq t_0 = CM(x) \wedge CM(a)$. We prove now Condition 4 of Definition 4. Let $x, y \in R$ and $z \in x \cdot y$. By setting $t_1 = CM(x)$ and $t_2 = CM(y)$, we get that $x \in CM_{t_1}$ and $y \in CM_{t_2}$. Having $CM_{t_1} \cdot R \subseteq CM_{t_1}$ and $R \cdot CM_{t_2} \subseteq CM_{t_2}$ implies that $z \in CM_{t_1}$ and $z \in CM_{t_2}$. Thus, $CM(z) \geq t_1 \vee t_2 \geq CM(x) \vee CM(y)$.

Conversely, let A be a fuzzy multi-H_v-ideal of R and $CM_t \neq \emptyset$. We need to show that $CM_t = a + CM_t = CM_t + a$ for all $a \in CM_t$. We prove that $CM_t = a + CM_t$ and $CM_t = CM_t + a$ is done similarly. Let $x \in CM_t$. Then $CM(z) \geq CM(x) \wedge CM(a) \geq t$ for all $z \in a + x$. The latter implies that $z \in CM_t$. Thus, $a + CM_t \subseteq CM_t$. Let $x \in CM_t$. Since A is a fuzzy multi-H_v-ideal of R, it follows that there exist $y \in R$ such that $x \in a + y$ and $CM(y) \geq CM(x) \wedge CM(a) \geq t$. The latter implies that $y \in CM_t$ and hence, $CM_t \subseteq a + CM_t$. We prove now that $R \cdot CM_t \subseteq CM_t$ and $CM_t \cdot R \subseteq R$ is done similarly. Let $y \in CM_t$ and $x \in R$. For all $z \in x \cdot y$, Condition 4 of Definition 4 implies that $CM(z) \geq CM(x) \vee CM(y) \geq t$. Thus, $z \in CM_t$. □

Corollary 1. *Let $(R, +, \cdot)$ be an H_v-ring. If R has no proper H_v-ideals then every fuzzy multi-H_v-ideal of R is the constant fuzzy multi-H_v-ideal.*

Proof. Let A be a fuzzy multi-H_v-ideal of R and suppose, to get contradiction, that A is not the constant fuzzy multi-H_v-ideal. Then there exist $x, y \in R$ with $CM(x) \neq CM(y)$. We have three cases for $CM(x) \neq CM(y)$: $CM(x) < CM(y)$, $CM(x) > CM(y)$, and $CM(x)$ and $CM(y)$ are not comparable. If $CM(x) < CM(y)$ then $y \in CM_t$ and $x \notin CM_t$ for $t = CM(y)$. If $CM(x) > CM(y)$ or $CM(x)$ and $CM(y)$ are not comparable, then $x \in CM_t$ and $y \notin CM_t$ for $t = CM(x)$. Using Theorem 1, we get that $CM_t (\neq R)$ is an H_v-ideal of R. □

Proposition 4. *Let $(R, +, \cdot)$ be an H_v-ring and S be an H_v-ideal of R. Then $S = CM_t$ for some $t = (t_1, \ldots, t_k)$ where $t_i \in [0, 1]$ for $i = 1, \ldots, k$ and $t_1 \geq t_2 \geq \ldots \geq t_k$.*

Proof. Let $t = (t_1, \ldots, t_k)$ where $t_i \in [0, 1]$ for $i = 1, \ldots, k$ and define the fuzzy multiset A of R as follows:

$$CM(x) = \begin{cases} t & \text{if } x \in S \\ 0 & \text{otherwise.} \end{cases}$$

It is clear that $S = CM_t$. We still need to prove that CM is a fuzzy multi-H_v-ideal of R. Using Theorem 1, it suffices to show that $CM_\alpha \neq \emptyset$ is an H_v-ideal of R for all $\alpha = (a_1, \dots, a_s)$ with $a_i \in [0,1]$ and $a_1 \geq \dots \geq a_s$ for $i = 1, \dots, s$. One can easily see that

$$CM_\alpha = \begin{cases} R & \text{if } \alpha = 0 \\ S & \text{if } 0 < \alpha \leq t \\ \emptyset & \text{if } (\alpha > t) \text{ or } (\alpha \text{ and } t \text{ are not comparable}). \end{cases}$$

Thus, CM_α is either the empty set or an H_v-ideal of R. □

Next, we deal with some operations on fuzzy multi-H_v-ideals.

Definition 6. *Let $(R, +, \cdot)$ be an H_v-ring and A, B be fuzzy multisets of R. Then $A \circ B$ is defined by the following fuzzy count function.*

$$CM_{A \circ B}(x) = \vee\{CM_A(y) \wedge CM_B(z) : x \in y + z\}.$$

Theorem 2. *Let $(R, +, \cdot)$ be an H_v-ring and A be a fuzzy multiset of H. If A is a fuzzy multi-H_v-ideal of R then $A \circ A = A$.*

Proof. Let $z \in R$. Then $CM_A(z) \geq CM_A(x) \wedge CM_A(y)$ for all $z \in x + y$. The latter implies that $CM_A(z) \geq \vee\{CM_A(x) \wedge CM_B(y) : z \in x + y\} \geq CM_{A \circ A}(z)$. Thus, $A \circ A \subseteq A$. Having $(R, +, \cdot)$ an H_v-ring and A a fuzzy multi-H_v-idear of R implies that for every $x \in R$ there exist $y \in R$ such that $x \in x + y$ and $CM_A(y) \geq CM_A(x)$. Moreover, we have $CM_{A \circ A}(x) = \vee\{CM_A(y) \wedge CM_B(z) : x \in y + z\} \geq CM_A(x) \wedge CM_A(y) = CM_A(x)$. Thus, $A \subseteq A \circ A$. □

Definition 7. *Let R be a non-empty set and A be a fuzzy multiset of R. We define A', the complement of A, to be the fuzzy multiset defined as: For all $x \in R$,*

$$CM_{A'}(x) = \underline{1} - CM_A(x).$$

Example 9. *Let $R = \{a, b, c\}$ be a set and A be a fuzzy multiset with fuzzy count function CM defined as: $CM(a) = 0, CM(b) = (1,1,1), CM(c) = (0.5, 0.3, 0.1)$. Then $A' = \{(1,1,1)/a, (0.9, 0.7, 0.5)/c\}$.*

Remark 3. *Let $(R, +, \cdot)$ be an H_v-ring and A be the constant fuzzy multi-H_v-ideal of R defined in Example 3. Then A' is also a fuzzy multi-H_v-ideal of R.*

Remark 4. *Let $(R, +, \cdot)$ be an H_v-ring and A be a fuzzy multi-H_v-ideal of R. Then A' is not necessary a fuzzy multi-H_v-ideal of R.*

We illustrate Remark 4 by the following example.

Example 10. *Let the triple $(R_3, +_3, \cdot_3)$ be the H_v-ring defined in Example 6 and $B = \{(0.9, 0.8, 0.8, 0.1)/d\}$ be a fuzzy multi-H_v-ideals of R_3.*

Then $B' = \{(0.9, 0.2, 0.2, 0.1)/d, (1,1,1,1)/e, (1,1,1,1)/f\}$ is not a fuzzy multi-H_v-ideals of R_3. This is clear as $d \in d \cdot e$ and $CM_{B'}(d) \ngeq CM_{B'}(d) \vee CM_{B'}(e) = (1,1,1,1)$.

Proposition 5. *Let $(R_\alpha, +_\alpha, \cdot_\alpha)$ be an H_v-ring with a fuzzy multiset A_α for all $\alpha \in \Gamma$. If A_α is a fuzzy multi-H_v-ideal of R_α for all $\alpha \in \Gamma$ then $\prod_{\alpha \in \Gamma} A_\alpha$ is a fuzzy multi-H_v-ideal of the $\prod_{\alpha \in \Gamma} R_\alpha$. Where $CM_{\prod_{\alpha \in \Gamma} A_\alpha}(< x_\alpha >) = \inf_{\alpha \in \Gamma} CM_{A_\alpha}(x_\alpha)$.*

Proof. The proof is straightforward. □

We present an example when $|\Gamma| = 2$.

Example 11. *Let* $(R_1, +_1, \cdot_1)$ *be the* H_v*-ring presented in Example 4 and*

$$A = \{(0.8, 0.6, 0.6, 0.1)/0, (0.5, 0.4, 0.4)/1\}$$

be a fuzzy multi-H_v*-ideal of* R_1. *Then* $A \times A$ *given by:*

$$\{(0.8, 0.6, 0.6, 0.1)/(0,0), (0.5, 0.4, 0.4)/(0,1), (0.5, 0.4, 0.4)/(1,0), (0.5, 0.4, 0.4)/(1,1)\}$$

is a fuzzy multi-H_v*-ideal of* $R_1 \times R_1$.

The next two propositions discuss the strong homomorphic image and pre-image of a fuzzy multi-H_v-ideal.

Proposition 6. *Let* $(R_1, +_1, \cdot_1), (R_2, +_2, \cdot_2)$ *be* H_v*-rings,* A *be a fuzzy multiset of* R_1 *and* $f : R_1 \to R_2$ *be a surjective strong homomorphism. If* A *is a fuzzy multi-*H_v*-ideal of* R_1 *then* $f(A)$ *is a fuzzy multi-*H_v*-ideal of* R_2.

Proof. Let $y_1, y_2 \in R_2$ and $y_3 \in y_1 +_2 y_2$. Since $f^{-1}(y_1) \neq \emptyset$ and $f^{-1}(y_2) \neq \emptyset$, it follows that there exist $x_1, x_2 \in R_1$ such that $CM_A(x_1) = \bigvee_{f(x)=y_1} CM_A(x)$ and $CM_A(x_2) = \bigvee_{f(x)=y_2} CM_A(x)$. Having f a homomorphism implies that $y_3 \in f(x_1) +_2 f(x_2) = f(x_1 +_1 x_2)$. The latter implies that there exists $x_3 \in x_1 + x_2$ such that $y_3 = f(x_3)$. Since A is a fuzzy multi-H_v-ideal of R_1, it follows that $CM_{f(A)}(y_3) \geq CM_A(x_3) \geq CM_A(x_1) \wedge CM_A(x_2) = CM_{f(A)}(y_1) \wedge CM_{f(A)}(y_2)$. We prove now Condition 2 of Definition 4 and Condition 3 is done similarly. Let $y, b \in R_2$. Since $f^{-1}(y) \neq \emptyset$ and $f^{-1}(b) \neq \emptyset$ then there exist $x_1, a \in R_1$ such that $CM_A(x_1) = \bigvee_{f(x)=y} CM_A(x)$ and $CM_A(a) = \bigvee_{f(x)=b} CM_A(x)$. Having A a fuzzy multi-H_v-ideal of R_1 implies that there exist $x_2 \in R_1$ with $x_1 \in a +_1 x_2$ and $CM_A(x_2) \geq CM_A(x_1) \wedge CM_A(a)$. Since f is a strong homomorphism, it follows that $y = f(x_1) \in f(x_2) +_2 b$ and $CM_{f(A)}(f(x_2)) \geq CM_A(x_2) \geq CM_A(x_1) \wedge CM_A(a) = CM_{f(A)}(y) \wedge CM_{f(A)}(b)$. We prove now Condition 4 of Definition 4 for $f(A)$. Let $y_1, y_2 \in R_2$ and $y_3 \in y_1 \cdot_2 y_2$. Since $f^{-1}(y_1) \neq \emptyset$ and $f^{-1}(y_2) \neq \emptyset$, it follows that there exist $x_1, x_2 \in R_1$ such that $CM_A(x_1) = \bigvee_{f(x)=y_1} CM_A(x)$ and $CM_A(x_2) = \bigvee_{f(x)=y_2} CM_A(x)$. Having f a strong homomorphism implies that $y_3 \in f(x_1) \cdot_2 f(x_2) = f(x_1 \cdot_1 x_2)$. The latter implies that there exists $x_3 \in x_1 \cdot_1 x_2$ such that $y_3 = f(x_3)$. Since A is a fuzzy multi-H_v-ideal of R_1, it follows that $CM_{f(A)}(y_3) \geq CM_A(x_3) \geq CM_A(x_1) \vee CM_A(x_2) = CM_{f(A)}(y_1) \vee CM_{f(A)}(y_2)$. □

Proposition 7. *Let* $(R_1, +_1, \cdot_1), (R_2, +_2, \cdot_2)$ *be* H_v*-rings,* B *be a fuzzy multiset of* R_2 *and* $f : R_1 \to R_2$ *be a surjective strong homomorphism. If* B *is a fuzzy multi-*H_v*-ideal of* R_2 *then* $f^{-1}(B)$ *is a fuzzy multi-*H_v*-ideal of* R_1.

Proof. Let $x_1, x_2 \in R_1$ and $x_3 \in x_1 +_1 x_2$. Then $CM_{f^{-1}(B)}(x_3) = CM_B(f(x_3))$. Having $f(x_3) \in f(x_1 +_1 x_2) = f(x_1) +_2 f(x_2)$ implies that $CM_{f^{-1}(B)}(x_3) = CM_B(f(x_3)) \geq CM_B(f(x_1)) \wedge CM_B(f(x_2)) = CM_{f^{-1}(B)}(x_1) \wedge CM_{f^{-1}(B)}(x_2)$. We prove now Condition 2 of Definition 4 and Condition 3 is done similarly. Let $x, a \in R_1$. Having $y = f(x), b = f(a) \in R_2$ and B a fuzzy multi-hypergroup of R_2 implies that there exist $z \in R_2$ such that $y \in b +_2 z$ and $CM_B(z) \geq CM_B(y) \wedge CM_B(b)$. Since f is a surjective strong homomorphism, it follows that there exist $w \in R_1$ such that $f(w) = z$ and $x \in a +_1 w$. We get now that $CM_{f^{-1}(B)}(w) = CM_B(z) \geq CM_B(y) \wedge CM_B(b) = CM_{f^{-1}(B)}(x) \wedge CM_{f^{-1}(B)}(a)$. To prove Condition 4 for $f^{-1}(B)$, let $x_3 \in x_1 \cdot_1 x_2$. Then $f(x_3) \in f(x_1) \cdot_2 f(x_2)$. Having $CM_{f^{-1}(B)}(x_3) = CM_B(f(x_3)) \geq CM_B(f(x_1)) \vee CM_B(f(x_2)) = CM_{f^{-1}(B)}(x_1) \vee CM_{f^{-1}(B)}(x_2)$ completes the proof. □

Corollary 2. *Let $(R,+,\cdot)$ be an H_v-ring with fundamental relation $\gamma^\star$ and A be a fuzzy multiset of R. If A is a fuzzy multi-H_v-ideal of R then B is a fuzzy multi-H_v-ideal of $(R/\gamma^\star,\oplus,\odot)$. Where*

$$CM_B(\gamma^\star(x)) = \bigvee_{\alpha\in\gamma^\star(x)} CM_A(\alpha).$$

Proof. Let A be a fuzzy multi-H_v-ideal of R and $f: R\to R/\gamma^\star$ be the map defined by $f(x)=\gamma^\star(x)$. Then f is a surjective homomorphism. Proposition 6 asserts that $f(A)$ is a fuzzy multi-H_v-ideal of $R/\gamma^\star$ where

$$CM_{f(A)}(\gamma^\star(x)) = \bigvee_{f(\alpha)=\gamma^\star(x)} CM_A(\alpha) = \bigvee_{\alpha\in\gamma^\star(x)} CM_A(\alpha) = CM_B(\gamma^\star(x)).$$

Therefore, B is a fuzzy multi-H_v-ideal of $(R/\gamma^\star,\oplus,\odot)$. □

Definition 8. *Let $(R,+,\cdot)$ be a ring. A fuzzy multiset A (with fuzzy count function CM_A) over R is a fuzzy multi-ideal of R if for all $x,y\in R$, the following conditions hold.*

1. $CM_A(x)\wedge CM_A(y)\leq CM_A(x+y)$ *for all* $x,y\in R$;
2. $CM_A(-x)\geq CM_A(x)$ *for all* $x\in R$;
3. $CM_A(x)\vee CM_A(y)\leq CM_A(x\cdot y)$ *for all* $x,y\in R$.

Proposition 8. *Let $(R,+,\cdot)$ be an H_v-ring with fundamental relation $\gamma^\star$ and A be a fuzzy multiset of R. If A is a fuzzy multi-H_v-ideal of R then B is a fuzzy multi-ideal of the ring $(R/\gamma^\star,\oplus,\odot)$. Where*

$$CM_B(\gamma^\star(x)) = \bigvee_{\alpha\in\gamma^\star(x)} CM_A(\alpha).$$

Proof. Corollary 2 asserts that Conditions 1 and 3 of Definition 8 are satisfied. We need to prove Condition 2. Having $(R/\gamma^\star,\oplus,\odot)$ a ring implies that there exist a zero element, say $\overline{0}$ such that $\overline{0}\oplus\gamma^\star(x)=\gamma^\star(x)\oplus\overline{0}=\gamma^\star(x)$ and $\overline{0}\odot\gamma^\star(x)=\gamma^\star(x)\odot\overline{0}=\overline{0}$ for all $\gamma^\star(x)\in R/\gamma^\star$. Having B a fuzzy multi-H_v-ideal of $(R/\gamma^\star,\oplus,\odot)$ implies that $CM_B(\overline{0})\geq CM_B(\gamma^\star(x))$ for all $\gamma^\star(x)\in R/\gamma^\star$. Since $(R/\gamma^\star,\oplus,\odot)$ a ring, it follows that for every $\gamma^\star(x)\in R/\gamma^\star$ there exist $-\gamma^\star(x)\in R/\gamma^\star$ with $-\gamma^\star(x)\oplus\gamma^\star(x)=\overline{0}$. Having B a fuzzy multi-H_v-ideal of $(R/\gamma^\star,\oplus,\odot)$ and using Condition 2 of Definition 4 implies that for $\gamma^\star(x)$ and $\overline{0}$ there exists $\gamma^\star(y)$ such that $\overline{0}\in\gamma^\star(x)\oplus\gamma^\star(y)$ and $CM_B(\gamma^\star(y))\geq CM_B(\overline{0})\wedge CM_B(\gamma^\star(x))=CM_B(\gamma^\star(x))$. It is clear that $\gamma^\star(y)=-\gamma^\star(x)$. □

Example 12. *Let $(R_3,+_3,\cdot_3)$ be the H_v-ring presented in Example 6. One can easily see that the fundamental ring $R_3/\gamma^\star=\{\gamma^\star(d),\gamma^\star(e)\}$ and is isomorphic to the ring of integers under standard addition and multiplication modulo 2. Using Proposition 8, we get that $\{(0.9,0.7,0.6,0.6,0.1)/\gamma^\star(d)\}$ is a fuzzy multi-ideal of $R_3/\gamma^\star$.*

4. Generalized Fuzzy Multi-H_v-Ideal

In this section, we generalize the notion of fuzzy multi-H_v-ideal defined in Section 3 to generalized fuzzy multi-H_v-ideal, investigate its properties, and present some examples.

Notation 3. *Let A be a fuzzy multiset of a non-empty set R with a fuzzy count function CM. We say that:*

1. $x_t\in CM$ *when* $CM(x)\geq t$,
2. $x_t\in qCM$ *when* $CM(x)+t\geq 1$,
3. $x_t\in\vee qCM$ *when* $x_t\in CM$ *or* $x_t\in qCM$,

4. $\underline{0.5} = \underbrace{(0.5, \ldots, 0.5)}_{s\ times}$ *where*

$$s = \max\{k \in \mathbb{N} : CM_A(y) = (\mu_A^1(y), \mu_A^2(y), \ldots, \mu_A^k(y)), \mu_A^k(y) \neq 0, y \in R\}.$$

Definition 9. *Let $(R, +, \cdot)$ be an H_v-ring. A fuzzy multiset A (with fuzzy count function CM) over R is an $(\in, \in \vee q)$-fuzzy multi-H_v-ideal of R if for all $x, y \in R$, $0 \leq t, r \leq \underline{1}$, the following conditions hold.*

1. $x_t \in CM$, $y_r \in CM$ *implies* $z_{t \wedge r} \in \vee q CM$ *for all* $z \in x + y$;
2. $x_t \in CM$, $a_r \in CM$ *implies* $y_{t \wedge r} \in \vee q CM$ *for some* $y \in R$ *with* $x \in a + y$;
3. $x_t \in CM$, $a_r \in CM$ *implies* $z_{t \wedge r} \in \vee q CM$ *for some* $z \in R$ *with* $x \in z + a$;
4. $y_t \in CM$, $x \in R$ *implies* $z_t \in \vee q CM$ *for all* $z \in x \cdot y$
   ($x_t \in CM$, $y \in R$ *implies* $z_t \in \vee q CM$ *for all* $z \in x \cdot y$).

Remark 5. *Let $(R, +, \cdot)$ be an H_v-ring and A a fuzzy multiset of R. If A is a fuzzy multi-H_v-ideal of R then A is an $(\in, \in \vee q)$-fuzzy multi-H_v-ideal of R.*

Example 13. *Let $(R, +, \cdot)$ be any H_v-ring. Then the constant fuzzy multiset of R is an $(\in, \in \vee q)$-fuzzy multi-H_v-ideal of R.*

Example 14. *Let $(R_1, +_1, \cdot_1)$ be the H_v-ring presented in Example 4. Having $A = \{(0.8, 0.6, 0.6, 0.1)/0, (0.5, 0.4, 0.4)/1\}$ is a fuzzy multi-H_v-ideal of R_1 implies that $A = \{(0.8, 0.6, 0.6, 0.1)/0, (0.5, 0.4, 0.4)/1\}$ is an $(\in, \in \vee q)$-fuzzy multi-H_v-ideal of R_1.*

The converse of Remark 5 does not always hold. We illustrate this idea by the following example.

Example 15. *Let $(R, +, \cdot)$ be the H_v-ring defined by the following tables:*

$+$	a	b	c	d
a	a	b	c	d
b	b	$\{a, b\}$	d	c
c	c	d	$\{a, c\}$	b
d	d	c	b	$\{a, d\}$

$\cdot$	a	b	c	d
a	a	a	a	a
b	a	b	b	b
c	a	c	c	c
d	a	d	d	d

One can easily see that

$$A = \{(0.7, 0.6, 0.5)/a, (0.9, 0.8, 0.8)/b, (0.9, 0.8, 0.8)/c, (0.9, 0.8, 0.8)/d\}$$

is an $(\in, \in \vee q)$-fuzzy multi-H_v-ideal of R but not a fuzzy multi-H_v-ideal of R. This is clear as $a \in a \cdot b$ but $CM_A(a) \not\geq CM_A(a) \vee CM_A(b)$.

Proposition 9. *Let $t = (t_1, \ldots, t_k)$, $s = (s_1, \ldots, s_p)$ with $t_1 \geq \ldots \geq t_k$ and $s_1 \geq \ldots \geq s_p$. If $t < s$ then there exists $r = (r_1, \ldots, r_m)$ such that $t < r < s$.*

Proof. We have the following cases:
Case $k < p$. Take $r = (s_1, \ldots, s_p, \frac{s_{p+1}}{2})$.
Case $k = p$. Then there exists $i \in \{1, \ldots, k\}$ with $t_i < s_i$. Since s_i, t_i are real numbers, it follows that there exists a real number r_i with $t_i < r_i < s_i$. By taking $r = (t_1, \ldots, t_{i-1}, r_i, r_i \wedge s_{i+1}, \ldots, r_i \wedge s_k)$, we get that $t < r < s$. $\square$

Theorem 3. *Let $(R, +, \cdot)$ be an H_v-ring, A a fuzzy multiset of R with fuzzy count function CM, and for all $x \in R$, $CM(x)$ and $\underline{0.5}$ are comparable. If A is an $(\in, \in \vee q)$-fuzzy multi-H_v-ideal of R then the following conditions hold:*

(a) $CM(x) \wedge CM(y) \wedge \underline{0.5} \leq CM(z)$ for all $z \in x + y$;

(b) For all $x, a \in R$ there exists $y \in R$ such that $x \in a + y$ and

$$CM(a) \wedge CM(x) \wedge \underline{0.5} \leq CM(y).$$

(c) For all $x, a \in R$ there exists $z \in R$ such that $x \in z + a$ and

$$CM(a) \wedge CM(x) \wedge \underline{0.5} \leq CM(z).$$

(d) For all $z \in x \cdot y$, $CM(y) \wedge \underline{0.5} \leq CM(z)$ and $CM(x) \wedge \underline{0.5} \leq CM(z)$.

Proof. It suffices to show that $(1) \to (a)$, $(2) \to (b)$, $(3) \to (c)$, and $(4) \to (d)$.

(1) → (a): Let $x, y \in R$. Since each of $CM(x), CM(y)$ are comparable with $\underline{0.5}$, we can consider the cases: $CM(x) \wedge CM(y) < \underline{0.5}$ and $CM(x) \wedge CM(y) \geq \underline{0.5}$.

For the case $CM(x) \wedge CM(y) < \underline{0.5}$, suppose that there exists $z \in x + y$ with $CM(z) < CM(x) \wedge CM(y) \wedge \underline{0.5}$. We get that $CM(z) < CM(x) \wedge CM(y)$. Proposition 9 asserts that there exists r with $CM(z) < r < CM(x) \wedge CM(y)$. The latter implies that $x_r, y_r \in CM$ and $z \notin CM$. Moreover, having $CM(z) + r < \underline{0.5} + r \leq \underline{1}$ implies that $z \notin qCM_r$. We get that $z_r \notin \vee qCM$ which contradicts (1).

For the case $CM(x) \wedge CM(y) \geq \underline{0.5}$, suppose that there exists $z \in x + y$ with $CM(z) < CM(x) \wedge CM(y) \wedge \underline{0.5}$. We get that $x_{\underline{0.5}}, y_{\underline{0.5}} \in CM$ and $CM(z) < \underline{0.5}$. It is clear that $z_{0.5} \notin \vee qCM$ which contradicts (1).

(2) → (b): Let $x, a \in R$. Since each of $CM(x), CM(a)$ are comparable with $\underline{0.5}$, we can consider the cases: $CM(x) \wedge CM(a) < \underline{0.5}$ and $CM(x) \wedge CM(a) \geq \underline{0.5}$.

For the case $CM(x) \wedge CM(a) < \underline{0.5}$, suppose that for all $y \in R$ with $x \in a + y$ we have $CM(x) \wedge CM(a) = CM(x) \wedge CM(a) \wedge \underline{0.5} > CM(y)$. Proposition 9 asserts that there exists r with $CM(y) < r < CM(x) \wedge CM(a)$. It is clear that $x_r, y_r \in CM$ and $y_r \notin \vee qCM$. The latter contradicts (2).

(3) → (c): This case is done in a similar manner to that of **(2) → (b)**.

(4) → (d): Let $x, y \in R$. Since $CM(y)$ is comparable with $\underline{0.5}$, we can consider the cases: $CM(y) < \underline{0.5}$ and $CM(y) \geq \underline{0.5}$.

For the case $CM(y) < \underline{0.5}$, suppose that there exists $z \in x \cdot y$ with $CM(z) < CM(y) \wedge \underline{0.5} < CM(y)$. Proposition 9 asserts that there exists r with $CM(z) < r < CM(y)$. Then $y_r \in CM$ and $z_r \notin \vee qCM$ which contradicts (4).

For the case $CM(y) \geq \underline{0.5}$, suppose that there exists $z \in x \cdot y$ with $CM(z) < CM(y) \wedge \underline{0.5} \leq \underline{0.5}$. Then $y_{\underline{0.5}} \in CM$ and $z_{\underline{0.5}} \notin \vee qCM$ which contradicts (4). $\square$

Remark 6. *Theorem 3 can be used only when $CM(x)$ and $\underline{0.5}$ are comparable. Otherwise, we should use Definition 9.*

Note that according to Remark 6, we can not use Theorem 3 to the fuzzy multiset given in Example 5 as $CM_A(0) = (0.9, 0.7, 0.6, 0.6, 0.1)$ is not comparable with $\underline{0.5} = (0.5, 0.5, 0.5, 0.5, 0.5)$.

Remark 7. *In case of fuzzy H_v-ideal of an H_v-ring, the conditions of Theorem 3 are necessary and sufficient for a fuzzy set to be a fuzzy H_v-ideal (see [16]). Whereas in our case (fuzzy multiset), the converse of Theorem 3 is not always true. (See Example 16.)*

Example 16. *Let $(R, +, \cdot)$ be the H_v-ring defined in Example 15 and let A be the fuzzy multiset of R with count function CM defined by: $CM(a) = (0.7, 0.6, 0.5)$, $CM(b) = (0.7, 0.5, 0.5)$, $CM(c) = CM(d) = (0.6, 0.6, 0.6)$. Having $\underline{0.5} = (0.5, 0.5, 0.5)$, it is easy to see that Conditions (a), (b), (c), and (d) of Theorem 3 are satisfied. But A is not an $(\in, \in \vee q)$-fuzzy multi-H_v-ideal of R. By taking $t = (0.6, 0.6, 0.3)$, we get that $c_t \in CM$. Having $b \in b \cdot c$, $CM(b) \ngeq t$ and $CM(b) + t = (1.3, 1.1, 0.8) \ngeq \underline{1}$ implies that Condition 4 of Definition 9 is not satisfied.*

5. Conclusions

This paper has introduced algebraic hyperstructures of fuzzy multisets, for the first time, in the forms of fuzzy multi-H_v-ideals and generalized fuzzy multi-H_v-ideals. Several interesting properties related to the new defined notions were investigated and operations on fuzzy multi-H_v-ideals were defined and discussed. It is well known that the concept of fuzzy multiset is well established in dealing with many real life problems. As a result, we can deal with real life problems involving the concept of fuzzy multiset with a different perspective.

Author Contributions: Conceptualization, M.A.T., S.H.-M. and B.D.; methodology, M.A.T. and S.H.-M.; formal analysis and investigation, M.A.T. and S.H.-M.; resources, M.A.T. and S.H.-M.; writing–original draft preparation, M.A.T.; writing–review and editing, M.A.T. and S.H.-M.; supervision, B.D.; project administration and funding acquisition, S.H.-M.

Funding: This research was supported and the APC was funded within the project DZRO K217, supported by the Ministry of Defence in the Czech Republic.

Conflicts of Interest: The authors declare no conflict of interest. The funders had no role in the design of the study; in the collection, analyses, or interpretation of data; in the writing of the manuscript, or in the decision to publish the results.

References

1. Marty, F. Sur une Generalization de la notion de Group. Available online: https://www.scienceopen.com/document?vid=037b45a2-5350-43d4-86e1-39673e906fb5 (accessed on 10 October 2019)
2. Corsini, P. *Prolegomena of Hypergroup Theory*, 2nd ed.; Aviani Editore: Udine, Tricesimo, Italy, 1993.
3. Corsini, P.; Leoreanu, V. Applications of hyperstructures theory. In *Advances in Mathematics*; Kluwer Academic Publisher: Dordrecht, The Netherlands, 2003.
4. Davvaz, B. *Semihypergroup Theory*; Elsevier-Academic Press: London, UK, 2016; 156p.
5. Davvaz, B. *Polygroup Theory and Related Systems*; World Scientific Publishing Co. Pte. Ltd.: Hackensack, NJ, USA, 2013; 200p.
6. Davvaz, B.; Leoreanu-Fotea, V. *Hyperring Theory and Applications*; International Academic Press: Palm Harbor, FL, USA, 2007.
7. Davvaz, B.; Cristea, I. *Fuzzy Algebraic Hyperstructures*; Studies in Fuzziness and Soft Computing 321; Springer International Publishing: Cham, Switzerland, 2015. [CrossRef]
8. Vougiouklis, T. *Hyperstructures and Their Representations*; Hadronic Press Monographs: Palm Harbour, FL, USA, 1994, 180p.
9. Hoskova-Mayerova, S.; Chvalina, J. A survey of investigations of the Brno research group in the hyperstructure theory since the last AHA Congress. In Proceedings of the AHA, Brno, Czech Republic, 8–12 November 2008; pp. 71–84.
10. Zadeh, L.A. Fuzzy sets. *Inf. Control* **1965**, *8*, 338–353. [CrossRef]
11. Yager, R.R. On the theory of bags. *Int. J. Gen. Syst.* **1987**, *13*, 23–37. [CrossRef]
12. Onasanya, B.O.; Hoskova-Mayerova, S. Multi-fuzzy group induced by multisets. *Ital. J. Pure Appl. Maths* **2019**, *41*, 597–604.
13. Al-Tahan, M.; Hoskova-Mayerova, S.; Davvaz, B. Fuzzy multi-polygroups. *J. Intell. Fuzzy Syst.* **2019**, submitted.

14. Al-Tahan, M.; Hoskova-Mayerova, S.; Davvaz, B. Fuzzy multi-hypergroups. *J. Intell. Fuzzy Syst.* **2019**, submitted.
15. Davvaz, B. On H_v-rings and fuzzy H_v-ideals. *J. Fuzzy Math.* **1998**, *6*, 33–42.
16. Davvaz, B.; Zhan, J.; Shum, K.P. Generalized fuzzy H_v-ideals of H_v-rings. *Int. J. Gen. Syst.* **2008**, *37*, 329–346. [CrossRef]
17. Dresher, M.; Ore, O. Theory of multigroups. *Am. J. Math.* **1938**, *60*, 705–733. [CrossRef]
18. Miyamoto, S. *Fuzzy Multisets and Their Generalizations, Multiset Processing;* Lecture Notes in Computer Science 2235; Springer: Berlin, Germany, 2001; pp. 225–235.
19. Vougiouklis, T. On Hv-rings and Hv-representations. *Discret. Math.* **1999**, *208/209*, 615–620. [CrossRef]
20. Vougiouklis, T. On the Hyperstructure Theory. *Southeast Asian Bull. Math.* **2016**, *40*, 603–620.
21. Koskas, M. Groupoids, demi-hypergroupes et hypergroupes. *J. Math. Pures Appl.* **1970**, *49*, 155–192.
22. Antampoufis, N.; Hoskova-Mayerova, S. A Brief Survey on the two Different Approaches of Fundamental Equivalence Relations on Hyperstructures. *Ratio Math.* **2017**, *33* 47–60. [CrossRef]
23. Norouzi, M.; Cristea, I. A new type of fuzzy subsemihypermodules. *J. Intell. Fuzzy Syst.* **2017**, *32*, 1711–1717. [CrossRef]
24. Norouzi, M.; Cristea, I. Transitivity of the $\in$ m-relation on (m-idempotent) hyperrings. *Open Math.* **2018**, *16*, 1012–1021. [CrossRef]
25. Norouzi, M.; Cristea, I. Fundamental relation on m-idempotent hyperrings. *Open Math.* **2017**, *15*, 1558–1567. [CrossRef]
26. Cristea, I.; Ştefănescu, M.; Angheluţă, C. About the fundamental relations defined on the hypergroupoids associated with binary relations. *Eur. J. Comb.* **2011**, *32*, 72–81. [CrossRef]
27. Freni, D. Hypergroupoids and fundamental relations. In *Proceedings of AHA*; Stefanescu, M., Ed.; Hadronic Press: Palm Harbour, FL, USA, 1994; pp. 81–92.
28. Jena, S.P.; Ghosh, S.K.; Tripathi B.K. On theory of bags and lists. *Inf. Sci.* **2011**, *132*, 241–254. [CrossRef]
29. Syropoulos, A. Mathematics of multisets, Multiset processing. *Lecture Notes in Comput. Sci.* **2001**, *2235*, 347–358.
30. Shinoj, T.K.; John, S.J. Intutionistic fuzzy multisets. *Int. J. Eng. Sci. Innov. Technol. (IJESIT)* **2013**, *2*, 1–24.
31. Shinoj, T.K.; Baby, A.; John S.J. On some algebraic structures of fuzzy multisets. *Ann. Fuzzy Math. Inform.* **2015**, *9*, 77–90.

symmetry

MDPI

Article

Results on Functions on Dedekind Multisets

Šarka Hoškova-Mayerova [1,*,†] and Babatunde Oluwaseun Onasanya [2,†]

1 Department of Mathematics and Physics, University of Defence Brno, Kounicova 65, 66210 Brno, Czech Republic
2 Department of Mathematics, Faculty of Science, University of Ibadan, Ibadan 200284, Oyo State, Nigeria
* Correspondence: sarka.mayerova@unob.cz; Tel.: +420-973-44-2225
† These authors contributed equally to this work.

Received: 16 July 2019; Accepted: 2 September 2019; Published: 4 September 2019

Abstract: Many real-life problems are well represented only by sets which allow repetition(s), such as the multiset. Although not limited to the following, such cases may arise in a database query, chemical structures and computer programming. The set of roots of a polynomial, say $f(x)$, has been found to correspond to a multiset, say F. If $f(x)$ and $g(x)$ are polynomials whose sets of roots respectively correspond to the multisets $F(x)$ and $G(x)$, the set of roots of their product, $f(x)g(x)$, corresponds to the multiset $F \uplus G$, which is the sum of multisets F and G. In this paper, some properties of the algebraic sum of multisets $\uplus$ and some results on selection are established. Also, the count function of the image of any function on Dedekind multisets is defined and some of its properties are established. Some applications of these multisets are also given.

Keywords: multisets; functions on multiset; selection operation; submultiset

MSC: 03B70

1. Introduction

Let us start with the words of Prof. Irina Cristea that motivated us for this study; in many cases *Symmetry plays a fundamental role in our daily lives and in the study of the structure of different objects in physics, chemistry, biology, mathematics, architecture, arts, sociology, linguistics, etc. For example, the structure of molecules is well explained by their symmetry properties, described by symmetry elements and symmetry operations. A symmetry operation is a change, a transformation after which certain objects remain invariant, such as rotations, reflections, inversions, or permutation operations. Until now, the most efficient way to better describe symmetry, is using mathematical tools offered by group theory.*

The notion of multiset can be traced to as far back as 1888 where Dedekind in [1] said that an element of a set may belongs to it more than once. In 1989, Blizard in his paper [2] developed a first-order two-sorted multi set theory for multisets that "contains" classical set theory. Later on, in 1993 [3], he identified a kind of multiset which is based on the function of the root set. This idea was primarily from the work of Dedekind in [1].

However, Syropoulos in [4] also studied various operations on multisets and extended his work to category of multisets. Wildberger in [5] considered the use of multisets in data structure and also related it to tropical mathematics and gave some applications of it responding to meet catalogue of orders emanating from various customers from a set of inventory of a sales company. Knuth [6] related multisets to various aspects of computer programming.

Multisets are furthermore studied in the form of, and substituted with, numerous concepts such as bag, fireset (finitely repeated element set), heap, bunches, etc. These concepts have all been studied by various mathematicians with different applications.

To be specific, Nazmul et al. [7] extended the study of multisets to multigroup and other related algebraic properties as in the classical group. Concepts such as multicosets, [8,9] symmetric multigroup, and many others have all been studied [10–12]. Yohanna and Simon studied symmetric groups under multiset in [13]. Congruences of Multialgebra were studie by Ameri and Rosenberg in [14]. Even if the list of related papers is long we would like to draw the readers attention only to some of them, e.g., [4,5,15,16].

In this paper, we present some of the operations on multisets and some applications to real-life problems, number theory and management [17].

2. Preliminaries

In this paper, we shall use X to denote a non-empty set.

Definition 1 ([7,10]). *A* multiset *M drawn from a set X is denoted by the count function $C_M\colon X \longrightarrow N$ defined by $C_M(x) = n \in N$, the multiplicity or number of occurrence of x in M, where N is the set of non-negative integers.*

Definition 2 ([7]). *Let multisets A and B be drawn from X. A is said to be a* submultiset *of B and is denoted $A \subseteq B$ if $C_A(x) \leq C_B(x)\ \forall x \in X$.*

Definition 3 ([7]). *The* root set *or* support *of a multiset M, which is denoted by M^*, is the set which contains the distinct elements in the multiset. Hence, M^* is the set of $x \in M$ such that $C_M(x) > 0$.*

A multiset M is called a *regular multiset* if $C_M(x) = C_M(y)\ \forall x, y \in M$. The count function of the intersection of two multisets A and B both drawn from X is denoted by $C_A(x) \cap C_B(x) = \min\{C_A(x), C_B(x)\}$ and that of their union is denoted $C_A(x) \cup C_B(x) = \max\{C_A(x), C_B(x)\}$.

Multisets A and B are said to be equal if and only if $C_A(x) = C_B(x)$. Denote by $[X]^\alpha$, all the multisets whose elements have the multiplicity not more than α and $MS(X)$ the *set of all multisets drawn from* X. An empty multiset ϕ is such that $C_\phi(x) = 0,\ \forall x \in X$. *Cardinality* of a multiset M is denoted by $|M| = \sum C_M(x),\ \forall x \in M$. The *peak element* $x \in M$ is such that $C_M(x) \geq C_M(y),\ \forall y \in M$.

Definition 4 ([18]). *i.* *Consider $A \in MS(X)$. The insertion of x into A results into a multiset denoted by $\mathcal{C} = x \uplus A$ which has the count function*

$$C_{\mathcal{C}}(y) = \begin{cases} C_A(y), & y \neq x \\ C_A(x) + 1, & y = x. \end{cases}$$

ii. *Consider $A, B \in MS(X)$. The insertion of A into B or of B into A results into a multiset $\mathcal{C}$ which has the count function denoted by $C_{\mathcal{C}}(x) = C_A(x) + C_B(x)$.*

It should be noted that the operation of insertion ($\uplus$) on the set of all multisets drawn from X, that is $MS(X)$, is commutative and associative.

Definition 5 ([18]). *i.* *Consider $A \in MS(X)$. The removal of x from A results into a multiset denoted by $\mathcal{D} = A \ominus x$ which has the count function*

$$C_{\mathcal{D}}(y) = \begin{cases} \max\{C_A(y) - 1, 0\}, & y = x \\ C_A(y), & y \neq x. \end{cases}$$

ii. *Consider $A, B \in MS(X)$. The removal of B from A results into a multiset which has the count function denoted by $C_{\mathcal{D}}(x) = \max\{C_A(x) - C_B(x), 0\}$.*

It should be noted that the removal operation is neither commutative nor associative. Besides, it is also possible to make some kind of selection in multisets using the following operations.

Definition 6 ([10,18]). *Consider $A \in MS(X)$ and $B \subseteq X$.*

i. *The multiset $\mathcal{E} = A \otimes B$ is such that $\mathcal{E}$ only contains elements of A which also occur in B. The count function of $\mathcal{E}$ is denoted by*

$$C_{\mathcal{E}}(x) = \begin{cases} C_A(x), & x \in B \\ 0, & x \notin B. \end{cases}$$

ii. *The multiset $\mathcal{F} = A \odot B$ is such that $\mathcal{F}$ only contains elements of A which do not occur in B. The count function of $\mathcal{F}$ is denoted by*

$$C_{\mathcal{F}}(x) = \begin{cases} C_A(x), & x \notin B \\ 0, & x \in B. \end{cases}$$

Operations defined above "$\otimes$" or "$\odot$" are called selection operations.

Definition 7 ([7]). *Let X be a group and $e \in X$ its identity. Then, $\forall x, y \in X$, a multiset M drawn from X is called a* multigroup *if*

i. $C_M(xy) \geq C_M(x) \wedge C_M(y)$,
ii. $C_M(x^{-1}) \geq C_M(x)$.

Remark 1. *The immediate consequenceof this is that $C_M(e) \geq C_M(x)$ and $C_M(x^{-1}) = C_M(x)$, for any $x \in X$. We shall call $MG(X)$ the* set of all multigroups drawn from X.

Example 1. *Let $X = \{1,2,3,4,5\}$. Let the multiset $M = \{1,1,1,1,2,2,3,3,3,4,4,5,5\}$, $N = \{1,1,2,2,3,4,4,4,5\}$ and $W = \{1,1,1,2,3,3\}$ be drawn from X. It is also justifiable to say that X is the support of M and N. Furthermore, the root set of W is $W^* = \{1,2,3\}$.*

i. $C_N(1) = 2$, $C_N(2) = 2$, $C_N(3) = 1$, $C_N(4) = 3$ *and* $C_N(5) = 1$
ii. $M \cap N = \{1,1,2,2,3,4,4,5\}$. *When two or more multisets are intersected, the minimum multiplicity of the common elements is taken.*
iii. $M \cup N = \{1,1,1,1,2,2,3,3,3,4,4,4,5,5\}$. *When the union of two or more multisets is taken, the maximum multiplicity of the common elements is taken.*
iv. *Neither is $M \subseteq N$, since $C_M(x) \leq C_N(x)$ for all $x \in X$, nor $M \supseteq N$, since $C_M(x) \geq C_N(x)$ for all $x \in X$, but $W \subseteq M$, since $C_W(x) \leq C_M(x)$ for all $x \in X$.*

Example 2. *i.* *Let $G = \{1,-1,i,-i\}$ be a group with the usual multiplication. $M = \{1,1,1,-1,-1,-1,i,i,-i,-i\}$ is a multigroup since it satisfies Definition 7 and Remark 1.*
ii. *Consider the group $Z_3 = \{0,1,2\}$ with the modulo addition. $M = \{0,0,0,1,1,2,2\}$ is a multigroup.*

Definition 8 ([10]). *Let $A \in MS(X)$, where X is a group.*

i. $A_n = \{x : C_A(x) \geq n\}$;
ii. *We denote a multiset containing only one element x with multiplicity n as $[n]_x$–*a simple multiset;
iii. *The* complement of the multiset *$M \in [X]^\alpha$ denoted by M' is such that $C_{M'}(x) = \alpha - C_M(x)$;*
iv. $nA = \{x^n, \forall x \in A$ *, n is the multiplicity of each element that appears in* $nA\}$.

Example 3. *Let $X = \{0,1,2,3\}$ which is a group with respect to addition modulo 4 and $A = \{0,0,0,1,1,2,2\}$ a multiset.*

i. $2A = \{0,0,0,0,0,0,1,1,1,1,2,2,2,2\}$, *the multiplicities of each element in A is doubled;*

ii. $[2]_0 = \{0,0\}$, *this multiset comprises only 0 and its multiplicity is 2;*
iii. $A_2 = A_1 = A_0 = \{0,1,2\}$ *and* $A_3 = \{0\}$. *These are the sets of elements of A respectively with multiplicities 2, 1, 0 and 3.*

Remark 2. *For a multigroup A drawn from a group X,* A_n *is a group, indeed the subgroup of X* [7].

Proposition 1 ([7], p. 645). *Let* $A, B \in MS(X)$ *and* $m, n \in \mathbf{N}$.

i. *If* $A \subseteq B$, *then* $A_n \subseteq B_n$;
ii. *If* $m \leq n$, *then* $A_m \supseteq A_n$;
iii. $(A \cap B)_n = A_n \cap B_n$;
iv. $(A \cup B)_n = A_n \cup B_n$;
v. $A = B$ *if and only if* $A_n = B_n$, $\forall n \in \mathbf{N}$.

Definition 9 ([7]). *Let X and Y be two nonempty sets such that* $f\colon X \longrightarrow Y$ *is a mapping. Consider the multisets* $M \in [X]^\alpha$ *and* $N \in [Y]^\alpha$. *Then,*

i. *the image of M under f denoted* $f(M)$ *has the count function*

$$C_{f(M)}(y) = \begin{cases} \bigvee_{f(x)=y} C_M(x), & \text{if } f^{-1}(y) \neq \emptyset \\ 0, & \text{otherwise;} \end{cases}$$

ii. *the inverse image of N under f denoted* $f^{-1}(N)$ *has the count function* $C_{f^{-1}(N)}(x) = C_N[f(x)]$.

The following Propositions were proved in [7]. But we shall later show that the items (iv), (v) and (vii) are not true and that the Proposition 2 needs to be restated.

Proposition 2 ([7]). *Let X, Y and Z be three nonempty sets such that* $f\colon X \longrightarrow Y$ *and* $g : Y \longrightarrow Z$ *are mappings. If* $M_i \in [X]^\alpha$, $N_i \in [Y]^\alpha$, $i \in I$ *then*

i. $M_1 \subseteq M_2 \Rightarrow f(M_1) \subseteq f(M_2)$;
ii. $f(\cup_{i\in I} M_i) = \cup_{i \in I} f(M_i)$;
iii. $N_1 \subseteq N_2 \Rightarrow f^{-1}(N_1) \subseteq f^{-1}(N_2)$;
iv. $f^{-1}(\cup_{i\in I} M_i) = \cup_{i\in I} f^{-1}(M_i)$;
v. $f^{-1}(\cap_{i\in I} M_i) = \cap_{i\in I} f^{-1}(M_i)$;
vi. $f(M_i) \subseteq N_j \Rightarrow M_j \subseteq f^{-1}(N_j)$;
vii. $g[f(M_i)] = [gf](M_i)$ *and* $f^{-1}[g^{-1}(N_j)] = [gf]^{-1}(N_j)$.

3. Some Illustrations of Properties of Operations on Multisets

In the following section we now introduce some new results and properties of defined operations.

Proposition 3. *The operation* $\uplus$ *in Definition 4(ii) is such that:*

i. *Let A be a multiset drawn a nonempty set X. The n insertion of A into itself denoted* $\uplus_n A = nA$;
ii. $A \uplus A = A \Leftrightarrow A = \phi$.

Proof. i. By Definition 4(ii),

$$C_{\uplus_n A}(x) = \underbrace{C_A(x) + C_A(x) + \cdots + C_A(x)}_{n \text{ times}} = nC_A(x)\ \forall x \in A;$$

ii. Assume that $A \uplus A = A$ but $A \neq \emptyset$. By (i), $A \uplus A = 2A$. Then, $2A = A$. This is not possible by Definition 4(ii). Thus, $A = \emptyset$.

Conversely, assume that $A = \emptyset$ but $A \uplus A \neq A = \emptyset$. Then $A \uplus A = 2A$. But $C_A(x) = 0$; $\forall x \in X$. Then, $C_{A \uplus A}(x) = C_A(x) + C_A(x) = 0$. This implies that $A \uplus A = \emptyset$. This is a contradiction. Hence, $A \uplus A = A$. □

Remark 3. *Some properties of the selection operation $\bigotimes$ in Definition 6(i) and $\bigodot$ in Definition 6(ii) will be illustrated here. Let X be a nonempty set, $A \in MS(X)$, $B \subseteq X$ and $E = A \bigodot B$.*

i. Let $D = A \bigotimes B$. If $B \subset A^*$ then $D \subseteq A$. If $A^* \subseteq B$ then $D = A$.
ii. If $B \subset A^*$ then $E \subseteq A$. If $A^* \subseteq B$ then $E = \emptyset$.
iii. $D^* \cup E^* = A^*$.

This will be illustrated by the following Examples.

Example 4. *Let $X = \{1,2,3,4,5\}$, $B = \{2,3,5\}$.*

i. If $A = \{1,1,1,2,3,3,3,4,4,5,5\}$. Then, $D = \{2,3,3,3,5,5\}$, $E = \{1,1,1,4,4\}$ and $A^* = \{1,2,3,4,5\}$. Obviously, $B \subseteq A^*$, $D \subseteq A$ and $E \subseteq A$.
ii. If, on the other hand, $A = \{3,3,3,5,5\}$, $A^* = \{3,5\}$. Then, $A^* \subseteq B$, $D = \{3,3,3,5,5\} = A$ and $E = \emptyset$.
iii. In [i], $D^* = \{2,3,5\}$ and $E^* = \{1,4\}$. Hence, $D^* \cup E^* = A^*$. Also, in [ii] $D^* = \{3,5\}$ and $E^* = \emptyset$. Hence, $D^* \cup E^* = A^*$.

Proposition 4. *Let A be a multiset drawn from X and $B \subseteq X$. Then, $A^* \cap B = (A \bigotimes B)^*$.*

Proof. Let $x \in (A^* \cap B)$. Then $x \in A^*$ and $x \in B$. Thus, $C_A(x) \neq 0$. Hence, $x \in A \bigotimes B$. It can be concluded that $x \in (A \bigotimes B)^*$, in which case $(A^* \cap B) \subseteq (A \bigotimes B)^*$.

Also, let $x \in (A \bigotimes B)^*$, then $C_{(A \bigotimes B)}(x) \neq 0$. This implies that $x \in B$ and $C_A(x) \neq 0$ in which case $x \in A^*$. Then $x \in (A^* \cap B)$. Thus, $(A \bigotimes B)^* \subseteq (A^* \cap B)$. □

Proposition 5. *Let A and B be multisets drawn from X.*

i. $A \cup B \subseteq A \uplus B$;
ii. $A \cup B = A \uplus B$ if $A \cap B = \emptyset$.

Proof. Note that for any non-negative integers n and m, $\max(n,m) \leq n + m$ and that $\max(n,m) = n + m$ if either m or n is 0. Let $C_A(x) = n$ and $C_B(x) = m$.

i. $C_{A \cup B}(x) = C_A(x) \vee C_B(x) = \max(n,m) \leq n + m = C_A(x) + C_B(x) = C_{A \uplus B}(x)$;
ii. Since $A \cap B = \emptyset$, if $x \in A$, $C_A(x) = n$ and $C_B(x) = 0$. On the other hand, if $x \in B$, $C_B(x) = m$ and $C_A(x) = 0$. $C_{A \cup B}(x) = C_A(x) \vee C_B(x) = \max(n,m) = n + m = C_A(x) + C_B(x) = C_{A \uplus B}(x)$. □

Recall the Definition of A_n and the complement of a multiset M denoted M' by [7]. For a non-empty set Y, define a *characteristic function*

$$\mu_Y(y) = \begin{cases} 1, \text{ if } y \in Y; \\ 0, \text{ if } y \notin Y. \end{cases}$$

We now introduce the complement of A_n denoted by A'_n.

Definition 10. *Let A be a multiset drawn from X and A_n as defined in Definition 8(iii). $A'_n = \{x \in X : \mu_{A_n}(x) = 0$ and $C_M(x) < n\}$.*

The following results shows that A'_n is well-defined.

Proposition 6. *i.* *If $m \leq n$, then $A'_m \subseteq A'_n$*
ii. $A \subseteq B \Rightarrow B'_n \subseteq A'_n$.
iii. $A'_n \cup B'_n = (A \cap B)'_n$;
iv. $A'_n \cap B'_n = (A \cup B)'_n$.

Proof. i. The proof is evident.
ii. Since $A \subseteq B, C_B(x) \geq C_A(x)$. Let $x \in A_n, C_A(x) \geq n$. But $C_B(x) \geq C_A(x) \geq n$. Then $x \in B_n$ and that implies that $A_n \subseteq B_n$. From elementary set theory, $B'_n \subseteq A'_n$.
iii. Let $x \in (A \cap B)'_n$ then $C_{(A\cap B)}(x) < n$. Thus, $\min\{C_A(x), C_B(x)\} < n$ in which case, $C_A(x) < n$ or $C_B(x) < n$. We conclude that $x \in A'_n$ or $x \in B'_n$. Hence, $x \in (A'_n \cup B'_n)$ and $(A \cap B)'_n \subseteq (A'_n \cup B'_n)$.
Now let $x \in (A'_n \cup B'_n)$. $x \in A'_n$ or $x \in B'_n$ or both. Then, $x \notin A_n$ or $x \notin B_n$ or not in both. $C_A(x) < n$ and $C_B(x) < n$. Therefore, $C_A(x) \wedge C_B(x) = C_{(A\cap B)}(x) < n$. We conclude that $x \in (A \cap B)'_n$ and $(A'_n \cup B'_n) \subseteq (A \cap B)'_n$.
iv. Let $x \in (A'_n \cap B'_n)$. Then, $x \in A'_n$ and $x \in B'_n$, which implies that $x \notin A_n$ and $x \notin B_n$. Thus, $C_A(x) < n$ and $C_B(x) < n$. As a result, $C_A(x) \vee C_B(x) = C_{A\cup B}(x) < n$. Consequently, $x \in (A \cup B)'_n \Rightarrow A'_n \cap B'_n \subseteq (A \cup B)'_n$.
On the other hand, let $x \in (A \cup B)'_n$. Then, $x \notin (A \cup B)_n$ and $C_{A\cup B}(x) < n$. Furthermore, if $C_{A\cup B}(x) = C_A(x) \vee C_B(x) < n$, then $C_A(x) < n$ and $C_B(x) < n$. The consequence is that $x \in A'_n$ and $x \in B'_n$. Hence, $x \in (A'_n \cap B'_n) \Rightarrow (A \cup B)'_n \subseteq (A'_n \cap B'_n)$. □

Then the complement of A_n denoted A'_n is well-defined.

4. Results on Function on Multisets

Dedekind, in his paper "Was sind und was sollen die Zahlen?" had said that "the frequency-number of an image is the number of its preimages" [1]. Hence, if there are n elements in a domain X (of a function f mapping X to Y) which are mapped to an element $y \in Y$, then y has frequency n so that it is an n-fold element of Y. This defines a kind of multiset (Dedekind's multiset) denoted by M_f and $|dom(f)| = |M_f|$. This fact will be illustrated by the following Example.

Definition 11. *Let $f : X \to Y$ be a mapping on two non-empty sets and $f : M \to N$ be a mapping on multisets M and N respectively drawn from X and Y. If $C_N(y) = C_M(x)$ for all $y \in Y$ such that $f(x) = y$, M and N are Dedekind's multisets.*

Example 5. *Let $X = \{-1, -1, -1, 1, 1, 2, 2\}$ and $f(x) = x^2$. Then, the Dedekind's multiset $M_f = \{1, 1, 1, 1, 1, 4, 4\}$.*

But, following the Definition by Nazmul et al. [7] which is stated in this article as Definition 2(i), if a multiset $M = \{-1, -1, -1, 1, 1, 2, 2\}$ is mapped by $f(x) = x^2, f(M) = \{1, 1, 1, 4, 4\}$. This is because $\emptyset \neq f^{-1}(1) = \pm 1$ and $\emptyset \neq f^{-1}(4) = \pm 2$; $\max\{C_M(1), C_M(-1)\} = 3$ and $\max\{C_M(2), C_M(-2)\} = 2$; thus $|M| \neq |f(M)|$. Hence, Nazmul et al.'s Definition of function on multisets fails for Dedekind's multisets.

Moreover, in Proposition 2, since $M_i \in [X]^\alpha$ and $N_i \in [Y]^\alpha$, and f maps X to Y, $f^{-1}(M_i)$ is undefined but $f^{-1}(N_i)$ is. Also, $f^{-1}g^{-1}[N_j]$ is not defined. There should be a multiset say $W_i \in [Z]^\alpha$ so that $f^{-1}g^{-1}[W_i]$ is defined. Hence, properties (iv), (v) and (vii) are not true. Against these backgrounds, there is the need to redefine Definition 2(i) and state some properties of this new Definition.

Definition 12. *Let X and Y be two non-empty sets and $f : X \longrightarrow Y$ a mapping such that $M \in [X]^\alpha$. Then,* $C_{f(M)}(y) = \sum_{f^{-1}(y)\neq\emptyset f(x)=y} C_M(x)$.

Proposition 7. *Let X, Y and Z be three nonempty sets such that $f : X \longrightarrow Y$ and $g : Y \longrightarrow Z$ are mappings. If $M_i \in [X]^\alpha$, $N_i \in [Y]^\alpha$ and $W_i \in [Z]^\alpha$ with $i \in I$, then*

i. $M_1 \subseteq M_2 \Rightarrow f(M_1) \subseteq f(M_2)$;
ii. $f(\cup_{i\in I} M_i) \supseteq \cup_{i\in I} f(M_i)$;
iii. $N_1 \subseteq N_2 \Rightarrow f^{-1}(N_1) \subseteq f^{-1}(N_2)$;
iv. $f^{-1}(\cup_{i\in I} N_i) = \cup_{i\in I} f^{-1}(N_i)$;
v. $f^{-1}(\cap_{i\in I} N_i) = \cap_{i\in I} f^{-1}(N_i)$;
vi. $f(M_i) \subseteq N_j \Rightarrow M_i \subseteq f^{-1}(N_j)$;
vii. $g[f(M_i)] = [gf](M_i)$ *and* $f^{-1}[g^{-1}(W_j)] = [gf]^{-1}(W_j)$.

Proof. i. Suppose $M_1 \subseteq M_2$ and let $y \in f(M_1)$. $C_{f(M_1)}(y) = \sum C_{M_1}(x) \le \sum C_{M_2}(x) = C_{f(M_2)}(y)$;
ii. Note that $C_{\cup M_i}(x) \ge C_{M_i}(x) \Rightarrow \sum C_{\cup M_i}(x) \ge \sum C_{M_i}(x)$. Then, $C_{f(\cup M_i)}(y) = \sum C_{\cup M_i}(x) = \sum \vee C_{M_i}(x) \ge \vee \sum C_{M_i}(x) = \vee C_{f(M_i)}(y) = C_{\cup f(M_i)}(y)$;
iii. Let $M_1 \subseteq M_2$. Then, $f(M_1) = N_1 \subseteq N_2 = f(M_2)$. Let $x \in f^{-1}(N_1)$. $C_{f^{-1}(N_1)}(x) = C_{N_1}(f(x)) = \sum C_{M_1}(x) \le \sum C_{M_2}(x) = C_{N_2}(f(x)) = C_{f^{-1}(N_2)}(x)$.
iv. $C_{f^{-1}\cup N_i}(x) = C_{\cup N_i}(f(x)) = \vee C_{N_i}(f(x)) = \vee C_{f^{-1}(N_i)}(x) = C_{\cup f^{-1}(N_i)}(x)$.
v. $C_{f^{-1}\cap N_i}(x) = C_{\cap N_i}(f(x)) = \wedge C_{N_i}(f(x)) = \wedge C_{f^{-1}(N_i)}(x) = C_{\cap f^{-1}(N_i)}(x)$.
vi. Let $f(M_i) \subseteq N_j$ and $x \in M_i$ such that $f(x) = y$. $C_{M_i}(x) = C_{f(M_i)}(y) \le C_{N_j}(y) = C_{f^{-1}(N_j)}(x)$.
vii. $C_{g[f(M_i)]}(z) = \sum_{g^{-1}(z)\neq\emptyset g(y)=z} C_{f(M_i)}(y) = \sum_{g^{-1}(z)\neq\emptyset g(y)=z} \sum_{f^{-1}(y)\neq\emptyset f(x)=y} C_{M_i}(x) =$
$= \sum_{gf(x)=z} C_{M_i}(x) = C_{gf(M_i)}(z)$. □

The following Examples will illustrate some of the properties in Proposition 7.

Example 6. *Let* $X = \{-1, 1, 2\}$ *and* $f(x) = x^2$. *Let* $M_1 = \{-1, -1, 1, 12, 2\}$, $M_2 = \{1, 1, -2, -2, -2\}$, $f(M_i) = N_i; i = 1, 2$. *Then,*
$f(M_1 \cup M_2) = \{1, 1, 1, 1, 4, 4, 4, 4\}$ *and* $f(M_1) \cup f(M_2) = \{1, 1, 1, 1, 4, 4, 4\}$.

5. Some Applications of Operations on Multiset

This chapter offers some possible applications: we are going to draw our attention to applying "removal and selection" operations defined in Definitions 5 and 6, respectively. Those may be and are used for mathematical sorting and blacklisting.

Informatics understands the expressions blacklist or blocklist as lists containing something forbidden; conversely, the so-called whitelist is used to create a list of entities, which are to be allowed. Server lists (more precisely, their IP addresses) are among of the most common uses; it is unsuitable to receive e-mails from them because they send spams. The blacklist is also used in programs for instant messaging to create a list of users from which the information sent is not to be received; they can also serve for the same purpose on social networks. Similarly, a list of e-mail addresses from which messages are not to be received can be created in an e-mail client or at an e-mail provider.

Both blacklists and whitelists are sometimes used at the same time. A typical case is a situation when a program uses its algorithm; an antispam filter can be used as an example of such a program. It decides according to the e-mail content, whether it is spam. If a user needs to inform the filter that only spams and no useful e-mails are received from a particular e-mail address, he/she places this address on the blacklist. Then, the filter will automatically classify e-mails sent from that address as spams. On the other hand, if a user needs to inform the filter that he/she receives no spams from a particular address or he/she does not want to miss any important e-mail, he/she places this address on the whitelist. Then, the filter will treat all messages from that particular address as useful and will not classify them as spam.

We also used a so-called yellow list containing for instance the IP address list of e-mail servers, which predominantly send non-spam e-mails, however, sometimes there are some spam as well, e.g., yahoo, Hotmail and Gmail. The yellow list comprises servers, which should never appear on the blacklist (for Example because of mistakes or misprints). The yellow list is checked as the first one and if a server is listed there, blacklist tests are ignored afterwards.

Operations such as *removal* "$\ominus$" and *selection* "$\odot$" and "$\otimes$" are useful for the above mentioned applications: see previous sections of this paper. See also [6]. Let us give several Examples:

Example 7. *Consider the polynomial* $f(x) = (x-1)^2(x-2)(x-3)^3$. *The associated multiset of roots is* $F = \{1,1,2,3,3,3\}$. *If it is intended to get a polynomial* $g(x) = \frac{f(x)}{x-2}$, *the associated multiset of roots is* $G = F \ominus 2 = \{1,1,3,3,3\}$. *If it is intended to find a factor of the polynomial* $f(x)$ *which does not contain the linear factor* $x-3$, *the multiset* $B = \{3\}_{C_F(3)} = \{3,3,3\}$ *is defined and the removal operation* $F \ominus B$ *is performed.*

Example 8. *Consider a unique factorisation of a positive integer* N *into its primes with the associated multiset of primes*

$$M = \{x_1^{m_1}, x_2^{m_2}, x_3^{m_3}, \cdots, x_n^{m_n}\},$$

where $m_i's$ *are the multiplicities of* $x_i's$ *for* $1 \le i \le n$. *To have a number which is not a multiple of* x_i, *construct a multiset* $B = \{x_i\}_{m_i}$ *and perform* $M \ominus B$; *to have a number which is just* $\frac{N}{x_i}$, *perform* $M \ominus x_i$.

Example 9. *Again, consider the polynomial* $f(x) = (x-1)^2(x-2)(x-3)^3$ *with the associated multiset of roots* $F = \{1,1,2,3,3,3\}$. *If it is intended to get a polynomial in which all linear factors* $(x-3)$ *is eliminated, define a subset* $B = \{1,2\}$ *of* $X = \{1,2,3\}$ *and perform* $F \otimes B$ *and the new polynomial has the associated multiset of roots as* $G = F \otimes B = \{1,1,2\}$. *It is another way of getting a polynomial whose only factors are* $x-1$ *and* $x-2$.

Example 10. *If tag numbers were to be given to a set of 20 workers in a manufacturing company from the set of numbers* $X = \{1,2,3,4\}$ *with which they are allocated into four different work stations, and if the allocation of these personnel is the multiset* $A = \{1^6, 2^4, 3^4, 4^6\}$, *the code to blacklist any of the group would consist the selection operation* $A \odot B$, *where* B *is the subset of* X *consisting the tag number of the group to be blacklisted. For instance, if the company wishes to blacklist or remove from their payroll everyone carrying the tag number 1 and 2,* $B = \{1,2\}$ *and* $A \odot B = \{3^4, 4^6\}$.

6. Conclusions

The theory of multisets is an important generalization of the classical set theory which has emerged by violating a basic property of classical sets that an element can belong to a set just once. It can be used in many applications, e.g., data encryption, data mining, coding theory, decision making or to write a mathematical programme which could do some sorting of data. The algebraic structure of such data could be studied by applying group theory to a multiset.

Author Contributions: On conceptualization, investigation, resources, writing—original draft preparation, writing–review and editing worked both authors equally. Project administration and funding acquisition: Š.H.-M.

Funding: The work presented in this paper was supported within the project for Development of basic and applied research developed in the long term by the departments of theoretical and applied bases FMT (Project code: DZRO K-217) supported by the Ministry of Defence the Czech Republic.

Acknowledgments: The authors also thank the Department of Mathematics of University of Ibadan, and also to Foundation of Chongqing Municipal Key Laboratory of Institutions of Higher Education ([2017]3), Foundation of Chongqing Development and Reform Commission (2017[1007]) and Foundation of Chongqing Three Gorges University.

Conflicts of Interest: The authors declare no conflict of interest. The funders had no role in the design of the study; in the collection, analyses, or interpretation of data; in the writing of the manuscript, or in the decision to publish the results.

References

1. Dedekind, R. *Essays on the Theory of Numbers*; Dover: New York, NY, USA, 1963.
2. Blizard, W.D. Multiset theory. *Notre Dame Form. Log.* **1989**, *30*, 36–66. [CrossRef]
3. Blizard, W.D. Dedekind multisets and function shells. *Theor. Comput. Sci.* **1993**, *110*, 79–98. [CrossRef]
4. Syropoulos, A. Mathematics of Multisets. In *WMC 2000: Multiset Processing*; Lecture Notes in Computer Science; Springer: Berlin/Hidelberg, Germany, 2001; Volume 2235, pp. 347–385.
5. Wildberger, N.J. *A New Look at Multisets*; School of Mathematics, UNSW: Sydney, Australia, 2003; pp. 1–21.
6. Knuth, D. *The Art of Computer Programming, Volume 2: Seminumerical Algorithms*; Addison Wesley: Reading, MA, USA, 1981; pp. 453–636.
7. Nazmul, S.K.; Majumdar, P.; Samanta, S.K. On multisets and multigroups. *Ann. Fuzzy Math. Inform.* **2013**, *6*, 643–656.
8. Lake, J. Sets, fuzzy sets, multisets and functions. *J. Lond. Math. Soc.* **1976**, *12*, 211–212. [CrossRef]
9. Onasanya, B.O.; Feng, Y. Multigroups and multicosets. *Ital. J. Pure Appl. Math.* **2019**, *41*, 251–261.
10. Onasanya, B.O.; Hošková-Mayerovxax, Š. Multi-fuzzy group induced by multisets. *Ital. J. Pure Appl. Math.* **2019**, *41*, 597–604.
11. Singh, D.; Ibrahim, A.M.; Yohanna, T.; Singh, J.N. An overview of the applications of multisets. *Novi Sad J. Math.* **2007**, *37*, 73–92.
12. Singh, D.; Ibrahim, A.M.; Yohanna, T.; Singh, J.N. A systemisation of fundamenals of multisets. *Lect. Mat.* **2008**, *29*, 33–48.
13. Yohanna, T.; Simon, D. Symmetric groups under multiset perspective. *IOSR J. Math.* **2013**, *7*, 47–52.
14. Ameri, R.; Rosenberg, I.G. Congruences of Multialgebras. *Multivalued Log. Soft Comput.* **2009**, *15*, 525–536.
15. Anusuya Ilamathi, V.S.; Vimala, J.; Davvaz, B. Multiset filters of residuated lattices and its application in medical diagnosis. *J. Intell. Fuzzy Syst.* **2019**, *36*, 2297–2305. [CrossRef]
16. Tella, Y.; Daniel, S. Computer representation of multisets. *Sci. World J.* **2011**, *6*, 21–22.
17. Ameri, R.; Hoskova-Mayerova, S. Fuzzy Continuous Polygroups. In Proceedings of the Aplimat—15th Conference on Applied Mathematics, Bratislava, Slovakia, 2–4 February 2016; pp. 13–19.
18. Yager, R.R. On the theory of bags. *Int. J. Gen. Syst.* **1986**, *13*, 23–37. [CrossRef]

symmetry

MDPI

Article

Primeness of Relative Annihilators in BCK-Algebra

Hashem Bordbar [1,*], G. Muhiuddin [2] and Abdulaziz M. Alanazi [2]

1 Center for Information Technologies and Applied Mathematics, University of Nova Gorica, Vipavska 13, 5000 Nova Gorica, Slovenia
2 Department of Mathematics, University of Tabuk, Tabuk 71491, Saudi Arabia; chishtygm@gmail.com (G.M.); am.alenezi@ut.edu.sa (A.M.A.)
* Correspondence: Hashem.bordbar@ung.si

Received: 15 January 2020; Accepted: 11 February 2020; Published: 15 February 2020

check for updates

Abstract: Conditions that are necessary for the relative annihilator in lower BCK-semilattices to be a prime ideal are discussed. Given the minimal prime decomposition of an ideal A, a condition for any prime ideal to be one of the minimal prime factors of A is provided. Homomorphic image and pre-image of the minimal prime decomposition of an ideal are considered. Using a semi-prime closure operation "cl", we show that every minimal prime factor of a cl-closed ideal A is also cl-closed.

Keywords: lower BCK-semilattice; relative annihilator; semi-prime closure operation; minimal prime decomposition; minimal prime factor

1. Introduction

For the first time, Aslam et al. in [1] discussed the concept of annihilators for a subset in BCK-algebras, and after that many researchers generalized it in different research articles (see [2–5]). Except these, the notion related to annihilator in BCK-algebras is investigated in the papers [6–8]. In [4], Bordbar et al. introduced the notion of the relative annihilator in a lower BCK-semilattice for a subset with respect to another subset as a logical extension of annihilator, and they obtained some properties related to this notion. They provide the conditions that the relative annihilator of an ideal with respect to an ideal needs to be ideal, and discussed conditions for the relative annihilator ideal to be an implicative (resp., positive implicative, commutative) ideal. Moreover, in some articles, different properties of ideals in logical algebras and ordered algebraic structures were concerned (see [9–18]). In order to investigate these kinds of properties for an arbitrary ideal in BCI/BCK-algebra, we need to know about the decomposition of an ideal. With this motivation, this article is the first try, as far as we know, to decompose an ideal in a BCI/BCK-algebra.

In this paper, we prove that the relative annihilator of a subset with respect to a prime ideal is also a prime ideal. Given the minimal prime decomposition of an ideal A, we provide a condition for any prime ideal to be one of minimal prime factors of A by using the relative annihilator. We consider homomorphic image and preimage of the minimal prime decomposition of an ideal. Using a semi-prime closure operation "cl", we show that, if an ideal A is cl-closed, then every minimal prime factor of A is also cl-closed.

2. Preliminaries

In this section, gather some results related to BCI/BCK-algebra and ideals, which will be used in the next section. For more details, the readers are refereed to [19].

The study of BCI/BCK-algebras was initiated by Imai and Iseki in 1966 as a generalization of the concept of set-theoretic difference and propositional calculi.

Suppose that X is a set and $(X;*,0)$ of type $(2,0)$ is an algebra. The set X is called a *BCI-algebra* if it satisfies the following conditions:

(I) $(\forall x,y,z \in X)\ (((x*y)*(x*z))*(z*y)=0)$,
(II) $(\forall x,y \in X)\ ((x*(x*y))*y=0)$,
(III) $(\forall x \in X)\ (x*x=0)$,
(IV) $(\forall x,y \in X)\ (x*y=0,\ y*x=0\ \Rightarrow\ x=y)$.

Every BCI-algebra X with the following condition

$$(\forall x \in X)\ (0*x=0)$$

is called a *BCK-algebra*.

Proposition 1. *Let X be a BCI/BCK-algebra. Then, the following statements are satisfied in every BCI/BCK-algebra:*

(1) $(\forall x \in X)\ (x*0=x)$,
(2) $(\forall x,y,z \in X)\ (x \leq y\ \Rightarrow\ x*z \leq y*z,\ z*y \leq z*x)$,
(3) $(\forall x,y,z \in X)\ ((x*y)*z=(x*z)*y)$,
(4) $(\forall x,y,z \in X)\ ((x*z)*(y*z) \leq x*y)$

where $x \leq y$ if and only if $x*y=0$.

Definition 1. *A BCK-algebra X is called a lower BCK-semilattice (see [19]) if X is a lower semilattice with respect to the BCK-order.*

Definition 2 ([19])**.** *Let X be a a BCI/BCK-algebra. An arbitrary subset A of X is called an ideal of X if it satisfies*

$$0 \in A, \tag{1}$$
$$(\forall x \in X)\ (\forall y \in A)\ (x*y \in A\ \Rightarrow\ x \in A)\,. \tag{2}$$

Remark 1 ([19])**.** *For every ideal A of a BCK-algebra X and for all $x,y \in X$, the following implication is satisfied:*

$$(x \leq y,\ y \in A\ \Rightarrow\ x \in A)\,. \tag{3}$$

Definition 3 ([19])**.** *Let P be a proper ideal of a lower BCK-semilattice X. Then, P is a prime ideal if, for $a,b \in X$ such that $a \wedge b \in P$, we conclude that $a \in P$ or $b \in P$, where $a \wedge b$ is the greatest lower bound of a and b.*

For an ideal A of a BCK-algebra X, the ideal B of X is called *minimal prime* associated with A if B is minimal in the set of all prime ideals containing A.

Lemma 1 ([20])**.** *If $\varphi : X \to Y$ is an epimorphism of lower BCK-semilattices, then*

$$(\forall x,y \in X)\ (\varphi(x \wedge_X y) = \varphi(x) \wedge_Y \varphi(y))\,. \tag{4}$$

Lemma 2 ([20])**.** 1. *Let $\varphi : X \to Y$ be an epimorphism of BCK-algebras. If A is an ideal of X, then $\varphi(A)$ is an ideal of Y.*
2. *Let $\varphi : X \to Y$ be an homomorphism of BCK-algebras. If B is an ideal of Y, then $\varphi^{-1}(B)$ is an ideal of X.*

Lemma 3 ([20])**.** *Let $\varphi : X \to Y$ be a homomorphism of BCK-algebras X and Y and let A be an ideal of X such that $Ker(\varphi) \subseteq A$. Then, $\varphi^{-1}(A') = A$ where $A' = \varphi(A)$.*

3. Primeness of Relative Annihilators

In this section, we use the notations X as a lower BCK-semilattice, $x \wedge y$ as the g.l.b.(greatest lower bound) of $x, y \in X$ and

$$A \wedge B := \{a \wedge b \mid a \in A, b \in B\}$$

for any two arbitrary subsets A, B of X, unless otherwise.
In a case that, $A = \{a\}$, then we use $a \wedge B$ instead of $\{a\} \wedge B$.

Definition 4 ([4]). *Let A and B be two arbitrary subsets of X. A set $(A :_\wedge B)$ is defined as follows:*

$$(A :_\wedge B) := \{x \in X \mid x \wedge B \subseteq A\} \tag{5}$$

and it is called the relative annihilator of B with respect to A.

Remark 2. *If $A = \{a\}$, then $(\{a\} :_\wedge B)$ is denoted by $(a :_\wedge B)$. Similarly, we use $(A :_\wedge b)$ instead of $(A :_\wedge \{b\})$, when $B = \{b\}$.*

The next two Lemmas are from [4].

Lemma 4. *For any ideal A and a nonempty subset B of X, the following implication*

$$A \subseteq (A :_\wedge B)$$

is satisfied.

Lemma 5. *Let B be an arbitrary nonempty subset of X in which the following statement is valid for all $x, y \in X$*

$$(\forall b \in B)\left((x \wedge b) * (y \wedge b) \leq (x * y) \wedge b\right). \tag{6}$$

Consider the relative annihilator $(A :_\wedge B)$. If A is an ideal of X, then the the relative annihilator $(A :_\wedge B)$ is an ideal of X.

Theorem 1. *Let B be an arbitrary subset of X such that the condition (6) is satisfied for B. If A is a prime ideal of X, then the relative annihilator $(A :_\wedge B)$ of B with respect to A is X itself or a prime ideal of X.*

Proof. Suppose that $(A :_\wedge B) \neq X$. Then $(A :_\wedge B)$ is a proper ideal of X by Lemma 5. Now, let $x \wedge y \in (A :_\wedge B)$ and $x \notin (A :_\wedge B)$ for elements $x, y \in X$. Then, $(x \wedge y) \wedge B \subseteq A$ and $x \wedge b \notin A$ for some $b \in B$. Thus,

$$(x \wedge b) \wedge y = (x \wedge y) \wedge b \in A.$$

Since A is a prime ideal of X, it follows from Definition 3 and Lemma 4 that $y \in A \subseteq (A :_\wedge B)$. Therefore, $(A :_\wedge B)$ is a prime ideal of X. $\square$

Corollary 1. *Suppose that X is a commutative BCK-algebra. If A is a prime ideal of X and B is a nonempty subset of X, then the relative annihilator $(A :_\wedge B)$ of B with respect to A is X itself or a prime ideal of X.*

Lemma 6 ([21]). *If A and B are ideals of X, then the relative annihilator $(A :_\wedge B)$ of B with respect to A is an ideal of X.*

Theorem 2. *If A is a prime ideal and B is an ideal of X, then the relative annihilator $(A :_\wedge B)$ of B with respect to A is X itself or a prime ideal of X.*

Proof. Suppose that $(A :_\wedge B) \neq X$. By using Lemma 6, $(A :_\wedge B)$ is a proper ideal of X. The primeness of $(A :_\wedge B)$ can be proved by a similar way as in the proof of Theorem 1. □

By changing the role of A and B in Theorem 2, the $(A :_\wedge B)$ may not be a prime ideal of X. The following example shows that it is not true in general case.

Example 1. *Let $X = \{0,1,2,3,4\}$ with the following Cayley table.*

$*$	0	1	2	3	4
0	0	0	0	0	0
1	1	0	0	1	1
2	2	2	0	2	2
3	3	3	3	0	3
4	4	4	4	4	0

Then, by routine calculation, X is a lower BCK-semilattice. Consider ideals $A = \{0,1\}$ and $B = \{0,1,2,4\}$ of X. It is easy to show that B is a prime ideal. Then,

$$(A :_\wedge B) = \{x \in X \mid x \wedge B \subseteq A\} = \{0,1,3\},$$

and it is not a prime ideal of X because $2 \wedge 4 = 0 \in (A :_\wedge B)$ but $2 \notin (A :_\wedge B)$ and $4 \notin (A :_\wedge B)$.

For any ideal I of X and any $x \in X$, we know that

$$I \subseteq (I :_\wedge x) \subseteq X. \tag{7}$$

Lemma 7. *For any ideal P of X and any $a \in X$, the following statements are satisfied:*

$$(a \in P \Rightarrow (P :_\wedge a) = X). \tag{8}$$

$$(a \notin P \text{ and } P \text{ is prime } \Rightarrow (P :_\wedge a) = P). \tag{9}$$

Proof. Let $a \in P$. Then, for arbitrary element $x \in X$, $x \wedge a \in P$. Hence, $x \in (P :_\wedge a)$. Therefore, (8) is valid. Let $a \notin P$ and P be a prime ideal of X. Obviously, $P \subseteq (P :_\wedge a)$. If $x \in (P :_\wedge a)$, then $x \wedge a \in P$ and so $x \in P$. Consequently, $(P :_\wedge a) = P$. □

Theorem 3. *Let A_1 and A_2 be ideals of X. For any prime ideal P of X, the following assertions are equivalent:*

(i) $A_1 \subseteq P$ *or* $A_2 \subseteq P$.
(ii) $A_1 \cap A_2 \subseteq P$.
(iii) $A_1 \wedge A_2 \subseteq P$.

Proof. The implications (i) ⇒ (ii) ⇒ (iii) are clear.
(iii) ⇒ (i) Suppose $A_1 \nsubseteq P$ and $A_2 \nsubseteq P$. Then, there exist $a_1 \in A_1$ and $a_2 \in A_2$ such that $a_1, a_2 \notin P$. Since P is a prime ideal, we have $a_1 \wedge a_2 \notin P$. This is a contradiction, and so $A_1 \subseteq P$ or $A_2 \subseteq P$. □

By using induction on n, the following theorem can be considered as an extension of Theorem 3.

Theorem 4. *Let $A_1, A_2, \cdots, A_n$ be ideals of X. For a prime ideal P of X, the following assertions are equivalent:*

(i) $A_j \subseteq P$ *for some* $j \in \{1,2,\cdots,n\}$.
(ii) $\bigcap_{i=1}^n A_i \subseteq P$.
(iii) $\bigwedge_{i=1}^n A_i \subseteq P$.

Theorem 5. *Let A_1 and A_2 be ideals of X. For any prime ideal P of X, if $P = A_1 \cap A_2$, then $P = A_1$ or $P = A_2$.*

Proof. It is straightforward by Theorem 3. □

Inductively, the following theorem can be proved as an extension of Theorem 5.

Theorem 6. *Let $A_1, A_2, \cdots, A_n$ be ideals of X. For a prime ideal P of X, if $P = \cap_{i=1}^{n} A_i$, then $P = A_j$ for some $j \in \{1, 2, \cdots, n\}$.*

Definition 5. *Letting A be an ideal of a lower BCK-semilattice X, we say that A has a minimal prime decomposition if there exist prime ideals $Q_1, Q_2, \cdots, Q_n$ of X such that*

(1) $A = \bigcap_{i \in \{1,2,\cdots,n\}} Q_i$,

(2) $\bigcap_{\substack{i \in \{1,2,\cdots,n\} \\ i \neq j}} Q_i \nsubseteq Q_j$.

The class $\{Q_1, Q_2, \cdots, Q_n\}$ is called a minimal prime decomposition of A, and each Q_i is called a minimal prime factor of A.

Lemma 8 ([22]). *Let A, B, and C be non-empty subsets of X. Then, we have*

$$(A \cap B :_\wedge C) = (A :_\wedge C) \cap (B :_\wedge C).$$

Given the minimal prime decomposition of an ideal A, we provide a condition for any prime ideal to be one of minimal prime factors of A by using the relative annihilator.

Theorem 7. *Let A be an ideal of X and $\{P_1, P_2\}$ be a minimal prime decomposition of A. For a prime ideal P of X, the following statements are equivalent:*

(i) *$P = P_1$ or $P = P_2$.*
(ii) *There exists $a \in X$ such that $(A :_\wedge a) = P$.*

Proof. (i) ⇒ (ii). Since $\{P_1, P_2\}$ is a minimal prime decomposition of A, there exist $a_1 \in P_1 \setminus P_2$ and $a_2 \in P_2 \setminus P_1$. If $P = P_2$, then Lemmas 7 and 8 imply that

$$(A :_\wedge a_1) = (P_1 \cap P_2 :_\wedge a_1) = (P_1 :_\wedge a_1) \cap (P_2 :_\wedge a_1) = X \cap P_2 = P_2 = P.$$

Similarly, if $P = P_1$, then $(A :_\wedge a_2) = P$.

Conversely, suppose that, for an element $a \in X$, we have $(A :_\wedge a) = P$. Then, we have

$$(A :_\wedge a) = (P_1 \cap P_2 :_\wedge a) = (P_1 :_\wedge a) \cap (P_2 :_\wedge a).$$

If $a \in P_1$, then $(P_1 :_\wedge a) = X$, and if $a \notin P_1$, then $(P_1 :_\wedge a) = P_1$ by Lemma 7. Similarly, $(P_2 :_\wedge a) = X$ or $(P_2 :_\wedge a) = P_2$. Thus,

$$P = (A :_\wedge a) = (P_1 \cap P_2 :_\wedge a) = (P_1 :_\wedge a) \cap (P_2 :_\wedge a)$$

is one of P_1, P_2, $P_1 \cap P_2$ and X. We know that $P \neq X$ since P is proper. If $P = P_1 \cap P_2$, then $P = P_1$ or $P = P_2$ by Theorem 5. □

Using an inductive method, the following theorem is satisfied.

Theorem 8. *Let $\{P_1, P_2, \cdots, P_n\}$ be a minimal prime decomposition of an ideal A in X. If P is a prime ideal of X, then the following statements are equivalent:*

(i) *$P = P_i$ for some $i \in \{1, 2, \cdots, n\}$.*
(ii) *There exists $a \in X$ such that $(A :_\wedge a) = P$.*

Theorem 9. *Suppose that $\varphi : X \to Y$ is an epimorphism of lower BCK-semilattices. Then,*

(i) *If P is a prime ideal of X such that $Ker\varphi \subseteq P$, then $\varphi(P)$ is a prime ideal of Y.*
(ii) *For prime ideals $P_1, P_2, \cdots, P_n$ of X, the following equation is satisfied:*

$$\varphi(P_1 \cap P_2 \cap \cdots \cap P_n) = \varphi(P_1) \cap \varphi(P_2) \cap \cdots \cap \varphi(P_n).$$

Proof. (i) Suppose that P is a prime ideal of X and $Ker\varphi \subseteq P$. Then, $\varphi(P)$ is an ideal of Y by using Lemma 2. Now, let $a \wedge_Y b \in \varphi(P)$ for any $a, b \in Y$. Then, there exist x and y in X such that $\varphi(x) = a$ and $\varphi(y) = b$. Using Lemma 1, we have the following:

$$\varphi(x \wedge_X y) = \varphi(x) \wedge_Y \varphi(y) = a \wedge_Y b \in \varphi(P).$$

Hence, there exists $q \in P$ such that $\varphi(x \wedge_X y) = \varphi(q)$. In addition, since φ is a homomorphism, it follows that

$$\varphi((x \wedge_X y) *_X q) = \varphi(x \wedge_X y) *_Y \varphi(q) = 0.$$

Thus, $(x \wedge_X y) *_X q \in Ker\varphi \subseteq P$. Since $q \in P$, we conclude that $x \wedge_X y \in P$. It follows from the primeness of P that

$$a = \varphi(x) \in \varphi(P) \text{ or } b = \varphi(y) \in \varphi(P).$$

Therefore, $\varphi(P)$ is a prime ideal of Y.

(ii) Let $x \in \varphi(P_1 \cap P_2 \cap \cdots \cap P_n)$. Then, there exists $a \in P_1 \cap P_2 \cap \cdots \cap P_n$ such that $x = \varphi(a)$. Since $a \in P_1 \cap P_2 \cap \cdots \cap P_n$, we have $a \in P_i$ and so $\varphi(a) \in \varphi(P_i)$ for all $i \in \{1, 2, \cdots, n\}$. Hence,

$$x = \varphi(a) \in \varphi(P_1) \cap \varphi(P_2) \cap \cdots \cap \varphi(P_n).$$

Therefore, $\varphi(P_1 \cap P_2 \cap \cdots \cap P_n) \subseteq \varphi(P_1) \cap \varphi(P_2) \cap \cdots \cap \varphi(P_n)$.

Assume that $x \in \varphi(P_1) \cap \varphi(P_2) \cap \cdots \cap \varphi(P_n)$. Then, $x \in \varphi(P_i)$, and thus there exists $a_i \in P_i$ such that $x = \varphi(a_i)$ for all $i \in \{1, 2, \cdots, n\}$. Note that $a_1 \wedge_X a_2 \wedge_X \cdots \wedge_X a_n \leq a_i$ for all $i \in \{1, 2, \cdots, n\}$. Since $a_i \in P_i$ and P_i is an ideal, we conclude that $a_1 \wedge_X a_2 \wedge_X \cdots \wedge_X a_n \in P_i$ for all $i \in \{1, 2, \cdots, n\}$. Therefore,

$$a_1 \wedge_X a_2 \wedge_X \cdots \wedge_X a_n \in P_1 \cap P_2 \cap \ldots \cap P_n$$

and so

$$\begin{aligned} x &= x \wedge_Y x \wedge_Y \cdots \wedge_Y x \\ &= \varphi(a_1) \wedge_Y \varphi(a_2) \wedge_Y \cdots \wedge_Y \varphi(a_n) \\ &= \varphi(a_1 \wedge_X a_2 \wedge_X \cdots \wedge_X a_n) \in \varphi(P_1 \cap P_2 \cap \cdots \cap P_n). \end{aligned}$$

Hence, $\varphi(P_1) \cap \varphi(P_2) \cap \cdots \cap \varphi(P_n) \subseteq \varphi(P_1 \cap P_2 \cap \cdots \cap P_n)$, and therefore the proof is completed. □

Lemma 9. *Let $\{P_1, P_2, \cdots, P_n\}$ be a minimal prime decomposition of an ideal A in X. If P is a prime ideal of X, then $A \subseteq P$ if and only if there exists $i \in \{1, 2, \cdots, n\}$ such that $P_i \subseteq P$.*

Proof. Straightforward. □

Theorem 10. *Let $\varphi : X \to Y$ be an epimorphism of lower BCK-semilattices. Let A be an ideal of X such that $Ker(\varphi) \subseteq A$. If $\{P_1, P_2, \cdots, P_n\}$ is a minimal prime decomposition of A in X, then $\{\varphi(P_1), \varphi(P_2), \cdots, \varphi(P_n)\}$ is a minimal prime decomposition of $\varphi(A)$ in Y.*

Proof. Note that $\varphi(A)$ is an ideal of Y (Lemma 1). If $\{P_1, P_2, \cdots, P_n\}$ is a minimal prime decomposition of A in X, then

$$A = \bigcap_{i\in\{1,2,\cdots,n\}} P_i$$

and so $\varphi(A) = \varphi\left(\bigcap_{i\in\{1,2,\cdots,n\}} P_i\right) = \bigcap_{i\in\{1,2,\cdots,n\}} \varphi(P_i)$. Suppose that

$$\bigcap_{\substack{i\in\{1,2,\cdots,n\}\\ i\neq j}} \varphi(P_i) \subseteq \varphi(P_j).$$

Since $Ker(\varphi) \subseteq P_i$, we conclude that $\varphi^{-1}(\varphi(P_i)) = P_i$ for all $i \in \{1,2,\cdots,n\}$ by using Lemma 3. Hence,

$$\begin{aligned}\bigcap_{\substack{i\in\{1,2,\cdots,n\}\\ i\neq j}} P_i &= \bigcap_{\substack{i\in\{1,2,\cdots,n\}\\ i\neq j}} \varphi^{-1}(\varphi(P_i))\\ &= \varphi^{-1}\left(\bigcap_{\substack{i\in\{1,2,\cdots,n\}\\ i\neq j}} \varphi(P_i)\right)\\ &\subseteq \varphi^{-1}(\varphi(P_j)) = P_j.\end{aligned}$$

This is a contradiction, so $\{\varphi(P_1), \varphi(P_2), \cdots, \varphi(P_n)\}$ is a minimal prime decomposition of $\varphi(A)$ in Y. □

Corollary 2. *Suppose that $\varphi : X \to Y$ is an isomorphism of lower BCK-semilattices. Let A be an ideal of X. If $\{P_1, P_2, \cdots, P_n\}$ is a minimal prime decomposition of A in X, then $\{\varphi(P_1), \varphi(P_2), \cdots, \varphi(P_n)\}$ is a minimal prime decomposition of $\varphi(A)$ in Y.*

Theorem 11. *Suppose that $\varphi : X \to Y$ is an epimorphism of lower BCK-semilattices. Let B be an ideal of Y. If $\{Q_1, Q_2, \cdots, Q_n\}$ is a minimal prime decomposition of B in Y, then $\{\varphi^{-1}(Q_1), \varphi^{-1}(Q_2), \cdots, \varphi^{-1}(Q_n)\}$ is a minimal prime decomposition of $\varphi^{-1}(B)$ in X.*

Proof. Obviously, $\varphi^{-1}(B)$ is an ideal of X. If $\{Q_1, Q_2, \cdots, Q_n\}$ is a minimal prime decomposition of B in Y, then

$$B = \bigcap_{i\in\{1,2,\cdots,n\}} Q_i.$$

Thus,

$$\varphi^{-1}(B) = \varphi^{-1}\left(\bigcap_{i\in\{1,2,\cdots,n\}} Q_i\right) = \bigcap_{i\in\{1,2,\cdots,n\}} \varphi^{-1}(Q_i).$$

Suppose that

$$\bigcap_{\substack{i\in\{1,2,\cdots,n\}\\ i\neq j}} \varphi^{-1}(Q_i) \subseteq \varphi^{-1}(Q_j).$$

Since φ is onto, $\varphi\left(\varphi^{-1}(Q_i)\right) = Q_i$ for all $i \in \{1,2,\cdots,n\}$. Hence,

$$\bigcap_{\substack{i\in\{1,2,\cdots,n\}\\ i\neq j}} Q_i = \bigcap_{\substack{i\in\{1,2,\cdots,n\}\\ i\neq j}} \varphi\left(\varphi^{-1}(Q_i)\right)$$
$$= \varphi\left(\bigcap_{\substack{i\in\{1,2,\cdots,n\}\\ i\neq j}} \varphi^{-1}(Q_i)\right)$$
$$\subseteq \varphi\left(\varphi^{-1}(Q_j)\right) = Q_j.$$

This is a contradiction, and so

$$\bigcap_{\substack{i\in\{1,2,\cdots,n\}\\ i\neq j}} \varphi^{-1}(Q_i) \nsubseteq \varphi^{-1}(Q_j).$$

Therefore, $\{\varphi^{-1}(Q_1), \varphi^{-1}(Q_2), \cdots, \varphi^{-1}(Q_n)\}$ is a minimal prime decomposition of $\varphi^{-1}(B)$ in X. □

Lemma 10 ([19]). *If X is Noetherian BCK-algebra, then each ideal of X has a unique minimal prime decomposition.*

Lemma 11 ([19]). *Every proper ideal of X is equal to the intersection of all minimal prime ideals associated with it.*

For an ideal A of X, consider the set $X \setminus A$. This set is not closed subset under the $\wedge$ operation in X in general. The following example shows it.

Example 2. *Let $X = \{0,1,2,3,4\}$ with the following Cayley table:*

$*$	0	1	2	3	4
0	0	0	0	0	0
1	1	0	0	0	0
2	2	1	0	1	1
3	3	3	3	0	3
4	4	4	4	4	0

Then, X is a lower BCK-semilattice. For an ideal $A = \{0,1,2\}$ of X, we have $X \setminus A = \{3,4\}$, which is not a $\wedge$-closed subset of X because $3,4 \in X \setminus A$, but $3 \wedge 4 = 1 \notin X \setminus A$.

For a subset A of X with $0 \notin A$, we can check that the set $X \setminus A$ may not be an ideal of X. In the following example, we check it.

Example 3. *Suppose that $X = \{0,1,2,3,4\}$ with the following Cayley table:*

$*$	0	1	2	3	4
0	0	0	0	0	0
1	1	0	1	0	1
2	2	2	0	2	0
3	3	1	3	0	3
4	4	4	4	4	0

Then, X is a lower BCK-semilattice. For a subset $A = \{3,4\}$ of X, we have $X \setminus A = \{0,1,2\}$. By routine verification, we can investigate that $X \setminus A$ is not an ideal of X.

The following theorem provided a characterization of a prime ideal.

Theorem 12. *For an arbitrary ideal P of X, the following assertions are equivalent:*

(i) *P is a prime ideal of X.*

(ii) $X \setminus P$ *is a closed subset under the* $\wedge$ *operation in* X*, that is,* $x \wedge y \in X \setminus P$ *for all* $x, y \in X \setminus P$.

Proof. (i) $\rightarrow$ (ii): Suppose that P is a prime ideal of X and $x, y \in X \setminus P$ are arbitrary elements. If $x \wedge y \notin X \setminus P$, then clearly $x \wedge y \in P$. Since P is a prime ideal, $x \in P$ or $y \in P$, which is contradictory because x and y were chosen from the set $X \setminus P$. Thus, $x \wedge y \in X \setminus P$ and $X \setminus P$ is the closed subset under the $\wedge$ operation.

(ii) $\rightarrow$ (i): Suppose that $x \wedge y \in P$. If $x \notin P$ and $y \notin P$, then clearly $x \in X \setminus P$ and also $y \in X \setminus P$. Using condition (ii), we conclude that $x \wedge y \in X \setminus P$, which is a contradiction from the first assumption $x \wedge y \in P$. Thus, $x \in P$ or $y \in P$ and P is a prime ideal of X. □

Definition 6. *Let* X *be a* BCK*-algebra. We defined in [2] the closure operation on* $\mathcal{I}_X$*, as the following function*

$$cl : \mathcal{I}(X) \rightarrow \mathcal{I}(X),\ A \mapsto A^{cl}$$

such that

$$(\forall A \in \mathcal{I}(X)) \left(A \subseteq A^{cl}\right), \tag{10}$$

$$(\forall A \in \mathcal{I}(X)) \left(A^{cl} = (A^{cl})^{cl}\right), \tag{11}$$

$$(\forall A, B \in \mathcal{I}(X)) \left(A \subseteq B \Rightarrow A^{cl} \subseteq B^{cl}\right), \tag{12}$$

where $\mathcal{I}(X)$ *is the set of all ideals of* X.

An ideal A in a BCK-algebra X is said to be *cl-closed* (see [2]) if $A = A^{cl}$.

Definition 7 ([3]). *For a closure operation "cl" on* X*, we have the following definitions:*

(i) *"cl" is a semi-prime closure operation if we have*

$$A \wedge B^{cl} \subseteq (A \wedge B)^{cl} \text{ and } A^{cl} \wedge B \subseteq (A \wedge B)^{cl}$$

for every $A, B \in \mathcal{I}(X)$.

(ii) *"cl" is a good semi-prime closure operation, if we have*

$$A \wedge B^{cl} = A^{cl} \wedge B = (A \wedge B)^{cl}$$

for every $A, B \in \mathcal{I}(X)$.

Theorem 13 ([3]). *Suppose that "cl" is a semi-prime closure operation on* X *and* S *is a closed subset of* X *under the* $\wedge$ *operation. If* X *is Noetherian and* A *is a cl-closed ideal of* X*, then the set*

$$B := \langle \{x \in X \mid x \wedge s \in A \text{ for some } s \in S\} \rangle$$

is a cl-closed ideal of X.

Lemma 12. *If* $\{P_1, P_2, \cdots, P_n\}$ *is a minimal prime decomposition of an ideal* A *of* X*, then*

$$(\forall i, j \in \{1, 2, \cdots, n\}) \left(i \neq j \Rightarrow P_i \cap (X \setminus P_j) \neq \emptyset\right). \tag{13}$$

Proof. Suppose that for $i, j \in \{1, 2, \cdots, n\}$ such that $i \neq j$, $P_i \cap (X \setminus P_j) = \emptyset$. Then, it follows that $P_i \subseteq P_j$ and this is a contradiction because $\{P_1, P_2, \cdots, P_n\}$ is a minimal prime decomposition of an ideal A of X. □

Theorem 14. *Suppose that* A *is an ideal of* X *with a minimal prime decomposition*

$$\{P_1, P_2, \cdots, P_n\}.$$

Assume that X is Noetherian and "cl" is a semi-prime closure operation on $\mathcal{I}(X)$. If A is cl-closed, then so is P_j for all $j \in \{1, 2, \cdots, n\}$.

Proof. For any $j \in \{1, 2, \cdots, n\}$, let

$$\Omega_j := \{x \in X \mid x \wedge s \in A \text{ for some } s \in X \setminus P_j\}. \tag{14}$$

Then, we will prove that $\Omega_j = P_j$. If $x \in \Omega_j$, then there exists $s \in X \setminus P_j$ such that $x \wedge s \in A$. It follows that $x \wedge s \in P_j$ and so $x \in P_j$. Thus, $\Omega_j \subseteq P_j$ for all $j \in \{1, 2, \cdots, n\}$. Now, assume that $y \in P_j$. Using Lemma 12, we can take an element $a \in P_i \cap (X \setminus P_j)$, and so $a \in P_i$ and $a \in X \setminus P_j$ for all $i \in \{1, 2, \cdots, n\}$ with $i \neq j$. Then, $y \wedge a \in P_j$ and $y \wedge a \in P_i$ for all $i \in \{1, 2, \cdots, n\}$ with $i \neq j$. Thus,

$$y \wedge a \in \bigcap_{i \in \{1,2,\cdots,n\}} P_i = A,$$

and so $y \in \Omega_j$. Therefore, $\Omega_j = P_j$, which implies that $\langle \Omega_j \rangle = \langle P_j \rangle = P_j$ for all $j \in \{1, 2, \cdots, n\}$. Since $X \setminus P_j$ is a $\wedge$-closed subset of X for all $j \in \{1, 2, \cdots, n\}$ by Theorem 12, we conclude from Theorem 13 that P_j is a cl-closed ideal of X for all $j \in \{1, 2, \cdots, n\}$. □

4. Conclusions

Necessary conditions for the relative annihilator in lower BCK-semilattices to be a prime ideal are discussed. In addition, we provided conditions for any prime ideal in the minimal prime decomposition of an ideal A, to be one of the minimal prime factors of A. Homomorphic image and pre-image of the minimal prime decomposition of an ideal are considered. Using a semi-prime closure operation "cl", we showed that every minimal prime factor of a cl-closed ideal A is also cl-closed.

These results can be applied to characterize the composable ideals in a BCK-algebra with their associated prime ideals. In our future research, we will focus on some properties of decomposable ideal such as intersections, unions, maximality, and height, and try to find the relations between these properties of ideals and the associated prime ideals. For instance, is the height of the arbitrary decomposable ideal, equal to the sum of the height of associated prime ideals? For information about the height of ideals, please refer to [23–25].

In addition, other kinds of closure operations such as meet, tender, nave, finite, prime, etc. can be checked for prime ideals in prime decompositions. For further information about other kinds of (weak) closure operation, please refer to [2,3,21,22].

In addition, for future research, we invite the researchers to join us and apply the results of this paper to new concepts in [26–28].

Author Contributions: Conceiving the idea, H.B.; literature review, H.B., G.M. and A.M.A.; writing—original draft preparation, H.B.; review and editing, G.M. and A.M.A. All authors conceived and designed the new definitions and results and read and approved the final manuscript for submission.

Funding: The authors extend their appreciation to the Deanship of Scientific Research at University of Tabuk for funding this work through Research Group no. RGP-0207-1440.

Conflicts of Interest: The authors declare no conflicts of interest.

References

1. Aslam, M.; Thaheem, A.B. On annihilators of BCK-algebras. *Czechoslovak Math. J.* **1995**, *45*, 727–735.
2. Bordbar, H.; Zahedi, M.M. A finite type of closure operations on BCK-algebra. *Appl. Math. Inf. Sci. Lett.* **2016**, *4*, 1–9. [CrossRef]
3. Bordbar, H.; Zahedi, M.M. Semi-prime closure operations on BCK-algebra. *Commun. Korean Math. Soc.* **2015**, *30*, 385–402. [CrossRef]

4. Bordbar, H.; Zahedi, M.M.; Jun, Y.B. Relative annihilators in lower BCK-semilattices. *Math. Sci. Lett.* **2017**, *6*, 1–7. [CrossRef]
5. Jun, Y.B.; Roh, E.H.; Meng, J. Annihilators in *BCI*-algebras. *Math. Jpn.* **1996**, *43*, 559–562.
6. Halas, R. Annihilators in *BCK*-algebras. *Czechoslov. Math. J.* **2003**, *153*, 1001–1007. [CrossRef]
7. Kondo, M. Annihilators in *BCK*-algebras. *Math. Jpn.* **1999**, *49*, 407–410.
8. Kondo, M. Annihilators in *BCK*-algebras II. *Mem. Fac. Sci. Eng. Shimane Univ. Ser. B Math. Sci.* **1998**, *31*, 21–25.
9. Jun, Y.B.; Smarandache, F.; Bordbar, H. Neutrosophic N-structures applied to BCK/BCI-algebras. *Information* **2017**, *8*, 128. [CrossRef]
10. Jun, Y.B.; Smarandache, F.; Song, S.Z.; Bordbar, H. Neutrosophic Permeable Values and Energetic Subsets with Applications in BCK/BCI-Algebras. *Mathematics* **2018**, *6*, 74. [CrossRef]
11. Jun, Y.B.; Song, S.Z.; Kim, S.J. Neutrosophic Quadruple BCI-Positive Implicative Ideals. *Mathematics* **2019**, *7*, 385. [CrossRef]
12. Jun, Y.B.; Song, S.Z.; Smarandache, F.; Bordbar, H. Neutrosophic Quadruple BCK/BCI-Algebras. *Axioms* **2018**, *7*, 41. [CrossRef]
13. Song, S.Z.; Bordbar, H.; Jun, Y.B. Quotient Structures of BCK/BCI-Algebras Induced by Quasi-Valuation Maps. *Axioms* **2018**, *7*, 26. [CrossRef]
14. Muhiuddin, G.; Al-Kenani, A.N.; Roh, E.H.; Jun, Y.B. Implicative neutrosophic quadruple BCK-algebras and ideals. *Symmetry* **2019**, *11*, 277. [CrossRef]
15. Muhiuddin, G.; Bordbar, H.; Smarandache, F.; Jun, Y.B. Further results on $(\in, \in)$-neutrosophic subalgebras and ideals in BCK/BCI-algebras. *Neutrosophic Sets Syst.* **2018**, *20*, 36–43.
16. Muhiuddin, G.; Jun, Y.B. p-semisimple neutrosophic quadruple BCI-algebras and neutrosophic quadruple p-ideals. *Ann. Commun. Math.* **2018**, *1*, 26–37.
17. Muhiuddin, G.; Kim, S.J.; Jun, Y.B. Implicative N-ideals of BCK-algebras based on neutrosophic N-structures. *Discrete Math. Algorithms Appl.* **2019**, *11*, 1950011. [CrossRef]
18. Muhiuddin, G.; Smarandache, F.; Jun, Y.B. Neutrosophic quadruple ideals in neutrosophic quadruple BCI-algebras. *Neutrosophic Sets Syst.* **2019**, *25*, 161–173.
19. Meng, J.; Jun, Y.B. *BCK-Algebras*; Kyung Moon Sa Co.: Seoul, Korea, 1994.
20. Huang, Y. *BCI-Algebra*; Science Press: Beijing, China, 2006.
21. Bordbar, H.; Zahedi, M.M.; Jun, Y.B. Semi-prime and meet weak closure operatons in lower *BCK*-semilattices. *Quasigroups Relat. Syst.* **2017**, *1*, 41–50.
22. Bordbar, H.; Ahn, S.S.; Song, S.Z.; Jun, Y.B. Tender and naive weak closure operations on lower BCK-semilattices. *J. Comput. Anal. Appl.* **2018**, *25*, 1354–1365.
23. Bordbar, H.; Cristea, I. Height of prime hyperideals in Krasner hyperrings. *Filomat* **2017**, *31*, 6153–6163. [CrossRef]
24. Bordbar, H.; Novak, M.; Cristea, I. A note on the support of a hypermodule. *J. Algebra Its Appl.* **2020**. [CrossRef]
25. Bordbar, H.; Cristea, I.; Novak, M. Height of hyperideals in Noetherian Krasner hyperrings. *Univ. Politeh. Buchar. Sci. Bull. Ser. A Appl. Math. Phys.* **2017**, *79*, 31–42.
26. Zhang, W.-R. G-CPT Symmetry of Quantum Emergence and Submergence—An Information Conservational Multiagent Cellular Automata Unification of CPT Symmetry and CP Violation for Equilibrium-Based Many-World Causal Analysis of Quantum Coherence and Decoherence. *J. Quantum Inf. Sci.* **2016**, *6*, 62–97. [CrossRef]
27. Zhang, W.-R. YinYang Bipolar Lattices and L-Sets for Bipolar Knowledge Fusion, Visualization, and Decision. *Int. J. Inf. Technol. Decis. Mak.* **2005**, *4*, 621–645. [CrossRef]
28. Zhang, W.; Zhang, L. YinYang Bipolar Logic and Bipolar Fuzzy Logic. *Inf. Sci.* **2004**, *165*, 265–287. [CrossRef]

symmetry

MDPI

Article

Intuitionistic Fuzzy Soft Hyper BCK Algebras

Xiaolong Xin [1,*], Rajab Ali Borzooei [2], Mahmood Bakhshi [3] and Young Bae Jun [2,4]

1. School of Mathematics, Northwest University, Xi'an 710127, China
2. Department of Mathematics, Shahid Beheshti University, Tehran 1983963113, Iran; borzooei@sbu.ac.ir (R.A.B.); skywine@gmail.com (Y.B.J.)
3. Department of Mathematics, University of Bojnord, P.O. Box 1339, Bojnord 9453155111, Iran; bakhshi@ub.ac.ir
4. Department of Mathematics Education, Gyeongsang National University, Jinju 52828, Korea

* Correspondence: xlxin@nwu.edu.cn

Received: 12 January 2019; Accepted: 25 February 2019; Published: 19 March 2019

Abstract: Maji et al. introduced the concept of fuzzy soft sets as a generalization of the standard soft sets, and presented an application of fuzzy soft sets in a decision-making problem. Maji et al. also introduced the notion of intuitionistic fuzzy soft sets in the paper [P.K. Maji, R. Biswas and A.R. Roy, Intuitionistic fuzzy soft sets, The Journal of Fuzzy Mathematics, **9** (2001), no. 3, 677–692]. The aim of this manuscript is to apply the notion of intuitionistic fuzzy soft set to hyper BCK algebras. The notions of intuitionistic fuzzy soft hyper BCK ideal, intuitionistic fuzzy soft weak hyper BCK ideal, intuitionistic fuzzy soft s-weak hyper BCK-ideal and intuitionistic fuzzy soft strong hyper BCK-ideal are introduced, and related properties and relations are investigated. Characterizations of intuitionistic fuzzy soft (weak) hyper BCK ideal are considered. Conditions for an intuitionistic fuzzy soft weak hyper BCK ideal to be an intuitionistic fuzzy soft s-weak hyper BCK ideal are provided. Conditions for an intuitionistic fuzzy soft set to be an intuitionistic fuzzy soft strong hyper BCK ideal are given.

Keywords: intuitionistic fuzzy soft hyper BCK ideal; intuitionistic fuzzy soft weak hyper BCK ideal; intuitionistic fuzzy soft s-weak hyper BCK-ideal; intuitionistic fuzzy soft strong hyper BCK-ideal

JEL Classification: 06F35; 03G25; 06D72

---

1. Introduction

Dealing with uncertainties is a major problem in many areas such as economics, engineering, environmental science, medical science, and social science etc. These problems cannot be dealt with by classical methods, because classical methods have inherent difficulties. To overcome these difficulties, Molodtsov [1] proposed a new approach, which was called soft set theory, for modeling uncertainty. In [2], Jun applied the notion of soft sets to the theory of BCK/BCI-algebras, and Jun et al. [3] studied ideal theory of BCK/BCI-algebras based on soft set theory. Maji et al. [4] extended the study of soft sets to fuzzy soft sets. They introduced the concept of fuzzy soft sets as a generalization of the standard soft sets, and presented an application of fuzzy soft sets in a decision-making problem. Maji et al. [5] also introduced the concept of intuitionistic fuzzy soft set which combines the advantage of soft set and Atanassov's intuitionistic fuzzy set. Jun et al. [6] applied fuzzy soft set to BCK/BCI-algebras.

Hyperstructure theory was born in 1934 when Marty defined hypergroups, began to analyze their properties, and applied them to groups and relational algebraic functions (see [7]). Algebraic hyperstructures represent a natural extension of classical algebraic structures. In a classical algebraic structure, the composition of two elements is an element, while in an algebraic hyperstructure, the composition of two elements is a set. Many papers and several books have been written on this topic. Presently, hyperstructures have a lot of applications in several branches of mathematics and computer sciences (see [8–19]). In [20], Jun et al. applied the hyperstructures to BCK-algebras, and introduced the concept of a hyper BCK-algebra which is a generalization of a BCK-algebra. Since then, Jun et al. studied more notions and results in [3,21,22]. Also, several fuzzy versions of hyper BCK-algebras have been considered in [23,24]. Recently Davvaz et al. summarize research progress of fuzzy hyperstructures in [25].

In this article, we introduce the notions of intuitionistic fuzzy soft hyper BCK ideal, intuitionistic fuzzy soft weak hyper BCK ideal, intuitionistic fuzzy soft s-weak hyper BCK-ideal and intuitionistic fuzzy soft strong hyper BCK-ideal, and investigate related properties and relations. We discuss characterizations of intuitionistic fuzzy soft (weak) hyper BCK ideal. We find conditions for an intuitionistic fuzzy soft weak hyper BCK ideal to be an intuitionistic fuzzy soft s-weak hyper BCK ideal. We provide conditions for an intuitionistic fuzzy soft set to be an intuitionistic fuzzy soft strong hyper BCK ideal.

2. Preliminaries

Let H be a nonempty set endowed with a hyper operation "$\circ$", that is, $\circ$ is a function from $H \times H$ to $\mathcal{P}^*(H) = \mathcal{P}(H) \setminus \{\emptyset\}$. For two subsets A and B of H, denote by $A \circ B$ the set $\cup\{a \circ b \mid a \in A,\, b \in B\}$. We shall use $x \circ y$ instead of $x \circ \{y\}$, $\{x\} \circ y$, or $\{x\} \circ \{y\}$.

By a *hyper BCK algebra* (see [20]) we mean a nonempty set H endowed with a hyper operation "$\circ$" and a constant 0 satisfying the following axioms:

(H1) $(x \circ z) \circ (y \circ z) \ll x \circ y$,
(H2) $(x \circ y) \circ z = (x \circ z) \circ y$,
(H3) $x \circ H \ll \{x\}$,
(H4) $x \ll y$ and $y \ll x$ imply $x = y$,

for all $x, y, z \in H$, where $x \ll y$ is defined by $0 \in x \circ y$ and for every $A, B \subseteq H$, $A \ll B$ is defined by $\forall a \in A, \exists b \in B$ such that $a \ll b$.

In a hyper BCK algebra H, the condition (H3) is equivalent to the condition:

$$x \circ y \ll \{x\} \text{ for all } x, y \in H.$$

In any hyper BCK algebra H, the following hold (see [20]):

$$x \circ 0 \ll \{x\},\ 0 \circ x \ll \{0\},\ 0 \circ 0 \ll \{0\}, \tag{1}$$
$$(A \circ B) \circ C = (A \circ C) \circ B, A \circ B \ll A, 0 \circ A \ll \{0\}, \tag{2}$$
$$0 \ll x,\ x \ll x,\ A \ll A, \tag{3}$$
$$A \subseteq B \ \Rightarrow\ A \ll B, \tag{4}$$
$$0 \circ x = \{0\},\ 0 \circ A = \{0\}, \tag{5}$$
$$A \ll \{0\} \ \Rightarrow\ A = \{0\}, \tag{6}$$
$$x \in x \circ 0, \tag{7}$$

for all $x, y, z \in H$ and for all non-empty subsets A, B and C of H.

A non-empty subset A of a hyper BCK algebra H is called a

- *hyper BCK ideal* of H (see [20]) if it satisfies

$$0 \in A, \tag{8}$$
$$(\forall x, y \in H)\,(x \circ y \ll A,\ y \in A \ \Rightarrow\ x \in A). \tag{9}$$

- *strong hyper BCK ideal* of H (see [22]) if it satisfies (8) and

$$(\forall x, y \in H)\,((x \circ y) \cap A \neq \emptyset,\ y \in A \ \Rightarrow\ x \in A). \tag{10}$$

- *weak hyper BCK ideal* of H (see [20]) if it satisfies (8) and

$$(\forall x, y \in H)\,(x \circ y \subseteq A,\, y \in A \Rightarrow x \in A). \tag{11}$$

Recall that every strong hyper BCK ideal is a hyper BCK ideal (see [22]).

Molodtsov [1] defined the soft set in the following way: Let U be an initial universe set and E be a set of parameters. Let $\mathscr{P}(U)$ denote the power set of U and $A \subseteq E$.

A pair (λ, A) is called a *soft set* over U, where λ is a mapping given by

$$\lambda : A \to \mathscr{P}(U).$$

In other words, a soft set over U is a parameterized family of subsets of the universe U. For $\varepsilon \in A$, $\lambda(\varepsilon)$ may be considered as the set of ε-approximate elements of the soft set (λ, A) (see [1]).

Let U be an initial universe set and E be a set of parameters. Let $\mathcal{F}(U)$ denote the set of all fuzzy sets in U. Then (f, A) is called a *fuzzy soft set* over U (see [4]) where $A \subseteq E$ and f is a mapping given by $f : A \to \mathcal{F}(U)$.

In general, for every parameter u in A, $f[u]$ is a fuzzy set in U and it is called *fuzzy value set* of parameter u. If for every $u \in A$, $f[u]$ is a crisp subset of U, then (f, A) is degenerated to be the standard soft set. Thus, from the above definition, it is clear that fuzzy soft set is a generalization of standard soft set.

3. Intuitionistic Fuzzy Soft Hyper BCK Ideals

In what follows let H and E be a hyper BCK algebra and a set of parameters, respectively, unless otherwise specified.

Definition 1. *Let $\mathcal{F}_I(H)$ denote the set of all intuitionistic fuzzy sets in H and $A \subseteq E$. Then a pair $(\tilde{\lambda}, A)$ is called an intuitionistic fuzzy soft set over H, where $\tilde{\lambda}$ is a mapping given by*

$$\tilde{\lambda} : A \to \mathcal{F}_I(H). \tag{12}$$

For any parameter $e \in A$, $\tilde{\lambda}(e)$ is an intuitionistic fuzzy set in H and it is called the *intuitionistic fuzzy value set* of parameter e, which is of the form

$$\tilde{\lambda}(e) = \left\{ \langle x, \mu_{\tilde{\lambda}(e)}(x), \gamma_{\tilde{\lambda}(e)}(x) \rangle \mid x \in H \right\}. \tag{13}$$

Definition 2. *An intuitionistic fuzzy soft set $(\tilde{\lambda}, A)$ over H is called an intuitionistic fuzzy soft hyper BCK ideal based on a parameter $e \in A$ over H (briefly, e-intuitionistic fuzzy soft hyper BCK ideal of H) if the intuitionistic fuzzy value set $\tilde{\lambda}(e)$ of e satisfies the following conditions:*

$$(\forall x, y \in H)\left(x \ll y \Rightarrow \mu_{\tilde{\lambda}(e)}(x) \geq \mu_{\tilde{\lambda}(e)}(y),\ \gamma_{\tilde{\lambda}(e)}(x) \leq \gamma_{\tilde{\lambda}(e)}(y)\right), \tag{14}$$

$$(\forall x, y \in H)\left(\begin{array}{c} \mu_{\tilde{\lambda}(e)}(x) \geq \min\left\{ \inf\limits_{a \in x \circ y} \mu_{\tilde{\lambda}(e)}(a),\ \mu_{\tilde{\lambda}(e)}(y) \right\} \\ \gamma_{\tilde{\lambda}(e)}(x) \leq \max\left\{ \sup\limits_{a \in x \circ y} \gamma_{\tilde{\lambda}(e)}(a),\ \gamma_{\tilde{\lambda}(e)}(y) \right\} \end{array} \right). \tag{15}$$

If $(\tilde{\lambda}, A)$ is an e-intuitionistic fuzzy soft hyper BCK ideal based on H for all $e \in A$, we say that $(\tilde{\lambda}, A)$ is an *intuitionistic fuzzy soft hyper BCK ideal* of H.

Example 1. *Consider a hyper BCK algebra $H = \{0, a, b\}$ with the hyper operation "$\circ$" which is given in Table 1.*

Table 1. Tabular representation of the binary operation $\circ$.

$\circ$	0	a	b
0	$\{0\}$	$\{0\}$	$\{0\}$
a	$\{a\}$	$\{0,a\}$	$\{0,a\}$
b	$\{b\}$	$\{a,b\}$	$\{0,a,b\}$

Given a set $A = \{x, y\}$ of parameters, we define an intuitionistic fuzzy soft set $(\tilde{\lambda}, A)$ by Table 2.

Table 2. Tabular representation of $(\tilde{\lambda}, A)$.

$\tilde{\lambda}$	0	a	b
x	$(0.9, 0.05)$	$(0.5, 0.35)$	$(0.3, 0.55)$
y	$(0.8, 0.15)$	$(0.4, 0.45)$	$(0.6, 0.25)$

Then $\tilde{\lambda}(x)$ satisfy conditions (14) and (15). Hence $(\tilde{\lambda}, A)$ is an intuitionistic fuzzy soft hyper BCK ideal based on x over H. But $\tilde{\lambda}(y)$ does not satisfy the condition (14) since $a \ll b$ and $\mu_{\tilde{\lambda}(y)}(a) \leq \mu_{\tilde{\lambda}(y)}(b)$ and/or $\gamma_{\tilde{\lambda}(y)}(a) \geq \gamma_{\tilde{\lambda}(y)}(b)$, and so it is not an intuitionistic fuzzy soft hyper BCK ideal based on y over H.

Proposition 1. *For every intuitionistic fuzzy soft hyper BCK ideal $(\tilde{\lambda}, A)$ of H and any parameter $e \in A$, the following assertions are valid.*

(1) *$(\tilde{\lambda}, A)$ satisfies the condition*

$$(\forall x \in H)\left(\mu_{\tilde{\lambda}(e)}(0) \geq \mu_{\tilde{\lambda}(e)}(x),\ \gamma_{\tilde{\lambda}(e)}(0) \leq \gamma_{\tilde{\lambda}(e)}(x)\right) \tag{16}$$

(2) *If $(\tilde{\lambda}, A)$ satisfies the condition*

$$(\forall T, S \in 2^H)(\exists (x_0, y_0) \in T \times S)\left(\begin{array}{c} \mu_{\tilde{\lambda}(e)}(x_0) = \inf\limits_{a \in T} \mu_{\tilde{\lambda}(e)}(a) \\ \gamma_{\tilde{\lambda}(e)}(y_0) = \sup\limits_{b \in S} \gamma_{\tilde{\lambda}(e)}(b) \end{array}\right), \tag{17}$$

then the following assertion is valid.

$$(\forall x, y \in H)(\exists a, b \in x \circ y)\left(\begin{array}{c} \mu_{\tilde{\lambda}(e)}(x) \geq \min\{\mu_{\tilde{\lambda}(e)}(a), \mu_{\tilde{\lambda}(e)}(y)\} \\ \gamma_{\tilde{\lambda}(e)}(x) \leq \max\{\gamma_{\tilde{\lambda}(e)}(b), \gamma_{\tilde{\lambda}(e)}(y)\} \end{array}\right). \tag{18}$$

Proof. Since $0 \ll x$ for all $x \in H$, we have $\mu_{\tilde{\lambda}(e)}(0) \geq \mu_{\tilde{\lambda}(e)}(x)$ and $\gamma_{\tilde{\lambda}(e)}(0) \leq \gamma_{\tilde{\lambda}(e)}(x)$ by (14). For any $x, y \in H$, there exists $x_0, y_0 \in x \circ y$ such that $\mu_{\tilde{\lambda}(e)}(x_0) = \inf\limits_{a \in x \circ y} \mu_{\tilde{\lambda}(e)}(a)$ and $\gamma_{\tilde{\lambda}(e)}(y_0) = \sup\limits_{b \in x \circ y} \gamma_{\tilde{\lambda}(e)}(b)$ by (17). It follows from (15) that

$$\mu_{\tilde{\lambda}(e)}(x) \geq \min\left\{\inf_{a \in x \circ y} \mu_{\tilde{\lambda}(e)}(a), \mu_{\tilde{\lambda}(e)}(y)\right\} = \min\left\{\mu_{\tilde{\lambda}(e)}(x_0), \mu_{\tilde{\lambda}(e)}(y)\right\}$$

and

$$\gamma_{\tilde{\lambda}(e)}(x) \leq \max\left\{\sup_{b \in x \circ y} \gamma_{\tilde{\lambda}(e)}(b), \gamma_{\tilde{\lambda}(e)}(y)\right\} = \max\left\{\gamma_{\tilde{\lambda}(e)}(y_0), \gamma_{\tilde{\lambda}(e)}(y)\right\}$$

which is the desired result. □

Lemma 1 ([21]). *Let A be a subset of a hyper BCK algebra H. If I is a hyper BCK ideal of H such that $A \ll I$, then A is contained in I.*

Given an intuitionistic fuzzy soft set $(\tilde{\lambda}, A)$ over H and $(\varepsilon, \delta) \in [0,1] \times [0,1]$ with $\varepsilon + \delta \leq 1$, we consider the following sets.

$$\begin{aligned} U_\varepsilon &:= \left\{x \in H \mid \mu_{\tilde{\lambda}(e)}(x) \geq \varepsilon\right\} \\ L_\delta &:= \left\{x \in H \mid \gamma_{\tilde{\lambda}(e)}(x) \leq \delta\right\} \end{aligned} \tag{19}$$

where e is a parameter in A.

Theorem 1. *An intuitionistic fuzzy soft set $(\tilde{\lambda}, A)$ over H is an intuitionistic fuzzy soft hyper BCK ideal of H if and only if the nonempty sets U_ε and L_δ are hyper BCK ideals of H for all $(\varepsilon, \delta) \in [0,1] \times [0,1]$ with $\varepsilon + \delta \leq 1$.*

Proof. Let e be a parameter in A. Assume that $(\tilde{\lambda}, A)$ is an intuitionistic fuzzy soft hyper BCK ideal of H and U_ε and L_δ are nonempty for all $(\varepsilon, \delta) \in [0,1] \times [0,1]$ with $\varepsilon + \delta \leq 1$. Then there exist $a \in U_\varepsilon$ and $b \in L_\delta$, and so $\mu_{\tilde{\lambda}(e)}(a) \geq \varepsilon$ and $\gamma_{\tilde{\lambda}(e)}(b) \leq \delta$. It follows from (16) that

$$\mu_{\tilde{\lambda}(e)}(0) \geq \mu_{\tilde{\lambda}(e)}(a) \geq \varepsilon \text{ and } \gamma_{\tilde{\lambda}(e)}(0) \leq \gamma_{\tilde{\lambda}(e)}(b) \leq \delta.$$

Hence $0 \in U_\varepsilon \cap L_\delta$. Let $x, y \in H$ be such that $x \circ y \ll U_\varepsilon$ and $y \in U_\varepsilon$. Then for any $a \in x \circ y$ there exists $a_0 \in U_\varepsilon$ such that $a \ll a_0$. Thus $\mu_{\tilde{\lambda}(e)}(a) \geq \mu_{\tilde{\lambda}(e)}(a_0) \geq \varepsilon$ by (14), which implies from (15) that

$$\mu_{\tilde{\lambda}(e)}(x) \geq \min\left\{\inf_{a \in x \circ y} \mu_{\tilde{\lambda}(e)}(a), \mu_{\tilde{\lambda}(e)}(y)\right\} \geq \min\left\{\varepsilon, \mu_{\tilde{\lambda}(e)}(y)\right\} \geq \varepsilon.$$

Hence $x \in U_\varepsilon$, and therefore U_ε is a hyper BCK ideal of H. Now suppose that $a \circ b \ll L_\delta$ and $b \in L_\delta$ for all $a, b \in H$. Then for any $x \in a \circ b$ there exists $x_0 \in L_\delta$ such that $x \ll x_0$. Thus $\gamma_{\tilde{\lambda}(e)}(x) \leq \gamma_{\tilde{\lambda}(e)}(x_0) \leq \delta$ by (14), which implies from (15) that

$$\gamma_{\tilde{\lambda}(e)}(a) \leq \max\left\{\sup_{x \in a \circ b} \gamma_{\tilde{\lambda}(e)}(x), \gamma_{\tilde{\lambda}(e)}(b)\right\} \leq \max\left\{\delta, \gamma_{\tilde{\lambda}(e)}(b)\right\} \leq \delta.$$

Hence $a \in L_\delta$, and therefore L_δ is a hyper BCK ideal of H.

Conversely, suppose that the nonempty sets U_ε and L_δ are hyper BCK ideals of H for all $(\varepsilon, \delta) \in [0,1] \times [0,1]$ with $\varepsilon + \delta \leq 1$. Let $x, y, u, v \in H$ be such that $x \ll y$, $\mu_{\tilde{\lambda}(e)}(y) = \varepsilon$, $u \ll v$ and $\gamma_{\tilde{\lambda}(e)}(v) = \delta$. Then $y \in U_\varepsilon$ and $v \in L_\delta$, which imply that $x \ll U_\varepsilon$ and $u \ll L_\delta$. It follows from Lemma 1 that $x \in U_\varepsilon$ and $u \in L_\delta$. Thus $\mu_{\tilde{\lambda}(e)}(x) \geq \varepsilon = \mu_{\tilde{\lambda}(e)}(y)$ and $\gamma_{\tilde{\lambda}(e)}(u) \leq \delta = \gamma_{\tilde{\lambda}(e)}(v)$. Now, for any $x, y, u, v \in H$, let $\varepsilon := \min\left\{\inf_{a \in x \circ y} \mu_{\tilde{\lambda}(e)}(a), \mu_{\tilde{\lambda}(e)}(y)\right\}$ and $\delta := \max\left\{\sup_{b \in u \circ v} \gamma_{\tilde{\lambda}(e)}(b), \gamma_{\tilde{\lambda}(e)}(v)\right\}$. Then $y \in U_\varepsilon$ and $v \in L_\delta$, and for each $a \in x \circ y$ and $b \in u \circ v$ we have

$$\mu_{\tilde{\lambda}(e)}(a) \geq \inf_{a \in x \circ y} \mu_{\tilde{\lambda}(e)}(a) \geq \min\left\{\inf_{a \in x \circ y} \mu_{\tilde{\lambda}(e)}(a), \mu_{\tilde{\lambda}(e)}(y)\right\} = \varepsilon$$

and

$$\gamma_{\tilde{\lambda}(e)}(b) \leq \sup_{b \in u \circ v} \gamma_{\tilde{\lambda}(e)}(b) \leq \max\left\{\sup_{b \in u \circ v} \gamma_{\tilde{\lambda}(e)}(b), \gamma_{\tilde{\lambda}(e)}(v)\right\} = \delta.$$

Thus, $a \in U_\varepsilon$ and $b \in L_\delta$, and so $x \circ y \subseteq U_\varepsilon$ and $u \circ v \subseteq L_\delta$. Hence $x \circ y \ll U_\varepsilon$ and $u \circ v \ll L_\delta$ by (4). Since $y \in U_\varepsilon$, $v \in L_\delta$ and U_ε and L_δ are hyper BCK ideal of H, it follows that $x \in U_\varepsilon$ and $u \in L_\delta$. Therefore

$$\mu_{\tilde{\lambda}(e)}(x) \geq \varepsilon = \min\left\{\inf_{a\in x\circ y}\mu_{\tilde{\lambda}(e)}(a), \mu_{\tilde{\lambda}(e)}(y)\right\}$$

and

$$\gamma_{\tilde{\lambda}(e)}(u) \leq \delta = \max\left\{\sup_{b\in u\circ v}\gamma_{\tilde{\lambda}(e)}(b), \gamma_{\tilde{\lambda}(e)}(v)\right\}.$$

Consequently, $(\tilde{\lambda}, A)$ is an intuitionistic fuzzy soft hyper BCK ideal of H. □

Definition 3. *An intuitionistic fuzzy soft set $(\tilde{\lambda}, A)$ over H is called an*

- *intuitionistic fuzzy soft weak hyper BCK ideal based on a parameter $e \in A$ over H (briefly, e-intuitionistic fuzzy soft weak hyper BCK ideal of H) if the intuitionistic fuzzy value set $\tilde{\lambda}(e)$ of e satisfies conditions* (15) *and* (16).
- *intuitionistic fuzzy soft s-weak hyper BCK ideal based on a parameter $e \in A$ over H (briefly, e-intuitionistic fuzzy soft s-weak hyper BCK ideal of H) if the intuitionistic fuzzy value set $\tilde{\lambda}(e)$ of e satisfies conditions* (16) *and* (18).

If $(\tilde{\lambda}, A)$ is an intuitionistic fuzzy soft weak (resp., s-weak) hyper BCK ideal based on e over H for all $e \in A$, we say that $(\tilde{\lambda}, A)$ is an *intuitionistic fuzzy soft weak (resp., s-weak) hyper BCK ideal* of H.

Example 2. *The intuitionistic fuzzy soft set $(\tilde{\lambda}, A)$ in Example 1 is an intuitionistic fuzzy soft weak hyper BCK ideal of H.*

Obviously, every intuitionistic fuzzy soft hyper BCK ideal is an intuitionistic fuzzy soft weak hyper BCK ideal. However, the converse is not true in general. In fact, the intuitionistic fuzzy soft weak hyper BCK ideal of H in Example 2 is not an intuitionistic fuzzy soft hyper BCK ideal of H since it is not an intuitionistic fuzzy soft hyper BCK ideal based on parameter y over H.

Theorem 2. *An intuitionistic fuzzy soft set $(\tilde{\lambda}, A)$ over H is an intuitionistic fuzzy soft weak hyper BCK ideal of H if and only if the nonempty sets U_ε and L_δ are weak hyper BCK ideals of H for all $(\varepsilon, \delta) \in [0,1] \times [0,1]$ with $\varepsilon + \delta \leq 1$ where e is any parameter in A.*

Proof. It is similar to the proof of Theorem 1. □

Theorem 3. *Every intuitionistic fuzzy soft s-weak hyper BCK ideal is an intuitionistic fuzzy soft weak hyper BCK ideal.*

Proof. Let $(\tilde{\lambda}, A)$ be an intuitionistic fuzzy soft s-weak hyper BCK ideal of H. Let $x, y \in H$ and $e \in A$. Then there exists $a, b \in x \circ y$ such that $\mu_{\tilde{\lambda}(e)}(x) \geq \min\{\mu_{\tilde{\lambda}(e)}(a), \mu_{\tilde{\lambda}(e)}(y)\}$ and $\gamma_{\tilde{\lambda}(e)}(x) \leq \max\{\gamma_{\tilde{\lambda}(e)}(b), \gamma_{\tilde{\lambda}(e)}(y)\}$ by (18). Since $\mu_{\tilde{\lambda}(e)}(a) \geq \inf_{c\in x\circ y}\mu_{\tilde{\lambda}(e)}(c)$ and $\gamma_{\tilde{\lambda}(e)}(b) \leq \sup_{d\in x\circ y}\mu_{\tilde{\lambda}(e)}(d)$, it follows that

$$\mu_{\tilde{\lambda}(e)}(x) \geq \min\left\{\inf_{c\in x\circ y}\mu_{\tilde{\lambda}(e)}(c), \mu_{\tilde{\lambda}(e)}(y)\right\}$$

and

$$\gamma_{\tilde{\lambda}(e)}(x) \leq \max\left\{\sup_{d\in x\circ y}\gamma_{\tilde{\lambda}(e)}(d), \gamma_{\tilde{\lambda}(e)}(y)\right\}.$$

Therefore $(\tilde{\lambda}, A)$ is an intuitionistic fuzzy soft weak hyper BCK ideal of H. □

Question 1. *Is the converse of Theorem 3 true?*

It is not easy to find an example of an intuitionistic fuzzy soft weak hyper BCK ideal which is not an intuitionistic fuzzy soft s-weak hyper BCK ideal. However, we have the following theorem.

Theorem 4. *If an intuitionistic fuzzy soft weak hyper BCK ideal $(\tilde{\lambda}, A)$ of H satisfies the condition (17) then $(\tilde{\lambda}, A)$ is an intuitionistic fuzzy soft s-weak hyper BCK ideal of H.*

Proof. Let e be a parameter in A. For any $x, y \in H$, there exists $x_0, y_0 \in x \circ y$ such that $\mu_{\tilde{\lambda}(e)}(x_0) = \inf\limits_{a \in x \circ y} \mu_{\tilde{\lambda}(e)}(a)$ and $\gamma_{\tilde{\lambda}(e)}(y_0) = \sup\limits_{b \in x \circ y} \gamma_{\tilde{\lambda}(e)}(b)$ by (17). It follows from (15) that

$$\mu_{\tilde{\lambda}(e)}(x) \geq \min\left\{\inf_{a \subset x \circ y} \mu_{\tilde{\lambda}(e)}(a), \mu_{\tilde{\lambda}(e)}(y)\right\} = \min\left\{\mu_{\tilde{\lambda}(e)}(x_0), \mu_{\tilde{\lambda}(e)}(y)\right\}$$

and

$$\gamma_{\tilde{\lambda}(e)}(x) \leq \max\left\{\sup_{a \in x \circ y} \gamma_{\tilde{\lambda}(e)}(a), \gamma_{\tilde{\lambda}(e)}(y)\right\} = \max\left\{\gamma_{\tilde{\lambda}(e)}(y_0), \gamma_{\tilde{\lambda}(e)}(y)\right\}$$

Therefore $(\tilde{\lambda}, A)$ is an e-intuitionistic fuzzy soft s-weak hyper BCK ideal of H, and hence $(\tilde{\lambda}, A)$ is an intuitionistic fuzzy soft s-weak hyper BCK ideal of H since e is arbitrary. □

The condition (17) is always true in a finite hyper BCK algebra. Hence the notion of intuitionistic fuzzy soft s-weak hyper BCK ideal is in accord with the notion of intuitionistic fuzzy soft weak hyper BCK ideal in a finite hyper BCK algebra.

Definition 4. *An intuitionistic fuzzy soft set $(\tilde{\lambda}, A)$ over H is called an intuitionistic fuzzy soft strong hyper BCK ideal over H based on a parameter e in A (briefly, e-intuitionistic fuzzy soft strong hyper BCK ideal of H) if the intuitionistic fuzzy value set $\tilde{\lambda}(e) : H \to [0,1]$ of e satisfies the condition*

$$(\forall x, y \in H)\left(\begin{array}{l} \mu_{\tilde{\lambda}(e)}(x) \geq \min\left\{\sup\limits_{a \in x \circ y} \mu_{\tilde{\lambda}(e)}(a),\ \mu_{\tilde{\lambda}(e)}(y)\right\} \\ \gamma_{\tilde{\lambda}(e)}(x) \leq \max\left\{\inf\limits_{a \in x \circ y} \gamma_{\tilde{\lambda}(e)}(a),\ \gamma_{\tilde{\lambda}(e)}(y)\right\} \end{array}\right). \tag{20}$$

and

$$(\forall x \in H)\left(\inf_{a \in x \circ x} \mu_{\tilde{\lambda}(e)}(a) \geq \mu_{\tilde{\lambda}(e)}(x),\ \sup_{a \in x \circ x} \gamma_{\tilde{\lambda}(e)}(a) \leq \gamma_{\tilde{\lambda}(e)}(x)\right). \tag{21}$$

If $(\tilde{\lambda}, A)$ is an e-intuitionistic fuzzy soft strong hyper BCK ideal of H for all $e \in A$, we say that $(\tilde{\lambda}, A)$ is an intuitionistic fuzzy soft strong hyper BCK ideal of H.

Proposition 2. *Every intuitionistic fuzzy soft strong hyper BCK ideal $(\tilde{\lambda}, A)$ of H satisfies the following assertions.*

(1) *$(\tilde{\lambda}, A)$ satisfies the condition (16) for all $e \in A$.*

(2) $(\forall x, y \in H)(\forall e \in A)\left(x \ll y \Rightarrow \left\{\begin{array}{l} \mu_{\tilde{\lambda}(e)}(x) \geq \mu_{\tilde{\lambda}(e)}(y) \\ \gamma_{\tilde{\lambda}(e)}(x) \leq \gamma_{\tilde{\lambda}(e)}(y) \end{array}\right.\right)$.

(3) $(\forall a, x, y \in H)(\forall e \in A)\left(a \in x \circ y \Rightarrow \left\{\begin{array}{l} \mu_{\tilde{\lambda}(e)}(x) \geq \min\{\mu_{\tilde{\lambda}(e)}(a), \mu_{\tilde{\lambda}(e)}(y)\} \\ \gamma_{\tilde{\lambda}(e)}(x) \leq \max\{\gamma_{\tilde{\lambda}(e)}(a), \gamma_{\tilde{\lambda}(e)}(y)\} \end{array}\right.\right)$.

Proof. (1) Let $e \in A$. Since $x \ll x$, i.e., $0 \in x \circ x$ for all $x \in H$, we have

$$\mu_{\tilde{\lambda}(e)}(0) \geq \inf_{a \in x \circ x} \mu_{\tilde{\lambda}(e)}(a) \geq \mu_{\tilde{\lambda}(e)}(x)$$

and

$$\gamma_{\tilde{\lambda}(e)}(0) \leq \sup_{a \in x \circ x} \gamma_{\tilde{\lambda}(e)}(a) \leq \gamma_{\tilde{\lambda}(e)}(x)$$

for all $x \in H$ by (21).

(2) Let $e \in A$ and $x, y \in H$ be such that $x \ll y$. Then $0 \in x \circ y$, and so $\mu_{\tilde{\lambda}(e)}(0) \leq \sup_{a \in x \circ y} \mu_{\tilde{\lambda}(e)}(a)$ and $\gamma_{\tilde{\lambda}(e)}(0) \geq \inf_{a \in x \circ y} \gamma_{\tilde{\lambda}(e)}(a)$. It follows from (20) and (16) that

$$\mu_{\tilde{\lambda}(e)}(x) \geq \min\left\{\sup_{a \in x \circ y} \mu_{\tilde{\lambda}(e)}(a), \mu_{\tilde{\lambda}(e)}(y)\right\} \geq \min\left\{\mu_{\tilde{\lambda}(e)}(0), \mu_{\tilde{\lambda}(e)}(y)\right\} = \mu_{\tilde{\lambda}(e)}(y)$$

and

$$\gamma_{\tilde{\lambda}(e)}(x) \leq \max\left\{\inf_{a \in x \circ y} \gamma_{\tilde{\lambda}(e)}(a), \gamma_{\tilde{\lambda}(e)}(y)\right\} \leq \max\left\{\gamma_{\tilde{\lambda}(e)}(0), \gamma_{\tilde{\lambda}(e)}(y)\right\} = \gamma_{\tilde{\lambda}(e)}(y).$$

(3) Let $e \in A$ and $a, x, y \in H$ be such that $a \in x \circ y$. Then $\sup_{b \in x \circ y} \mu_{\tilde{\lambda}(e)}(b) \geq \mu_{\tilde{\lambda}(e)}(a)$ and $\inf_{c \in x \circ y} \gamma_{\tilde{\lambda}(e)}(c) \leq \gamma_{\tilde{\lambda}(e)}(a)$, which imply from (20) that

$$\mu_{\tilde{\lambda}(e)}(x) \geq \min\left\{\sup_{b \in x \circ y} \mu_{\tilde{\lambda}(e)}(b), \mu_{\tilde{\lambda}(e)}(y)\right\} \geq \min\left\{\mu_{\tilde{\lambda}(e)}(a), \mu_{\tilde{\lambda}(e)}(y)\right\}$$

and

$$\gamma_{\tilde{\lambda}(e)}(x) \leq \max\left\{\inf_{c \in x \circ y} \gamma_{\tilde{\lambda}(e)}(c), \gamma_{\tilde{\lambda}(e)}(y)\right\} \leq \max\left\{\gamma_{\tilde{\lambda}(e)}(a), \gamma_{\tilde{\lambda}(e)}(y)\right\}.$$

This proves (3). □

Please note that if $a \in x \circ y$ for all $a, x, y \in H$, then $\mu_{\tilde{\lambda}(e)}(a) \geq \inf_{b \in x \circ y} \mu_{\tilde{\lambda}(e)}(b)$ and $\gamma_{\tilde{\lambda}(e)}(a) \leq \sup_{b \in x \circ y} \gamma_{\tilde{\lambda}(e)}(b)$ for all $e \in A$. Hence, we have the following corollary.

Corollary 1. *Every intuitionistic fuzzy soft strong hyper BCK ideal $(\tilde{\lambda}, A)$ of H satisfies the following condition:*

$$(\forall e \in A)(\forall x, y \in H) \left(\begin{array}{l} \mu_{\tilde{\lambda}(e)}(x) \geq \min\left\{\inf_{a \in x \circ y} \mu_{\tilde{\lambda}(e)}(a), \mu_{\tilde{\lambda}(e)}(y)\right\} \\ \gamma_{\tilde{\lambda}(e)}(x) \leq \max\left\{\sup_{a \in x \circ y} \gamma_{\tilde{\lambda}(e)}(a), \gamma_{\tilde{\lambda}(e)}(y)\right\} \end{array} \right). \quad (22)$$

Corollary 2. *Every intuitionistic fuzzy soft strong hyper BCK ideal is both an intuitionistic fuzzy soft s-weak hyper BCK ideal and an intuitionistic fuzzy soft hyper BCK ideal.*

Proof. Straightforward. □

The following example shows that there is an intuitionistic fuzzy soft hyper BCK ideal (and hence an intuitionistic fuzzy soft weak hyper BCK ideal) which is not an intuitionistic fuzzy soft strong hyper BCK ideal.

Example 3. *Consider the hyper BCK algebra* $H = \{0, a, b\}$ *in Example 1. Given a set* $E = \{x, y, z\}$ *of parameters, let* $(\tilde{\lambda}, A)$ *be an intuitionistic fuzzy soft set over H defined by Table 3.*

Table 3. Tabular representation of $(\tilde{\lambda}, A)$.

$\tilde{\lambda}$	0	a	b
x	$(0.8, 0.15)$	$(0.7, 0.25)$	$(0.6, 0.35)$
y	$(0.5, 0.35)$	$(0.3, 0.45)$	$(0.2, 0.45)$
z	$(0.9, 0.05)$	$(0.6, 0.35)$	$(0.1, 0.65)$

Then $(\tilde{\lambda}, A)$ *is an intuitionistic fuzzy soft (weak) hyper BCK ideal of H. However, it is not an intuitionistic fuzzy soft strong hyper BCK ideal of H since*

$$\mu_{\tilde{\lambda}(y)}(b) = 0.2 < 0.3 = \min\left\{\sup_{c \in b \circ a} \mu_{\tilde{\lambda}(y)}(c), \mu_{\tilde{\lambda}(y)}(a)\right\}$$

and

$$\gamma_{\tilde{\lambda}(y)}(b) = 0.45 = 0.45 = \max\left\{\inf_{c \in b \circ a} \gamma_{\tilde{\lambda}(y)}(c), \gamma_{\tilde{\lambda}(y)}(a)\right\}.$$

Theorem 5. *If* $(\tilde{\lambda}, A)$ *is an intuitionistic fuzzy soft strong hyper BCK ideal of H, then the nonempty sets* U_ε *and* L_δ *are strong hyper BCK ideals of H for all* $(\varepsilon, \delta) \in [0, 1] \times [0, 1]$ *with* $\varepsilon + \delta \leq 1$.

Proof. Let $(\tilde{\lambda}, A)$ be an intuitionistic fuzzy soft strong hyper BCK ideal of H and $(\varepsilon, \delta) \in [0, 1] \times [0, 1]$ be such that $\varepsilon + \delta \leq 1$ and $U_\varepsilon \neq \emptyset \neq L_\delta$ where e is any parameter in A. Then there exist $a \in U_\varepsilon$ and $b \in L_\delta$, and thus $\mu_{\tilde{\lambda}(e)}(a) \geq \varepsilon$ and $\gamma_{\tilde{\lambda}(e)}(b) \leq \delta$. By Proposition 2(1), $\mu_{\tilde{\lambda}(e)}(0) \geq \mu_{\tilde{\lambda}(e)}(a) \geq \varepsilon$ and $\gamma_{\tilde{\lambda}(e)}(0) \leq \gamma_{\tilde{\lambda}(e)}(b) \leq \delta$, and thus $0 \in U_\varepsilon \cap L_\delta$. Let $x, y \in H$ be such that $(x \circ y) \cap U_\varepsilon \neq \emptyset$ and $y \in U_\varepsilon$. Then $\mu_{\tilde{\lambda}(e)}(y) \geq \varepsilon$ and there exists $a_0 \in (x \circ y) \cap U_\varepsilon$. It follows from (20) that

$$\mu_{\tilde{\lambda}(e)}(x) \geq \min\left\{\sup_{c \in x \circ y} \mu_{\tilde{\lambda}(e)}(c), \mu_{\tilde{\lambda}(e)}(y)\right\} \geq \min\left\{\mu_{\tilde{\lambda}(e)}(a_0), \mu_{\tilde{\lambda}(e)}(y)\right\} \geq \varepsilon.$$

Hence $x \in U_\varepsilon$, and therefore U_ε is a strong hyper BCK ideal of H. Now assume that $(x \circ y) \cap L_\delta \neq \emptyset$ and $y \in L_\delta$ for all $x, y \in H$. Then there exists $b_0 \in (x \circ y) \cap L_\delta$ and $\gamma_{\tilde{\lambda}(e)}(y) \leq \delta$. Using (20), we get

$$\gamma_{\tilde{\lambda}(e)}(x) \leq \max\left\{\inf_{c \in x \circ y} \gamma_{\tilde{\lambda}(e)}(c), \gamma_{\tilde{\lambda}(e)}(y)\right\} \leq \max\left\{\gamma_{\tilde{\lambda}(e)}(b_0), \gamma_{\tilde{\lambda}(e)}(y)\right\} \leq \delta.$$

Thus, $x \in L_\delta$, and so L_δ is a strong hyper BCK ideal of H. □

We provide conditions for an intuitionistic fuzzy soft set to be an intuitionistic fuzzy soft strong hyper BCK ideal.

Theorem 6. *Let* $(\tilde{\lambda}, A)$ *be an intuitionistic fuzzy soft set over H such that*

$$(\forall T \subseteq H)(\exists x_0, y_0 \in T)\left(\mu_{\tilde{\lambda}(e)}(x_0) = \sup_{a \in T} \mu_{\tilde{\lambda}(e)}(a),\ \gamma_{\tilde{\lambda}(e)}(y_0) = \inf_{b \in T} \gamma_{\tilde{\lambda}(e)}(b)\right) \tag{23}$$

where e is any parameter in A. If the sets U_ε *and* L_δ *in (19) are nonempty strong hyper BCK ideals of H for all* $(\varepsilon, \delta) \in [0, 1] \times [0, 1]$ *with* $\varepsilon + \delta \leq 1$*, then* $(\tilde{\lambda}, A)$ *is an intuitionistic fuzzy soft strong hyper BCK ideal of H.*

Proof. For any parameter e in A and $x \in H$, let $\mu_{\tilde{\lambda}(e)}(x) = \varepsilon$ and $\gamma_{\tilde{\lambda}(e)}(x) = \delta$. Then $x \in U_\varepsilon$ and $x \in L_\delta$. Since $x \circ x \ll x$ by (H3), it follows from Lemma 1 that $x \circ x \subseteq U_\varepsilon$. Hence $\mu_{\tilde{\lambda}(e)}(a) \geq \varepsilon$ and $\gamma_{\tilde{\lambda}(e)}(a) \leq \delta$ for all $a \in x \circ x$, and so $\inf_{a \in x \circ x} \mu_{\tilde{\lambda}(e)}(a) \geq \varepsilon = \mu_{\tilde{\lambda}(e)}(x)$ and $\sup_{a \in x \circ x} \gamma_{\tilde{\lambda}(e)}(a) \leq \delta = \gamma_{\tilde{\lambda}(e)}(x)$. For any $x, y \in H$, let $k = \min\left\{\sup_{a \in x \circ y} \mu_{\tilde{\lambda}(e)}(a), \mu_{\tilde{\lambda}(e)}(y)\right\}$ and $r = \max\left\{\inf_{a \in x \circ y} \gamma_{\tilde{\lambda}(e)}(a), \gamma_{\tilde{\lambda}(e)}(y)\right\}$. Then U_k and L_r are nonempty and are strong hyper BCK ideals of H by hypothesis. Using the condition (23) implies that $\mu_{\tilde{\lambda}(e)}(a_0) = \sup_{a \in x \circ y} \mu_{\tilde{\lambda}(e)}(a)$ and $\gamma_{\tilde{\lambda}(e)}(b_0) = \inf_{a \in x \circ y} \gamma_{\tilde{\lambda}(e)}(a)$ for some $a_0, b_0 \in x \circ y$. Hence

$$\mu_{\tilde{\lambda}(e)}(a_0) = \sup_{a \in x \circ y} \mu_{\tilde{\lambda}(e)}(a) \geq \min\left\{\sup_{a \in x \circ y} \mu_{\tilde{\lambda}(e)}(a), \mu_{\tilde{\lambda}(e)}(y)\right\} = k$$

and

$$\gamma_{\tilde{\lambda}(e)}(b_0) = \inf_{a \in x \circ y} \gamma_{\tilde{\lambda}(e)}(a) \leq \max\left\{\inf_{a \in x \circ y} \gamma_{\tilde{\lambda}(e)}(a), \gamma_{\tilde{\lambda}(e)}(y)\right\} = r,$$

which imply that $a_0 \in U_k$ and $b_0 \in L_r$. It follows that $(x \circ y) \cap U_k \neq \emptyset$ and $(x \circ y) \cap L_r \neq \emptyset$. Since U_k and L_r are strong hyper BCK ideals of H, we have $x \in U_k$ and $x \in L_r$. Thus

$$\mu_{\tilde{\lambda}(e)}(x) \geq k = \min\left\{\sup_{a \in x \circ y} \mu_{\tilde{\lambda}(e)}(a), \mu_{\tilde{\lambda}(e)}(y)\right\}$$

and

$$\gamma_{\tilde{\lambda}(e)}(x) \leq r = \max\left\{\inf_{a \in x \circ y} \gamma_{\tilde{\lambda}(e)}(a), \gamma_{\tilde{\lambda}(e)}(y)\right\}$$

Therefore $(\tilde{\lambda}, A)$ is an intuitionistic fuzzy soft strong hyper BCK ideal of H. □

Theorem 7. *Let H satisfy the following condition:*

$$(\forall x, y \in H)\,(|x \circ y| < \infty). \tag{24}$$

Given an intuitionistic fuzzy soft set $(\tilde{\lambda}, A)$ over H, if the nonempty sets U_ε and L_δ in (19) are strong hyper BCK ideals of H for all $(\varepsilon, \delta) \in [0,1] \times [0,1]$ with $\varepsilon + \delta \leq 1$, then $(\tilde{\lambda}, A)$ is an intuitionistic fuzzy soft strong hyper BCK ideal of H.

Proof. Assume that U_ε and L_δ in (19) are nonempty strong hyper BCK ideals of H for all $(\varepsilon, \delta) \in [0,1] \times [0,1]$ with $\varepsilon + \delta \leq 1$. Then U_ε and L_δ are hyper BCK ideals of H, and so $(\tilde{\lambda}, A)$ is an intuitionistic fuzzy soft hyper BCK ideal of H by Theorem 1. Please note that $x \circ x \subseteq x \circ H \ll \{x\}$ for all $x \in H$. Hence $a \ll x$ for every $a \in x \circ x$, which implies from (14) that $\mu_{\tilde{\lambda}(e)}(a) \geq \mu_{\tilde{\lambda}(e)}(x)$ and $\gamma_{\tilde{\lambda}(e)}(a) \leq \gamma_{\tilde{\lambda}(e)}(x)$ for all $a \in x \circ x$ and any parameter e in A. Thus $\mu_{\tilde{\lambda}(e)}(x) \leq \inf_{a \in x \circ x} \mu_{\tilde{\lambda}(e)}(a)$ and $\gamma_{\tilde{\lambda}(e)}(x) \geq \sup_{a \in x \circ x} \gamma_{\tilde{\lambda}(e)}(a)$. Let $\min\left\{\sup_{a \in x \circ y} \mu_{\tilde{\lambda}(e)}(a), \mu_{\tilde{\lambda}(e)}(y)\right\} = \varepsilon$ and $\max\left\{\inf_{a \in x \circ y} \gamma_{\tilde{\lambda}(e)}(a), \gamma_{\tilde{\lambda}(e)}(y)\right\} = \delta$. Then $\sup_{a \in x \circ y} \mu_{\tilde{\lambda}(e)}(a) \geq \varepsilon$, $\mu_{\tilde{\lambda}(e)}(y) \geq \varepsilon$, $\inf_{a \in x \circ y} \gamma_{\tilde{\lambda}(e)}(a) \leq \delta$ and $\gamma_{\tilde{\lambda}(e)}(y) \leq \delta$. Since $|x \circ y| < \infty$ for all $x, y \in H$, there exists $b \in x \circ y$ such that $\mu_{\tilde{\lambda}(e)}(b) \geq \varepsilon$, $\mu_{\tilde{\lambda}(e)}(y) \geq \varepsilon$, $\gamma_{\tilde{\lambda}(e)}(b) \leq \delta$ and $\gamma_{\tilde{\lambda}(e)}(y) \leq \delta$. It follows that $(x \circ y) \cap U_\varepsilon \neq \emptyset$, $y \in U_\varepsilon$, $(x \circ y) \cap L_\delta \neq \emptyset$ and $y \in L_\delta$. Since U_ε and L_δ are strong hyper BCK ideal of H, we have $x \in U_\varepsilon \cap L_\delta$. Consequently, $\mu_{\tilde{\lambda}(e)}(x) \geq \varepsilon = \min\left\{\sup_{a \in x \circ y} \mu_{\tilde{\lambda}(e)}(a), \mu_{\tilde{\lambda}(e)}(y)\right\}$ and

$\gamma_{\tilde{\lambda}(e)}(x) \leq \delta = \max\left\{\inf_{a \in x \circ y} \gamma_{\tilde{\lambda}(e)}(a), \gamma_{\tilde{\lambda}(e)}(y)\right\}$. Therefore $(\tilde{\lambda}, A)$ is an intuitionistic fuzzy soft strong hyper BCK ideal of H. $\square$

4. Conclusions

We have introduced the notions of intuitionistic fuzzy soft hyper BCK ideal, intuitionistic fuzzy soft weak hyper BCK ideal, intuitionistic fuzzy soft s-weak hyper BCK-ideal and intuitionistic fuzzy soft strong hyper BCK-ideal, and have investigated related properties and relations. We have discussed characterizations of intuitionistic fuzzy soft (weak) hyper BCK ideal, and have found conditions for an intuitionistic fuzzy soft weak hyper BCK ideal to be an intuitionistic fuzzy soft s-weak hyper BCK ideal. We have provided conditions for an intuitionistic fuzzy soft set to be an intuitionistic fuzzy soft strong hyper BCK ideal. In future work, different types of intuitionistic fuzzy soft hyper BCK ideals will be defined and discussed.

Author Contributions: Investigation, X.X. and R.A.B.; Methodology, M.B. and Y.B.J.

Funding: This research is partially supported by a grant of National Natural Science Foundation of China (11571281).

Acknowledgments: The authors wish to thank the anonymous reviewers for their valuable comments and suggestions.

Conflicts of Interest: The authors declare no conflict of interest.

References

1. Molodtsov, D. Soft set theory—First results. *Comput. Math. Appl.* **1999**, *37*, 19–31. [CrossRef]
2. Jun, Y.B. Soft *BCK/BCI*-algebras. *Comput. Math. Appl.* **2008**, *56*, 1408–1413. [CrossRef]
3. Jun, Y.B.; Park, C.H. Applications of soft sets in ideal theory of BCK/BCI-algebras. *Inform. Sci.* **2008**, *178*, 2466–2475. [CrossRef]
4. Maji, P.K.; Biswas, R.; Roy, A.R. Fuzzy soft sets. *J. Fuzzy Math.* **2001**, *9*, 589–602.
5. Maji, P.K.; Biswas, R.; Roy, A.R. Intuitionistic fuzzy soft sets. *J. Fuzzy Math.* **2001**, *9*, 677–692.
6. Jun, Y.B.; Lee, K.J.; Park, C.H. Fuzzy soft set theory applied to *BCK/BCI*-algebras. *Comput. Math. Appl.* **2010**, *59*, 3180–3192. [CrossRef]
7. Marty, F. Sur une generalization de la notion de groupe. In Proceedings of the 8th Congrès des Mathématiciens Scandinaves, Stockholm, Sweden, 2–7 September 1934; pp. 45–49.
8. Ameri, R. On categories of hypergroups and hypermodules. *Ital. J. Pure Appl. Math.* **2003**, *6*, 121–132. [CrossRef]
9. Ameri, R.; Rosenberg, I.G. Congruences of multialgebras. *Mult.-Valued Log. Soft Comput.* **2009**, *15*, 525–536.
10. Ameri, R.; Zahedi, M.M. Hyperalgebraic systems. *Ital. J. Pure Appl. Math.* **1999**, *6*, 21–32.
11. Corsini, P. *Prolegomena of Hypergroup Theory*; Aviani Editore: Tricesimo, Italy, 1993.
12. Corsini, P.; Leoreanu, V. *Applications of Hyperstructure Theory*; Kluwer: Dordrecht, The Netherlands, 2003.
13. Leoreanu-Fotea, V.; Davvaz, B. Join n-spaces and lattices. *Mult.-Valued Log. Soft Comput.* **2008**, *15*, 421–432.
14. Leoreanu-Fotea, V.; Davvaz, B. n-hypergroups and binary relations. *Eur. J. Combin.* **2008**, *29*, 1207–1218. [CrossRef]
15. Pelea, C. On the direct product of multialgebras. *Studia Univ. Babes-Bolyai Math.* **2003**, *XLVIII*, 93–98.
16. Pickett, H.E. Homomorphism and subalgebras of multialgebras. *Pac. J. Math.* **2001**, *10*, 141–146. [CrossRef]
17. Schweigert, D. Congruence relations of multialgebras. *Discret. Math.* **1985**, *53*, 249–253. [CrossRef]
18. Serafimidis, K.; Kehagias, A.; Konstantinidou, M. The L-fuzzy Corsini join hyperoperation. *Ital. J. Pure Appl. Math.* **2002**, *12*, 83–90.
19. Vougiouklis, T. *Hyperstructures and Their Representations*; Hadronic Press, Inc.: Palm Harbor, FL, USA, 1994.
20. Jun, Y.B.; Zahedi, M.M.; Xin, X.L.; Borzooei, R.A. On hyper *BCK*-algebras. *Ital. J. Pure Appl. Math.* **2000**, *8*, 127–136.
21. Jun, Y.B.; Xin, X.L. Scalar elements and hyper atoms of hyper *BCK*-algebras. *Sci. Math.* **1999**, *2*, 303–309.
22. Jun, Y.B.; Xin, X.L.; Zahedi, M.M.; Roh, E.H. Strong hyper *BCK*-ideals of hyper *BCK*-algebras. *Math. Jpn.* **2000**, *51*, 493–498.

23. Jun, Y.B.; Shim, W.H. Fuzzy implicative hyper *BCK*-ideals of hyper *BCK*-algebras. *Int. J. Math. Math. Sci.* **2002**, *29*, 63–70. [CrossRef]
24. Jun, Y.B.; Xin, X.L. Fuzzy hyper *BCK*-ideals of hyper *BCK*-algebras. *Sci. Math. Jpn.* **2001**, *53*, 353–360. [CrossRef]
25. Davvaz, B.; Cristea, I. *Fuzzy Algebraic Hyperstructures*; Springer: Cham, Switzerland, 2015.

symmetry

MDPI

Article

Series of Semihypergroups of Time-Varying Artificial Neurons and Related Hyperstructures

Jan Chvalina [1] **and Bedřich Smetana** [1,2,*]

1 Department of Mathematics, Faculty of Electrical Engineeering and Comunication, Brno University of Technology, Technická 8, 616 00 Brno, Czech Republic
2 Department of Quantitative Methods, University of Defence in Brno, Kounicova 65, 662 10 Brno, Czech Republic
* Correspondence: bedrich.smetana@unob.cz or xsmeta06@stud.feec.vutbr.cz; Tel.: +420-973-443-429

Received: 30 May 2019; Accepted: 2 July 2019; Published: 16 July 2019

check for updates

Abstract: Detailed analysis of the function of multilayer perceptron (MLP) and its neurons together with the use of time-varying neurons allowed the authors to find an analogy with the use of structures of linear differential operators. This procedure allowed the construction of a group and a hypergroup of artificial neurons. In this article, focusing on semihyperstructures and using the above described procedure, the authors bring new insights into structures and hyperstructures of artificial neurons and their possible symmetric relations.

Keywords: time-varying artificial neuron; ordered group; transposition hypergroup; linear differential operator

---

1. Introduction

As mentioned in the PhD thesis [1], neurons are the atoms of neural computation. Out of those simple computational units all neural networks are build up. The output computed by a neuron can be expressed using two functions $y = g(f(w,x))$. The details of computation consist in several steps: In a first step the input to the neuron, $x := \{x_i\}$, is associated with the weights of the neuron, $w := \{w_i\}$, by involving the so-called propagation function f. This can be thought as computing the activation potential from the pre-synaptic activities. Then from that result the so-called activation function g computes the output of the neuron. The weights, which mimic synaptic strength, constitute the adjustable internal parameters of the neuron. The process of adapting the weights is called learning [1–18].

From the biological point of view it is appropriate to use an integrative propagation function. Therefore, a convenient choice would be to use the weighted sum of the input $f(w,x) = \sum\limits_i w_i x_i$, that is the activation potential equal to the scalar product of input and weights. This is, in fact, the most popular propagation function since the dawn of neural computation. However, it is often used in a slightly different form:

$$f(w,x) = \sum_i w_i x_i + \Theta. \tag{1}$$

The special weight Θ is called bias. Applying $\Theta(x) = 1$ for $x > 0$ and $\Theta(x) = 0$ for $x < 0$ as the above activation function yields the famous perceptron of Rosenblatt. In that case the function Θ works as a threshold.

Let $F : \mathbb{R} \to \mathbb{R}$ be a general non-linear (or piece-wise linear) transfer function. Then the action of a neuron can be expressed by

$$y(k) = F\left(\sum_{i=1}^{m} w_i(k) x_i(k) + b\right),$$

where $x_i(k)$ is input value in discrete time k where $i = 0, \ldots, m$, $w_i(k)$ is weight value in discrete time where $i = 0, \ldots, m$, b is bias, $y_i(k)$ is output value in discrete time k.

Notice that in some very special cases the transfer function F can be also linear. Transfer function defines the properties of artificial neuron and this can be any mathematical function. Usually it is chosen on the basis of the problem that the artificial neuron (artificial neural network) needs to solve and in most cases it is taken (as mentioned above) from the following set of functions: step function, linear function and non-linear (sigmoid) function [1,2,5,7,9,12,16,19].

In what follows we will consider a certain generalization of classical artificial neurons mentioned above such that inputs x_i and weight w_i will be functions of an argument t belonging into a linearly ordered (tempus) set T with the least element 0. As the index set we use the set $\mathbb{C}(J)$ of all continuous functions defined on an open interval $J \subset \mathbb{R}$. So, denote by W the set of all non-negative functions $w : T \to \mathbb{R}$ forming a subsemiring of the ring of all real functions of one real variable $x : \mathbb{R} \to \mathbb{R}$. Denote by $Ne(\vec{w}_r) = Ne(w_{r1}, \ldots, w_{rn})$ for $r \in \mathbb{C}(J)$, $n \in \mathbb{N}$ the mapping

$$y_r(t) = \sum_{k=1}^{n} w_{r,k}(t) x_{r,k}(t) + b_r$$

which will be called the artificial neuron with the bias $b_r \in \mathbb{R}$. By $\mathbb{AN}(T)$ we denote the collection of all such artificial neurons.

Neurons are usually denoted by capital letters X, Y or X_i, Y_i, nevertheless we use also notation $Ne(\vec{w})$, where $\vec{w} = (w_1, \ldots, w_n)$ is the vector of weights [20–22].

We suppose - for the sake of simplicity - that transfer functions (activation functions) φ, $\sigma(or\, f)$ are the same for all neurons from the collection $\mathbb{AN}(T)$ and the role of this function plays the identity function $f(y) = y$.

Feedforward multilayer networks are architectures, where the neurons are assembled into layers, and the links between the layers go only into one direction, from the input layer to the output layer. There are no links between the neurons in the same layer. Also, there may be one or several hidden layers between the input and the output layer [5,9,16].

2. Preliminaries on Hyperstructures

From an algebraic point of view, it is useful to describe the terms and concepts used in the field of algebraic structures. A hypergroupoid is a pair $(H, \cdot)$, where H is a (nonempty) set and

$$\cdot : H \times H \to \mathcal{P}^*(H) (= \mathcal{P}(H) - \{\emptyset\})$$

is a binary hyperoperation on the set H. If $a \cdot (b \cdot c) = (a \cdot b) \cdot c$ for all $a, b, c \in H$ (the associativity axiom), the the hypergroupoid $(H, \cdot)$ is called a semihypergroup. A semihypergroup is said to be a hypergroup if the following axiom:

$$a \cdot H = H = H \cdot a$$

for all $a \in H$ (the reproduction axiom), is satisfied. Here, for sets $A, B \subseteq H$, $A \neq \emptyset \neq B$ we define as usually

$$A \cdot B = \bigcup \{a \cdot b; a \in A, b \in B\}.$$

Thus, hypergroups considered in this paper are hypergroups in the sense of F. Marty [23,24]. In some constructions it is useful to apply the following lemma (called also the Ends-lemma having many applications—cf. [25–29]). Recall, first that by a (quasi-)ordered semigroup we mean a triad $(S, \cdot, \leq)$, where $(S, \cdot)$ is a semigroup, $(S, \leq)$ is a (quasi-)ordered set, i.e., a set S endowed with a reflexive and transitive binary relation "$\leq$" and for all triads of elements $a, b, c \in S$ the implication $a \leq b \Rightarrow a \cdot c \leq b \cdot c$, $c \cdot a \leq c \cdot b$ holds.

Lemma 1 (Ends-Lemma). *Let $(S, \cdot, \leq)$ be a (quasi-)ordered semigroup. Define a binary hyperoperation*

$$* : S \times S \to \mathcal{P}^*(S) \text{ by } a * b = \{x \in S; a \cdot b \leq x\}.$$

Then $(S, *)$ is a semihypergroup. Moreover, if the semigroup $(S, \cdot)$ is commutative, then the semihypergroup $(S, *)$ is also commutative and if $(S, \cdot, \leq)$ is a (quasi-)ordered group then the semihypergroup $(S, *)$ is a hypergroup.

Notice, that if $(G, \cdot)$, $(H, \cdot)$ are (semi-)hypergroups, then a mapping $h : G \to H$ is said to be the homomorphism of $(G, \cdot)$ into $(H, \cdot)$ if for any pair $a, b \in G$ we have

$$h(a \cdot b) \subseteq h(a) \cdot h(b).$$

If for any pair $a, b \in G$ the equality $h(a \cdot b) = h(a) \cdot h(b)$ holds, the homomorphism h is called the good (or strong) homomorphism—cf. [30,31]. By $EndG$ we denote the endomorphism monoid of a semigroup (group) G.

Concerning the basics of the hypergroup theory see also [23,25–28,32–41].

Linear differential operators described in the article and used e.g., in [29,42] are of the following form:

Definition 1. *Let $J \subseteq \mathbb{R}$ be an open interval, $\mathbb{C}(J)$ be the ring of all continuous functions $\varphi : J \to \mathbb{R}$. For $p_k \in \mathbb{C}(J)$, $k = 0, \ldots, n-1, p_0 \neq 0$ we define*

$$L(p_{n-1}, \ldots, p_0)y(x) = y^{(n)}(x) + \sum_{k=0}^{n-1} p_k(x)y^{(k)}(x),\ y \in \mathbb{C}^n(J)$$

(the ring of all smooth functions up to order n, i.e., having derivatives up to order n defined on the interval $J \subseteq \mathbb{R}$).

Definition 2 ([41,49]). *Let $(G, \cdot)$ be a semigroup and $P \subset G$, $P \neq \emptyset$. A hyperoperation $*^P : G \times G \to \mathcal{P}(G)$ defined by $[x, y] \to xPy$, i.e., $x * y = xPy$ for any pair $[x, y] \in P \times P$ is said to be the P-hyperoperation in G. If*

$$x *^P (y *^P z) = xPyPz = (x *^P y) *^P z$$

holds for any triad $x, y, z \in G$, the P-hyperoperation is associative. If also the axiom of reproduction is satisfied, the hypergrupoid $(G, *^P)$ is said to be a P-hypergroup.

Evidently, if $(G, \cdot)$ is a group, then also $(G, *^P)$ is a P-hypergroup. If the set P is a singleton, then the P-operation $*^P$ is a usual single—valued operation.

Definition 3. *A subset $H \subset G$ is said to be a sub-P-hypergroup of $(G, *^P)$ if $P \subset H \subset G$ and $(H, *^P)$ is a hypergroup.*

Now, similarly as in the case of the collection of linear differential operators [29], we will construct a group and hypergroup of artificial neurons, cf. [29,32,42–44].

Denote by δ_{ij} Kronecker delta, $i, j \in \mathbb{N}$, i.e., $\delta_{ii} = \delta_{jj} = 1$ and $\delta_{ij} = 0$, whenever $i \neq j$.

Suppose $Ne(\vec{w}_r)$, $Ne(\vec{w}_s) \in \mathbb{AN}(T)$, $r, s \in \mathbb{C}(J)$, $\vec{w}_r = (w_{r1}, \ldots, w_{r,n})$, $\vec{w}_s = (w_{s1}, \ldots, w_{s,n})$, $n \in \mathbb{N}$. Let $m \in \mathbb{N}$, $1 \leq m \leq n$ be a such an integer that $w_{r,m} > 0$. We define

$$Ne(\vec{w}_r) \cdot_m Ne(\vec{w}_s) = Ne(\vec{w}_u),$$

where

$$\vec{w}_u = (w_{u,1}, \ldots, w_{u,n}) = (w_{u,1}(t), \ldots, w_{u,n}(t)),$$

$$\vec{w}_{u,k}(t) = w_{r,m}(t)w_{s,k}(t) + (1 - \delta_{m,k})w_{r,k}(t), t \in T$$

and, of course, the neuron $Ne(\vec{w}_u)$ is defined as the mapping $y_u(t) = \sum\limits_{k=1}^{n} w_k(t)x_k(t) + b_u,\ t \in T$, $b_u = b_r b_s$. Further, for a pair $Ne(\vec{w}_r)$, $Ne(\vec{w}_s)$ of neurons from $\mathbb{AN}(T)$ we put $Ne(\vec{w}_r) \leq_m Ne(\vec{w}_s)$, $w_r = (w_{r,1}(t), \ldots, w_{r,n}(t))$, $w_s = (w_{s,1}(t), \ldots, w_{s,n}(t))$ if $w_{r,k}(t) \leq w_{s,k}(t),\ k \in \mathbb{N},\ k \neq m$ and $w_{r,m}(t) = w_{s,m}(t),\ t \in T$ and with the same bias. Evidently $(\mathbb{AN}(T), \leq_m)$ is an ordered set. A relationship (compatibility) of the binary operation "$\cdot_m$" and the ordering $\leq_m$ on $\mathbb{AN}(T)$ is given by this assertion analogical to Lemma 2 in [29].

Lemma 2. *The triad $(\mathbb{AN}(T), \cdot_m, \leq_m)$ (algebraic structure with an ordering) is a non-commutative ordered group.*

Sketch of the proof was published in [21].

Denoting

$$\mathbb{AN}_1(T)_m = \{Ne(\vec{w}); \vec{w} = (w_1, \ldots, w_n),\ w_k \in \mathbb{C}(T),\ k = 1, \ldots, n,\ w_m(t) \equiv 1\},\ 1 \leq m \leq n,$$

we get the following assertion:

Proposition 1 (Prop. 1. [21], p. 239). *Let $T = \langle 0, t_0) \subset \mathbb{R},\ t_0 \in \mathbb{R} \cup \{\infty\}$. Then for any positive integer $n \in \mathbb{N},\ n \geq 2$ and for any integer m such that $1 \leq m \leq n$ the semigroup $(\mathbb{AN}_1(T)_m, \cdot_m)$ is an invariant subgroup of the group$(\mathbb{AN}(T)_m, \cdot_m)$.*

Proposition 2 (Prop. 2. [21], p. 240). *Let $t_0 \in \mathbb{R},\ t_0 > 0,\ T = \langle 0, t_0) \subset \mathbb{R}$ and $m, ,n \in \mathbb{N}$ are integers such that $1 \leq m \leq n-1$. Define a mapping $F : \mathbb{AN}_n(T)_m \to \mathbb{LA}_n(T)_{m+1}$ by this rule: For an arbitrary neuron $Ne(\vec{w}_r) \in \mathbb{AN}_n(T)_m$, where $\vec{w}_r = (w_{r,1}(t), \ldots, w_{r,n}(t)) \in [\mathbb{C}(T)]^n$ we put $F(Ne(\vec{w}_r)) = L(w_{r,1}, \ldots, w_{r,n}) \in \mathbb{LA}_n(T)_{m+1}$ with the action :*

$$L(w_{r,1}, \ldots, w_{r,n})y(t) = \frac{d^n y(t)}{dt^n} + \sum_{k=1}^{n} w_{r,k}(t)\frac{d^{k-1}y(t)}{dt^{k-1}},\ y \in \mathbb{C}^n(T).$$

Then the mapping $F : \mathbb{AN}_n(T)_m \to \mathbb{LA}_n(T)_{m+1}$ is a homomorphism of the group $(\mathbb{AN}_n(T)_m, \cdot_m)$ into the group $(\mathbb{LA}_n(T)_{m+1}, \circ_{m+1})$.

Now, using the construction described in the Lemma 1, we obtain the final transposition hypergroup (called also non-commutative join space). Denote by $\mathbb{P}(\mathbb{AN}(T)_m)^*$ the power set of $\mathbb{AN}(T)_m$ consisting of all nonempty subsets of the last set and define a binary hyperoperation

$$*_m : \mathbb{AN}(T)_m \times \mathbb{AN}(T)_m \to \mathbb{P}(\mathbb{AN}(T)_m)^*$$

by the rule

$$Ne(\vec{w}_r) *_m Ne(\vec{w}_s) = \{Ne(\vec{w}_u); Ne(\vec{w}_r) \cdot_m Ne(\vec{w}_s) \leq_m Ne(\vec{w}_u)\}$$

for all pairs $Ne(\vec{w}_r), Ne(\vec{w}_s) \in \mathbb{AN}(T)_m$. More in detail if $\vec{w}(u) = (w_{u,1}, \ldots, w_{u,n})$, $\vec{w}(r) = (w_{r,1}, \ldots, w_{r,n})$, $\vec{w}(s) = (w_{s,1}, \ldots, w_{s,n})$, then $w_{r,m}(t)w_{s,m}(t) = w_{u,m}(t)$, $w_{r,m}(t)w_{s,k}(t) + w_{r,k}(t) \leq w_{u,k}(t)$, if $k \neq m,\ t \in T$. Then we have that $(\mathbb{AN}(T)_m, *_m)$ is a non-commutative hypergroup. We say that this hypergroup is constructed by using the Ends Lemma (cf. e.g., [8,25,29]. These hypergroups can be called as EL-hypergroups. The above defined invariant (called also normal) subgroup $(\mathbb{AN}_1(T)_m, \cdot_m)$ of the group $(\mathbb{AN}(T)_m, \cdot_m)$ is the carrier set of a subhypergroup of the hypergroup $(\mathbb{AN}(T)_m, *_m)$ and it has certain significant properties.

Using certain generalization of methods from [42] (p. 283), we obtain, after we investigate the constructed structures, the following result:

Theorem 1. *Let $T = \langle 0, t_0) \subset \mathbb{R}$, $t_0 \in \mathbb{R} \cup \{\infty\}$. Then for any positive integer $n \in \mathbb{N}$, $n \geq 2$ and for any integer m such that $1 \leq m \leq n$ the hypergroup $(\mathbb{AN}(T)_m, *_m)$,where*

$$\mathbb{AN}(T)_m = \{Ne(\vec{w}_r);\ \vec{w}_r = (w_{r,1}(t), \ldots, w_{r,n}(t)) \in [\mathbb{C}(T)]^n,\ w_{r,m}(t) > 0,\ t \in T\},$$

is a transposition hypergroup (i.e., a non-commutative join space) such that $(\mathbb{AN}(T)_m, *_m)$ is its subhypergroup, which is

- *invertible (i.e., $Ne(\vec{w}_r)/Ne(\vec{w}_s) \cap \mathbb{AN}_1(T)_m \neq \emptyset$ implies $Ne(\vec{w}_s)/Ne(\vec{w}_r) \cap \mathbb{AN}_1(T)_m \neq \emptyset$ and $Ne(\vec{w}_r)\ Ne(\vec{w}_s) \cap \mathbb{AN}_1(T)_m \neq \emptyset$ implies $Ne(\vec{w}_s)\ Ne(\vec{w}_r) \cap \mathbb{AN}_1(T)_m \neq \emptyset$ for all pairs of neurons $Ne(\vec{w}_r), Ne(\vec{w}_s) \in \mathbb{AN}_1(T)_m$,*
- *closed (i.e., $Ne(\vec{w}_r)/Ne(\vec{w}_s) \subset \mathbb{AN}_1(T)_m$, $Ne(\vec{w}_r) \setminus Ne(\vec{w}_s) \subset \mathbb{AN}_1(T)_m$ for all pairs $Ne(\vec{w}_r), /, Ne(\vec{w}_s) \in \mathbb{AN}_1(T)_m$,*
- *reflexive (i.e., $Ne(\vec{w}_r)\ \mathbb{AN}_1(T)_m = \mathbb{AN}_1(T)_m / Ne(\vec{w}_r)$ for any neuron $Ne(\vec{w}_r) \in \mathbb{AN}(T)_m$ and*
- *normal (i.e.$Ne(\vec{w}_r) * \mathbb{AN}_1(T)_m = \mathbb{AN}_1(T)_m * Ne(\vec{w}_r)$ for any neuron $Ne(\vec{w}_r) \in \mathbb{AN}(T)_m$.*

Remark 1. *A certain generalization of the formal (artificial) neuron can be obtained from expression of a linear differential operator of the n-th order. Recall the expression of formal neuron with inner potential $y_{-in} = \sum\limits_{k=1}^{n} w_k(t)x_k(t)$, where $\vec{x}(t) = (x_1(t), \ldots, x_n(t))$ is the vector of inputs, $\vec{w}(t) = (w_1(t), \ldots, w_n(t))$ is the vector of weights. Using the bias b of the considered neuron and the transfer function σ we can expressed the output as $y(t) = \sigma\left(\sum\limits_{k=1}^{n} w_k(t)x_k(t) + b\right)$.*

Now consider a tribal function $u : J \to \mathbb{R}$, where $J \subseteq \mathbb{R}$ is an open interval; inputs are derived from $u \in \mathbb{C}^n(J)$ as follows: Inputs $x_1(t) = u(t), x_2 = \frac{du(t)}{dt}, \ldots, x_n(t) = \frac{d^{n-1}(t)}{dt^{n-1}}$, $n \in \mathbb{N}$. Further the bias $b = b_0\frac{d^n u(t)}{dt^n}$. As weights we use the continuous functions $w_k : J \to \mathbb{R}$, $k = 1, \ldots, n-1$.

Then formula

$$y(t) = \sigma\left(\sum_{k=1}^{n} w_k(t)\frac{d^{k-1}u(t)}{dt^{k-1}} + b_0\frac{d^n u(t)}{dt^n}\right)$$

is a description of the action of the neuron Dn which will be called a formal(artificial) differential neuron. This approach allows to use solution spaces of corresponding linear differential equations.

Proposition 3 ([41], p. 16). *Let $(G_1, \cdot)$, $(G_2, \cdot)$ be two groups $f \in Hom(G_1, G_2)$ and $P \subset G_1$. Then the homomorphism f is a good homomorphism between P-hypergroups $(G_1, *^P)$ and $(G_2, *^P)$.*

Concerning the discussed theme see [26–28,30,32,36,39,45]. Now denote by $S \subseteq \mathbb{C}(T)$ an arbitrary non/empty subset and let

$$P = \{Ne(\vec{w}_u(t)); u \in S\} \subseteq \mathbb{AN}(T).$$

Then defining

$$Ne(\vec{w}_p(t)) * Ne(\vec{w}_q(t)) = Ne(\vec{w}_p(t)) \cdot_m P \cdot_m Ne(\vec{w}_q(t)) =$$
$$\{Ne(\vec{w}_p(t)) \cdot_m Ne(\vec{w}_u(t)) \cdot_m Ne(\vec{w}_q(t)); u \in S\}$$

for any pair of neurons $Ne(\vec{w}_p(t)), Ne(\vec{w}_q(t)) \in \mathbb{AN}(T)$, we obtain a P-hypergroup of artificial time varying neurons. If S is a singleton, i.e., P is a one-element subset of $\mathbb{AN}(T)$, the obtained structure is a variant of $\mathbb{AN}(T)$. Notice, that any $f \in EndG$ for a group $(G, \cdot)$ induces a good homomorphism of the P-hypergroups $(G, *^P)$, $(G, *^{f(P)})$ and any automorphism creates an isomorphism beween the above P-hypergroups.

Let $(\mathbb{Z}, +)$ be the additive group of all integers. Let $Ne(\vec{w}_s(t)) \in \mathbb{AN}(T)$ be arbitrary but fixed chosen artificial neuron with the output function $y_s(t) = \sum\limits_{k=1}^{n} w_{s,k}(t)x_{s,k}(t) + b_s$. Denote by

$\lambda_s : \mathbb{AN}(T) \to \mathbb{AN}(T)$ the left translation within the group of time varying neurons determined by $Ne(\vec{w}_s(t)$, i.e.,

$$\lambda_s(Ne(\vec{w}_p(t)) = Ne(\vec{w}_s(t) \cdot_m Ne(\vec{w}_p(t))$$

for any neuron $Ne(\vec{w}_p(t)) \in \mathbb{AN}(T)$. Further, denote by λ_s^r the r-th iteration of λ_s for $r \in \mathbb{Z}$. Define the projection $\pi_s : \mathbb{AN}(T) \times \mathbb{Z} \to \mathbb{AN}(T)$ by

$$\pi_s(Ne(\vec{w}_p(t)), r) = \lambda_s^r(Ne(\vec{w}_p(t)).$$

It is easy to see that we get a usual (discrete) transformation group, i.e., the action of $(\mathbb{Z}, +)$ (as the phase group) on the group $\mathbb{AN}(T)$. Thus the following two requirements are satisfied:

1. $\pi_s(Ne(\vec{w}_p(t)), 0) = Ne(\vec{w}_p(t))$ for any neuron $Ne(\vec{w}_p(t)) \in \mathbb{AN}(T)$,
2. $\pi_s(Ne(\vec{w}_p(t)), r+u) = \pi_s(\pi_s(Ne(\vec{w}_p(t)), r), u)$ for any integers $r, u \in \mathbb{Z}$ and any artificial neuron $Ne(\vec{w}_p(t))$. Notice that, in the dynamical system theory this structure is called a cascade.

On the phase set we will define a binary hyperoperation. For any pair of neurons $Ne(\vec{w}_p(t))$, $Ne(\vec{w}_q(t))$ define

$$Ne(\vec{w}_p(t)) * Ne(\vec{w}_q(t)) = \pi_s(Ne(\vec{w}_p(t)), \mathbb{Z}) \cup \pi_s(Ne(\vec{w}_q(t)), \mathbb{Z}) =$$

$$\{\lambda_s^a(Ne(\vec{w}_p(t)); a \in \mathbb{Z}\} \cup \{\lambda_s^b(Ne(\vec{w}_q(t)); b \in \mathbb{Z}\}.$$

Then we have that $* : \mathbb{AN}(T) \times \mathbb{AN}(T) \to \mathcal{P}(\mathbb{AN}(T))$ is a commutative binary hyperoperation and since $Ne(\vec{w}_p(t)), Ne(\vec{w}_q(t)) \in Ne(\vec{w}_p(t)) * Ne(\vec{w}_q(t))$, we obtain that the hypergroupoid $(\mathbb{AN}(T), *)$ is a commutative, extensive hypergroup [20,27,29–31,34,35,38,43,46,47]. Using its properties we can characterize certain properties of the cascade $(\mathbb{AN}(T), \mathbb{Z}, \pi_s)$. The hypergroup $(\mathbb{AN}(T), *)$ can be called phase hypergroup of the given cascade.

Recall now the concept of invariant subsets of the phase set of a cascade $(X, \mathbb{Z}, \pi_s)$ and the concept of a critical point. A subset M of of a phase set X of the cascade $(X, \mathbb{Z}, \pi_s)$ is called invariant whenever $\pi(x, r) \in M$, for all $x \in M$ and all $r \in \mathbb{Z}$. A critical point of a cascade is an invariant singleton. It is evident that a subset M of neurons , i.e., $M \subseteq \mathbb{AN}(T)$ is invariant in the cascade $(\mathbb{AN}(T), \mathbb{Z}, \pi_s)$ whenever it is a carrier set of a subhypergroup of the hypergroup $(\mathbb{AN}(T), *)$, i.e., M is closed with respect to the hyperoperation $*$, which means $M * M = \bigcup\limits_{a,b \in M} a * b \subseteq M$. Moreover, union or intersection of an arbitrary non-empty system $\mathcal{M} \subseteq \mathbb{AN}(T)$ is also invariant.

3. Main Results

Now, we will construct series of groups and hypergroups of artificial neurons using certain analogy with series of groups of differential operators described in [29].

We denote by $\mathbb{LA}_n(J)$ (for an open interval $J \subseteq \mathbb{R}$) the set of all linear differential operators $L(p_{n-1}, \ldots, p_0)$, $p_0 \neq 0$, $p_k \in \mathbb{C}^n(J)$, i.e., the ring of all continuous functions defined on the interval J, acting as

$$L(p_{n-1}, \ldots, p_0)y(x) = y^n(x) + \sum_{k=0}^{n-1} p_k(x)y^k(x),\ y \in \mathbb{C}^n(J)$$

and endowed the binary operation

$$L(q_{n-1}, \ldots, q_0) \circ L(p_{n-1}, \ldots, p_0) = L(q_0 p_{n-1} + q_{n-1}, \ldots, q_0 p_1 + q_1, q_0 p_0).$$

Now denote by $\overline{\mathbb{LA}}_n(J)$ the set of all operators $\overline{L}(q_n, \ldots, q_0)$, $q_0 \neq 0$, $q_k \in \mathbb{C}(J)$ acting as

$$\overline{L}(q_n, \ldots, q_0)y(x) = \sum_{k=0}^{n} q_k(x)y^{(k)}(x),\ q_0 \neq 0,\ q_k \in \mathbb{C}(J)$$

with similarly defined binary operations such that $\mathbb{LA}_n(J)$, $\overline{\mathbb{LA}}_n(J)$ are noncommutative groups. Define mappings $F_n : \mathbb{LA}_n(J) \to \mathbb{LA}_{n-1}(J)$ by

$$F_n(L(p_{n-1}, \dots, p_0)) = L(p_{n-2}, \dots, p_0)$$

and $\phi_n : \mathbb{LA}(J) \to \overline{\mathbb{LA}}_{n-1}(J)$ by

$$\phi_n(L(p_{n-1}, \dots, p_0)) = \overline{L}(p_{n-2}, \dots, p_0).$$

It can be easily verified that both F_n and ϕ_n, for an arbitrary $n \in \mathbb{N}$, are group homomorphisms.

Evidently, $\mathbb{LA}_n(J) \subset \overline{\mathbb{LA}}_n(J)$, $\overline{\mathbb{LA}}_{n-1}(J) \subset \overline{\mathbb{LA}}_n(J)$ for all $n \in \mathbb{N}$. Thus we obtain complete sequences of ordinary linear differential operators with linking homomorphisms F_n, ϕ_n :

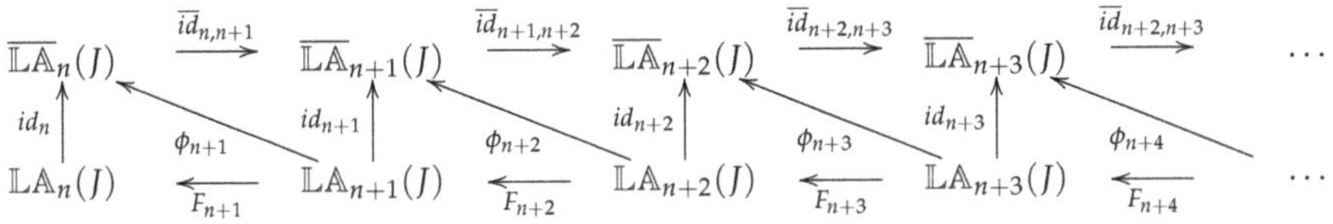
$\overline{\mathbb{LA}}_n(J)$
$\overline{id}_{n,n+1}$
$\overline{\mathbb{LA}}_{n+1}(J)$
$\overline{id}_{n+1,n+2}$
$\overline{\mathbb{LA}}_{n+2}(J)$
$\overline{id}_{n+2,n+3}$
$\overline{\mathbb{LA}}_{n+3}(J)$
$\overline{id}_{n+2,n+3}$
id_n
id_{n+1}
id_{n+2}
id_{n+3}
ϕ_{n+1}
ϕ_{n+2}
ϕ_{n+3}
ϕ_{n+4}
$\mathbb{LA}_n(J)$
$\mathbb{LA}_{n+1}(J)$
$\mathbb{LA}_{n+2}(J)$
$\mathbb{LA}_{n+3}(J)$
F_{n+1}
F_{n+2}
F_{n+3}
F_{n+4}

Now consider the groups of time-varying neurons $(\mathbb{AN}(T)_m, \cdot_m)$ from Proposition 3 and above defined homomorphism of the group $(\mathbb{AN}_n(T)_m, \cdot_m)$ into the group $(\mathbb{LA}_n(T)_{m+1}, \circ_{m+1})$. Then we can change the diagram in the following way:

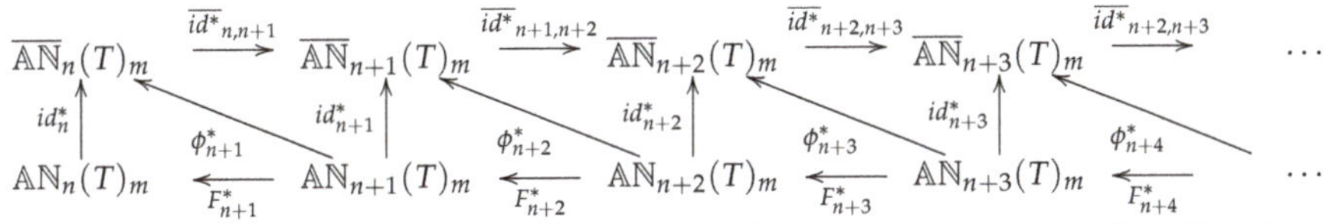
$\overline{\mathbb{AN}}_n(T)_m$
$\overline{id^*}_{n,n+1}$
$\overline{\mathbb{AN}}_{n+1}(T)_m$
$\overline{id^*}_{n+1,n+2}$
$\overline{\mathbb{AN}}_{n+2}(T)_m$
$\overline{id^*}_{n+2,n+3}$
$\overline{\mathbb{AN}}_{n+3}(T)_m$
$\overline{id^*}_{n+2,n+3}$
id^*_n
id^*_{n+1}
id^*_{n+2}
id^*_{n+3}
ϕ^*_{n+1}
ϕ^*_{n+2}
ϕ^*_{n+3}
ϕ^*_{n+4}
$\mathbb{AN}_n(T)_m$
$\mathbb{AN}_{n+1}(T)_m$
$\mathbb{AN}_{n+2}(T)_m$
$\mathbb{AN}_{n+3}(T)_m$
F^*_{n+1}
F^*_{n+2}
F^*_{n+3}
F^*_{n+4}

Using the *Ends lemma* and results the theory of linear operators we can describe also mapping morphisms in sequences groups of linear differential operators:

$$\mathbb{LA}_n(J) \xleftarrow[F_{n+1}]{} \mathbb{LA}_{n+1}(J) \xleftarrow[F_{n+2}]{} \mathbb{LA}_{n+2}(J) \xleftarrow[F_{n+3}]{} \mathbb{LA}_{n+3}(J) \xleftarrow[F_{n+4}]{} \cdots$$

as so analogy in sequences groups of time-varying neurons: (2)

$$\mathbb{AN}_n(T)_m \xleftarrow[F^*_{n+1}]{} \mathbb{AN}_{n+1}(T)_m \xleftarrow[F^*_{n+2}]{} \mathbb{AN}_{n+2}(T)_m \xleftarrow[F^*_{n+3}]{} \mathbb{AN}_{n+3}(T)_m \xleftarrow[F^*_{n+4}]{} \cdots$$

Theorem 2. *Let $T = \langle 0, t_0) \subset \mathbb{R}$, $t_0 \in \mathbb{R} \cup \{\infty\}$, $n \in \mathbb{N}$ such that $n \geq 2$, $m \in \mathbb{N}$ such that $m \leq n$. Let $(\mathbb{HAN}_n(T)_m, *_m)$ be the hypergroup obtained from the group $(\mathbb{AN}_n(T)_m, \circ_m)$ by Proposition 2. Suppose that $F_n : (\mathbb{AN}_n(T)_m, \circ_m) \to (\mathbb{AN}_{n-1}(T)_m, \circ_m)$ are the above defined surjective group homomorphisms. Then $F_n : (\mathbb{HAN}_n(T)_m, *_m) \to (\mathbb{HAN}_{n-1}(J)_m, *_m)$ are surjective homomorphisms of hypergroups.*

Remark 2. *The second sequence of (2) can thus be bijectively mapped onto sequence of hypergroups*

$$\mathbb{HAN}_n(T)_m \xleftarrow[F_{n+1}]{} \mathbb{HAN}_{n+1}(T)_m \xleftarrow[F_{n+2}]{} \mathbb{HAN}_{n+2}(T)_m \xleftarrow[F_{n+3}]{} \mathbb{HAN}_{n+3}(T)_m \xleftarrow[F_{n+4}]{} \cdots$$

wit the linking surjective homomorphisms F_n. Therefore, the bijective mapping of the above mentioned sequences is functorial.

Now, shift to the concept of an automaton. This was developed as a mathematical interpretation of real-life systems that work on a discrete time-scale. Using the binary operation of concatenation of chains of input symbols we obtain automata with input alphabets endowed with the structure of a semigroup or a group. Considering mainly the structure given by transition function and neglecting output functions with output sets we reach a very useful generalization of the concept of automaton called quasi—automaton [29,31,48,49]. Let us introduce the concept of automata as an action of time varying neurons. Moreover, let system (A, S, δ), consists of nonempty time-varying neuron set of states $A \subseteq \mathbb{AN}(T)_m$, arbitrary semigroup of their inputs S and let mapping $\delta : A \times S \to A$ fulfill the following condition:

$$\delta(\delta(a, r), s) = \delta(a, r\dot{s})$$

for arbitrary $a \in A$ and $r, s \in S$ can be understood as a analogy of concept of quasi-automaton, as a generalization of the Mealy-type automaton. The above condition is some times called Mixed Associativity Condition (MAC).

Definition 4. *Let A be a nonempty set, $(H, \cdot)$ a semihypergroup and $\delta : A \times H \to A$ a mapping satisfying the condition:*

$$\delta(\delta(s, a), b) \in \delta(s, a\dot{b}) \qquad (3)$$

for any triad $(s, a, b) \in A \times H \times H$, where $\delta(s, a\dot{b}) = \{\delta(s, x); x \in a \cdot b\}$. The triad (A, H, δ) is called a quasi-multiautomaton with the state set A and the input semihypergroup $(H, \cdot)$. The mapping $\delta : A \times H \to A$ is called transition function (or next-state function) of the quasi-multiautomaton (A, H, δ). Condition (3) is called Generalized Mixed Associativity Condition (or GMAC).

The just defined structures are also called as actions of semihypergroups $(H, \cdot)$ on sets A (called state sets).

Neuron $Ne(\vec{w})$ acts as described above:

$$y(t) = \sum_{i=1}^{n} w_i(t)x_i(t) + b,$$

where i goes from 0 to n, $w_i(t)$ is the weight value in continuous time, b is a bias and $y(t)$ is the output value in continuous time t. Here the transition function F is the identity function.

Now suppose that the input functions x_i are differentiable up to arbitrary order n.

We consider linear differential operators

$$L(m, w_n, \ldots, w_0) : \mathbb{C}^n(T) \times \cdots \times \mathbb{C}^n(T) \to \mathbb{C}^n(T), \text{ i. e. } \mathbb{C}^n(T) \times \cdots \times \mathbb{C}^n(T) = [\mathbb{C}^n(T)]^{n+1},$$

defined

$$L(m, w_n, \ldots, w_0)x(t) =$$

$$= mb + \sum_{k=1}^{n} w_k(t)\frac{d^k x_k(t)}{dt^k},\ x(t) = (x_0(t), x_1(t), \ldots, x_n(t)) \in \mathbb{C}^n(T) \times \cdots \times \mathbb{C}^n(T) = [\mathbb{C}^n(T)]^{n+1}.$$

Then we denote by $\mathbb{L}Ne_n(T)$ the additive Abelian group of linear differential operators $L(m, w_n, \ldots, w_0)$, where for $L(m, w_n, \ldots, w_0)$, $L(k, w_n^*, \ldots, w_0^*) \in \mathbb{L}Ne_n(T)$ with the bias b we define

$$L(m, w_n, \ldots, w_0) + L(s, w_n^* \ldots, w_0^*) = L(m + s, w_n + w_n^*, \ldots, w_0 + w_0^*),$$

where

$$L(m + s, w_n + w_n^*, \ldots, w_0 + w_0^*)x(t) =$$

$$= (m + s)b + \sum_{k=0}^{n}(w_k(t) + w_k^*(t))\frac{d^k x_k(t)}{dt^k},\ t \in T \text{ and } x(t) = (x_0(t), x_1(t), \ldots, x_n(t)) \in [\mathbb{C}^n(T)]^{n+1}.$$

Suppose that $w_k(t) \in \mathbb{C}^n(T)$ and define

$$\delta_n : \mathbb{C}^n(T) \times \mathbb{L}Ne_n(T) \to \mathbb{C}^n(T)$$

by

$$\delta_n(x(t), L(m, w_n, \dots, w_0)) = mb + x(t) + m + \sum_{k=0}^{n} w_k(t)\frac{d^k x(t)}{dt^k},\ x(t) \in \mathbb{C}^n(T), \text{ where}$$

$w_n, \dots, w_0$ are weights corresponding with inputs and b is the bias of a neuron corresponding to the operator $L(m, w_n, \dots, w_0) \in \mathbb{L}Ne_n(T)$.

Theorem 3. *Let $\mathbb{L}Ne_n(T)$, $\mathbb{C}^n(T)$ be the above defined structures and $\delta_n : \mathbb{C}^n(T) \times \mathbb{L}Ne_n(T) \to \mathbb{C}^n(T)$ be the above defined mapping. Then the triad $(\mathbb{C}^n(T), \mathbb{L}Ne_n(T), \delta_n)$ is an action of the group $\mathbb{L}Ne_n(T)$ on the group $\mathbb{C}^n(T)$, i.e., a quasi-automaton with the state space $\mathbb{C}^n(T)$ and with the alphabet $\mathbb{L}Ne_n(T)$ with the group structure of artificial neurons.*

Proof. We are going to verify the mixed associativity condition (MAC). Suppose $x \in \mathbb{C}^n(T)$ and $L(m, w_n, \dots, w_0)$, $L(k, u_n, \dots, u_0) \in \mathbb{L}Ne_n(T)$. Then we have

$$\delta_n(\delta_n(x(t), L(m, w_n, \dots, w_0), L(k, u_n, \dots, u_0)) =$$

$$= \delta_n(mb + x(t) + m + \sum_{k=0}^{n} w_k(t)\frac{d^k x(t)}{dt^k}, L(k, u_n, \dots, u_0)) =$$

$$= kb + mb + x(t) + m + k + \sum_{k=0}^{n} w_k(t)\frac{d^k x(t)}{dt^k} + \sum_{k=0}^{n} u_k(t)\frac{d^k x(t)}{dt^k} =$$

$$= (m+k)b + x(t) + m + k + \sum_{k=0}^{n}(w_k(t) + u_k(t))\frac{d^k x(t)}{dt^k} =$$

$$\delta_n(x(t), L(m+k, w_n(t) + u_n(t), \dots, w_0)(t) + u_0)(t)) =$$

$$\delta_n(x(t), L(m, w_n, \dots, w_0) + L(k, u_n, \dots, u_0)),$$

thus MAC is satisfied. □

Consider an interval $T \subseteq \mathbb{R}$ and the ring $\mathbb{C}(T)$ of all continuous functions defined on the interval. Let $\{\varphi_k; k \in \mathbb{N}\}$ be a sequence of ring-endomorphism of $\mathbb{C}(T)$. Denote $\mathbb{A}_{n+k}\mathbb{N}(T)_m$ the EL-hypergroup of artificial neurons constructed above, with vectors of weights of the dimension $n + k \in \mathbb{N} (= \{1, 2, 3 \dots\})$. Let $[\mathbb{C}(T)]^{n+k} = \mathbb{C}(T) \times \mathbb{C}(T) \times \dots \times \mathbb{C}(T)$ $(n + k - times)$ i.e., $\mathbb{C}(T)]^{n+k}$ is the $n + k$-dimensional cartesian cube. Denote by $\bar{\varphi}_k : [\mathbb{C}(T)]^{n+k} \to [\mathbb{C}(T)]^{n+k-1}$ the extension of φ_k such that $\bar{\varphi}_k(\vec{w}) = \bar{\varphi}_k((w_1, \dots, w_{n+k-1}, w_{n+k})) = (w_1, \dots, w_{n+k-1})$. Let us denote the mapping $F_k : \mathbb{A}_{n+k}\mathbb{N}(T)_m \to \mathbb{A}_{n+k-1}\mathbb{N}(T)_m$ defined by $F_k(Ne(\vec{w}) = Ne(\vec{w}_1)$ with $\vec{w}_1 = (w_1, \dots, w_{n+k})$. Consider underlying sets of hypergroups $\mathbb{A}_{n+k}\mathbb{N}(T)_m$ endowed with the above defined ordering relation:

$$\text{for } \vec{w} = (w_1, \dots, w_{n+k}),\ \vec{u} = (u_1, \dots, u_{n+k}) \in [\mathbb{C}(T)]^{n+k}$$

we have $\vec{w} \le \vec{u}$ if $w_r \le u_r$, $r = 1, 2, \dots, n + k$ and $w_m \le u_m$. Now, for $Ne(\vec{w}), Ne(\vec{u}) \in \mathbb{A}_{n+k}\mathbb{N}(T)_m$ such that $\vec{w} = (w_1, \dots, w_{n+k})$, $\vec{u} = (u_1, \dots, u_{n+k})$, $Ne(\vec{w}) \le Ne(\vec{u})$, which means $\vec{w} \le \vec{u}$ ($w_m = u_m$ and biases of corresponding neurons are the same) we have $\bar{\varphi}_k(\vec{w}) = (w_1, \dots, w_{n+k-1}) \le (u_1, \dots, u_{n+k-1}) = \bar{\varphi}_k(\vec{u})$,which implies $F_k(\vec{w}) \le F_k(\vec{u})$.

Consequently the mapping $F_k : (\mathbb{A}_{n+k}\mathbb{N}(T)_m, \le) \to (\mathbb{A}_{n+k-1}\mathbb{N}(T)_m, \le)$ is order-preserving, i.e., this is an order-homomorphism of hypergroups. The final result of our considerations is the following sequence of hypergroups of artificial neurons and linking homomorphisms:

$$\mathbb{A}_n\mathbb{N}(T)_m \xleftarrow{F_1} \mathbb{A}_{n+1}\mathbb{N}(T)_m \xleftarrow{F_2} \quad \dots \quad \xleftarrow{F_k} \mathbb{A}_{n+k}\mathbb{N}(T)_m \xleftarrow{F_{k+1}} \mathbb{A}_{n+k+1}\mathbb{N}(T)_m \dots$$

4. Conclusions

Artificial neural networks and structured systems of artificial neurons have been discussed by a great number of researchers. They are an important part of artificial intelligence with many useful applications in various branches of science and technical constructions. Our considerations are based on algebraic and analytic approach using certain formal similarity with classical structures and new hyperstructures of differential operators. We discussed a certain generalizations of classical artificial time-varying neurons and studied them using recently derived methods. The presented investigations allow further development.

Author Contributions: Investigation, J.C. and B.S.; Methodology, J.C. and B.S.; Supervision, J.C. and B.S.; Writing original draft, J.C. and B.S.; Writing review and editing, J.C. and B.S.

Funding: The first author was supported by the FEKT-S-17-4225 grant of Brno University of Technology.

Conflicts of Interest: The authors declare no conflict of interest.

References

1. Koskela, T. Neural Network Methods in Analysing and Modelling Time Varying Processes. Ph.D. Thesis, Helsinki University of Technology, Helsinki, Finland, 2003; pp. 1–113.
2. Behnke, S. Hierarchical Neural Networks for Image Interpretation. In *Lecture Notes in Computer Science*; Springer: Heidelberg, Germany, 2003; Volume 2766, p. 245.
3. Bishop, C.M. *Neural Networks for Pattern Recognition*; Oxford University Press: Oxford, UK, 1995; pp. 1–482.
4. Buchholz, S. *A Theory of Neural Computation With Clifford-Algebras*; Technical Report Number 0504; Christian-Albrechts-Universität zu Kiel, Institut für Informatik und Praktische Mathematik: Kiel, Germany, 2005; pp. 1–135.
5. Gardner, M.W.; Dorling, S.R. Artificial Neural Networks (the Multilayerperceptron)–a Review of Applications in Theatmospheric Sciences. *Atmos. Environ.* **1998**, *32*, 2627–2636. [CrossRef]
6. Koudelka, V.; Raida, Z.; Tobola, P. Simple electromagnetic modeling of small airplanes: Neural network approach. *Radioengineering* **2009**, *18*, 38–41.
7. Krenker, A.; Bešter, J.; Kos, A. Introduction to the artificial neural networks. In *Artificial Neural Networks—Methodological Advances and Biomedical Applications*; Suzuki, K., Ed.; Tech: Rijeka, Croatia, 2011; pp. 3–18.
8. Raida, Z.; Lukeš, Z.; Otevřel, V. Modeling broadband microwave structures by artificial neural networks. *Radioengineering* **2004**, *13*, 3–11.
9. Rosenblatt, F. The Perceptron: A probabilistic model for information storage and organiyation in the brain. *Psychol. Rev.* **1958**, *65*, 386–408.
10. Srivastava, N.; Hinton, G.; Krizhevsky, A.; Sutskever, I.; Salakhutdinov, R. Dropout: A simple way to prevent neural networks from overfitting. *J. Mach. Learn. Res.* **2014**, *15*, 1929–1958.
11. Tučková, J. Comparison of two Approaches in the Fundamental Frequency Control by Neural Nets. In Proceedings of the 6th Czech-German Workshop in Speech Processing Praha, Praha, Czech Republic, 2–4 September 1996; p. 37.
12. Tučková, J.; Boreš, P. The Neural Network Approach in Fundamental Frequency Control. In *Speech Processing: Forum Phoneticum*; Wodarz, H.-W., Ed.; Hector Verlag: Frankfurt am Main, Germany, 1997; pp. 143–154, ISSN 0341–3144, ISBN 3-930220-10-5.
13. Tučková, J.; Šebesta, V. Data Mining Approach for Prosody Modelling by ANN in Text-to-Speech Synthesis. In Proceedings of the IASTED Inernational Conference on "Artificial Intelligence and Applications (AIA 2001)", Marbella, Spain, 4–7 September 2001; Hamza, M.H., Ed.; ACTA Press Anaheim-Calgary-Zurich: Marbella, Spain, 2001; pp. 164–166.
14. Volná, E. *Neuronové sítě 1*, 2nd ed.; Ostravská Univerzita: Ostrava, Czech Republic, 2008; p. 86.
15. Waldron, M.B. Time varying neural networks. In Proceedings of the Annual International Conference of the IEEE Engineering in Medicine and Biology Society, New Orleans, LA, USA, 4–7 November 1988.
16. Widrow, B.; Lehr, A. 30years of adaptive networks: Perceptron, Madaline, and Backpropagation. *Proc. IEEE* **1990**, *78*, 1415–1442. [CrossRef]
17. Kremer, S. Spatiotemporal Connectionist Networks: A Taxonomy and Review. *Neural Comput.* **2001**, *13*, 249–306. [CrossRef]
18. Narendra, K.; Parthasarathy, K. Identification and control of dynamical systems using neural networks. *IEEE Trans. Neural Netw.* **1990**, *1*, 4–27. [CrossRef] [PubMed]

19. Hagan, M.; Demuth, H.; Beale, M. *Neural Network Design*; PWS Publishing: Boston, MA, USA, 1996.
20. Chvalina, J.; Smetana, B. Models of Iterated Artificial Neurons. In Proceedings of the 18th Conference on Aplied Mathematics Aplimat, Bratislava, Slovakia, 5–7 February 2019; pp. 203–212, ISBN 978-80-227-4884-1.
21. Chvalina, J.; Smetana, B. Groups and Hypergroups of Artificial Neurons. In Proceedings of the 17th Conference on Aplied Mathematics Aplimat, Bratislava, Slovakia, 6–8 February 2018; University of Technology in Bratislava, Faculty of Mechanical Engineering: Bratislava, Slovakia, 2018; pp. 232–243.
22. Chvalina, J.; Smetana, B. Solvability of certain groups of time varying artificial neurons. *Ital. J. Pure Appl. Math.* **2018**, submitted.
23. Chvalina, J. Commutative hypergroups in the sense of Marty and ordered sets. In Proceedings of the Summer School an General Algebra and Ordered Sets, Olomouc, Czech Republic, 4–12 September 1994; pp. 19–30.
24. Marty, F. Sur une généralisation de la notion de groupe. In Proceedings of the IV Congrès des Mathématiciens Scandinaves, Stockholm, Sweden, 14–18 August 1934; pp. 45–49.
25. Novák, M.; Cristea, I. Composition in *EL*–hyperstructures. *Hacet. J. Math. Stat.* **2019**, *48*, 45–58. [CrossRef]
26. Novák, M. n-ary hyperstructures constructed from binary quasi-ordered semigroups. *An. Stiintifice Ale Univ. Ovidius Constanta Ser. Mat.* **2014**, *22*, 147–168, ISSN 1224–1784. [CrossRef]
27. Novák, M. On EL—Semihypergroups. *Eur. J. Comb.* **2015**, *44*, 274–286, ISSN 0195–6698. [CrossRef]
28. Novák, M. Some basic properties of *EL*-hyperstructures. *Eur. J. Combin.* **2013**, *34*, 446–459. [CrossRef]
29. Chvalina, J.; Novák, M.; Staněk, D. Sequences of groups and hypergroups of linear ordinary differential operators. *Ital. J. Pure Appl. Math.* **2019**, accepted for publication.
30. Corsini, P. *Prolegomena of Hypergroup Theory*; Aviani Editore Tricesimo: Udine, Italy, 1993.
31. Corsini, P.; Leoreanu, V. *Applications of Hyperstructure Theory*; Kluwer Academic Publishers: Dordrecht, The Netherlands; Boston, MA, USA; London, UK, 2003.
32. Chvalina, J.; Hošková-Mayerová, Š.; Dehghan Nezhad, A. General actions of hypergroups and some applications. *An. Stiintifice Ale Univ. Ovidius Constanta* **2013**, *21*, 59–82.
33. Cristea, I. Several aspects on the hypergroups associated with n-ary relations. *An. Stiintifice Ale Univ. Ovidius Constanta* **2009**, *17*, 99–110.
34. Cristea, I.; Ştefănescu, M. Binary relations and reduced hypergroups. *Discret. Math.* **2008**, *308*, 3537–3544. [CrossRef]
35. Cristea, I.; Ştefănescu, M. Hypergroups and n-ary relations. *Eur. J. Combin.* **2010**, *31*, 780–789. [CrossRef]
36. Leoreanu-Fotea, V.; Ciurea, C.D. On a P-hypergroup. *J. Basic Sci.* **2008**, *4*, 75–79.
37. Pollock, D.; Waldron, M.B. Phase dependent output in a time varying neural net. *Proc. Ann Conf. EMBS* **1989**, *11*, 2054–2055.
38. Račková, P. Hypergroups of symmetric matrices. In Proceedings of the 10th International Congress of Algebraic Hyperstructures and Applications (AHA 2008), Brno, Czech Republic, 3–9 September 2008; pp. 267–272.
39. Vougiouklis, T. Generalization of P-hypergroups. In *Rendiconti del Circolo Matematico di Palermo*; Springer: Berlin, Germany, 1987; pp. 114–121.
40. Vougiouklis, T. Hyperstructures and their Representations. In *Monographs in Mathematics*; Hadronic Press: Palm Harbor, FL, USA, 1994.
41. Vougiouklis, T.; Konguetsof, L. P-hypergoups. *Acta Univ. C. Math. Phys.* **1987**, *28*, 15–20.
42. Chvalina, J.; Chvalinová, L. Modelling of join spaces by n-th order linear ordinary differential operators. In Proceedings of the Fourth International Conference APLIMAT 2005, Bratislava, Slovakia, 1–4 February 2005; pp. 279–284, ISBN 80-969264-2- X.
43. Chvalina, J.; Chvalinová, L. Action of centralizer hypergroups of n-th order linear differential operators on rings on smooth functions. *J. Appl. Math.* **2008**, *1*, 45–53.
44. Neuman, F. *Global Properties of Linear Ordinary Differential Equations*; Academia Praha—Kluwer Academic Publishers: Dordrecht, The Netherlands; Boston, MA, USA; London, UK, 1991; p. 320.
45. Vougiouklis, T. Cyclicity in a special class of hypergroups. *Acta Univ. C Math. Phys.* **1981**, *22*, 3–6.
46. Chvalina, J.; Svoboda, Z. Sandwich semigroups of solutions of certain functional equations and hyperstuctures determined by sandwiches of functions. *J. Appl. Math.* **2009**, *2*, 35–43.
47. Cristea, I.; Novák, M.; Křehlík, Š. A class of hyperlattices induced by quasi-ordered semigroups. In Proceedings of the 16th Conference on Aplied Mathematics Aplimat 2017, Bratislava, Slovakia, 31 January–2 February 2017; pp. 1124–1135.

48. Bavel, Z. The source as a tool in automata. *Inf. Control* **1971**, *18*, 140–155. [CrossRef]
49. Borzooei, R.A.; Varasteh, H.R.; Hasankhani, A. $\mathcal{F}$-Multiautomata on Join Spaces Induced by Differential Operators. *Appl. Math.* **2014**, *5*, 1386–1391. [CrossRef]

symmetry

MDPI

Article

Elements of Hyperstructure Theory in UWSN Design and Data Aggregation

Michal Novák [1,*], Štepán Křehlík [2] and Kyriakos Ovaliadis [3]

1 Faculty of Electrical Engineering and Communication, Brno University of Technology, Technická 8, 616 00 Brno, Czech Republic

2 Department of Applied Mathematics and Computer Science, Masaryk University, Lipová 41a, 602 00 Brno, Czech Republic; Stepan.Krehlik@econ.muni.cz

3 Department of Electrical Engineering, Eastern Macedonia and Thrace Institute of Technology, Agios Loukas, 654 04 Kavala, Greece; ovaliad@teiemt.gr

* Correspondence: novakm@feec.vutbr.cz; Tel.: +420-54114-6077

Received: 8 May 2019; Accepted: 27 May 2019; Published: 29 May 2019

check for updates

Abstract: In our paper we discuss how elements of algebraic hyperstructure theory can be used in the context of underwater wireless sensor networks (UWSN). We present a mathematical model which makes use of the fact that when deploying nodes or operating the network we, from the mathematical point of view, regard an operation (or a hyperoperation) and a binary relation. In this part of the paper we relate our context to already existing topics of the algebraic hyperstructure theory such as quasi-order hypergroups, EL-hyperstructures, or ordered hyperstructures. Furthermore, we make use of the theory of quasi-automata (or rather, semiautomata) to relate the process of UWSN data aggregation to the existing algebraic theory of quasi-automata and their hyperstructure generalization. We show that the process of data aggregation can be seen as an automaton, or rather its hyperstructure generalization, with states representing stages of the data aggregation process of cluster protocols and describing available/used memory capacity of the network.

Keywords: clustering protocols; quasi-automaton; quasi-multiautomaton; semihypergroup; UWSN

1. Introduction

Underwater wireless sensor networks (UWSN) are often used in environment monitoring where they review how human activities affect marine ecosystems, undersea explorations such as detecting oilfields, for disaster prevention, e.g., when monitoring ocean currents, in assisted navigation for the location of dangerous rocks in shallow waters, or for disturbed tactical surveillance for intrusion detection.

The fact that such wireless sensor networks are deployed underwater results in profound differences from terrestrial wireless sensor networks. The key aspects that are different include the communication method, i.e., radio waves vs acoustic signals, cost (while terrestrial networks experience decreasing prices of components, underwater sensors are still expensive devices), memory capacity (because water is a problematic medium resulting in the loss of large quantities of data), power limitations due to the nature of the signal and longer distances handled, as well as problems related to the deployment of the network, i.e., issues connected to static or dynamic deployment. In underwater sensor networks, we commonly face challenges of limited bandwith, high bit error rates, large propagation delays, and limited battery resources caused by the fact that in an underwater environment, sensor batteries are impossible to recharge especially because no solar energy is available underwater. The power losses, which cannot be avoided, result in the need to reconfigure the network topology in order to maintain network connectivity and communication between sensor nodes. Thus, size of the UWSN coverage area and efficiency of data aggregation are affected. Obviously, efficiency in battery use influences network lifetime without sacrificing system performances. These differences are shown in Table 1.

Table 1. Comparison of some features of terrestrial and underwater wireless sensor networks (UWSN).

	(Terrestrial) WSN	UWSN
Communication Media	RF Waves	Acoustic Waves
Frequency	High	Low
Node size	Small	Large
Deployment	Dense	Sparse
Power	Low	High
Energy consumption	Low	High
Propagation delay	Low	High
Bandwidth	High	Low
Path loss	Low	High
Cost	Inexpensive	Expensive
Memory	Sensor nodes have low capacity	Sensor nodes require large capacity

We use different protocols for discovering and maintaining routes between sensor nodes. As mentioned in Novák, Křehlík, and Ovaliadis [1], the most commonly used routing protocols are: Flooding, multipath, cluster, and miscellaneous protocols, see Wahid and Dongkyun [2]. In the flooding approach, the transmitters send a packet to all nodes within the transmission range. In the multipath approach, source sensor nodes establish more than one path towards sink nodes on the surface. Finally, in the clustering approach the sensor nodes are grouped together in a cluster. For an easy-to-follow reading on how UWSN's work and on advantages of clustering see Domingo and Prior [3], the basic idea is shown in Figures 1, 2.

Recent research shows that the cluster based protocols give a great contribution towards the concept of energy efficient networks, see Ayaz et al. [4], Ovaliadis and Savage [5], or Rault, Abdelmadjid, and Yacine [6]. A common cluster based network consists of a centralized station deployed at the surface of the sea called a sink (or surface station) and sensor nodes deployed at various tiers inside the sea environment. These are grouped into clusters. In this architecture, each cluster has a head sensor node called a cluster head (CH). The cluster head is assumed to be inside the transmission range of all sensor nodes that belong to its cluster. Every cluster head operates as a coordinator for its cluster, performing significant tasks such as cluster maintenance, transmission arrangements, data aggregation, and data routing (Figure 2).

Mathematical Background of the Model

In the UWSN topology, several aspects are important for successful data aggregation. First of all, there must exist a path linking every element of the network to the surface station. However, these paths need not be unique as there might be multiple possible paths which the data from a given element can use to reach the surface station. Second, there always exists a cetain kind of ordering of the set of the network elements. They can be ordered with respect to their physical depth, with respect to their importance, with respect to communication priority, remaining battery power, etc. Finally, as data are collected, they are combined in the "upwards" elements in order to be sent further on.

Thus one may employ techniques of algebra or graph theory in the description of the data aggregation process as has been recently done by Aboyamita et al., Domingo, or Jiang et al. [7–9]. However, given the multivalued nature of data aggregation (multiple paths, more than one possible links of elements, etc.), it seems relevant to make use of the elements of the algebraic hyperstructure theory. Notice that while in "classical" algebra, we regard operations, i.e., mappings $f : H^n \to H$, in the algebraic hyperstructure theory we work with hyperoperations, i.e., mappings $g : H^n \to \mathcal{P}^*(H)$, where $\mathcal{P}^*(H)$ is the power set of H with $\emptyset$ excluded (one need not consider this exclusion though). For the general introduction to the theory as well as definitions of concepts not explicitly defined further on, see Corsini and Leoreanu [10].

In the algebraic hyperstructure theory, there are several concepts which make use of the aspect of ordering. A small selection includes Comer, Corsini, Cristea, De Salvo et al. [11–15]. Further on we discuss three of these: EL–hyperstructures, quasi-order hypergroups, and ordered hyperstructures. Each of these concepts uses somewhat different background and assumptions:

EL–hyperstructures are constructed from pre- and partially-ordered semigroups, i.e., the hyperoperation is defined using an operation and a relation compatible with it;

Quasi-order hypergroups are constructed from pre-ordered sets, i.e., the hyperoperation is defined using a relation only;

Ordered hyperstructures are algebraic hyperstructures on which a relation compatible with the hyperoperation is defined.

All of these have been studied in depth and numerous results have been achieved in their respective theories. The idea of EL–hyperstructures has been implicitely present in a number of works since at least the 1960s, for example Pickett [16]. The definition and first results were given by Chvalina [17] and the theory has been elaborated by Novák (later jointly with Chvalina, Křehlík, and Cristea) in a series of papers including [18–22]. It is to be noted that, since the class of EL–hyperstructures is rather broad, the aim of many theorems included in some of those papers was to establish a common ground for some already existing ad hoc derived results. Recently, some examples concerning various types of cyclicity in hypergroups have been constructed using EL–hyperstructures, see Novák, Křehlík and Cristea [23].

The idea of quasi-order hypergroups was proposed by Chvalina in [17,24,25]. Some results achieved with the help of this concept are included in Corsini and Leoreanu [10]. Not to be missed are results concerning the theory of automata collected in Chvalina and Chvalinová [25]. It should be stressed that these results were motivated by Comer [26] and Massouros and Mittas [27].

Ordered hyperstructures were introduced by Heidari and Davvaz [28]. Numerous results have been published since, mainly by Iranian authors.

For the following set of basic definitions see Novák, Křehlík, and Ovaliadis [1].

Definition 1. *By an EL–semihypergroup we mean a semihypergroup, in which, for all $a, b \in H$, there is $a * b = \{x \in H \mid a \cdot b \leq x\}$, where $(H, \cdot, \leq)$ is a quasi-ordered semigroup.*

Proposition 1. *[20,22] If, for all $a, b \in H$, there is $\{a, b\} \in a * b$, then the EL–semihypergroup $(H, *)$ is a hypergroup. If $(H, \cdot, \leq)$ is a partially ordered group, then its EL–hypergroup $(H, *)$ is a join space.*

Definition 2. *Let $(H, *)$ be a hypergroupoid. We say that H is a quasi-order hypergroup, i.e., a hypergroup determined by a quasi-order, if, for all $a, b \in H$, $a \in a^3 = a^2$, and $a * b = a^2 \cup b^2$. Moreover, if $a^2 = b^2 \Rightarrow a = b$ holds for all $a, b \in H$, then $(H, *)$ is called an order hypergroup.*

Proposition 2. *[10] A hypergroupoid is a quasi-order hypergroup if and only if there exists a quasi-order "$\leq$" on the set H such that, for all $a, b \in H$, there is $a * b = [a)_\leq \cup [b)_\leq$.*

Definition 3. *An ordered semihypergroup $(H, *, \preceq)$ is a semihypergroup $(H, *)$ together with a partial ordering "$\preceq$" which is compatible with the hyperoperation, i.e., $x \preceq y \Rightarrow a * x \preceq a * y$ and $x * a \preceq y * a$ for all $a, x, y \in H$. By $a * x \preceq a * y$ we mean that for every $c \in a * x$ there exists $d \in a * y$ such that $c \preceq d$.*

Notation. Further on, for some $a \in H$, by $[a)_\leq$ means the set $\{x \in H \mid a \leq x\}$. For this reason, closed intervals will not be denoted by $[a, b]$ but by $\langle a; b \rangle$.

2. Mathematical Model

The mathematical model presented in this section was published as an extended abstract of the conference contribution Novák, Ovaliadis, and Křehlík [1] presented by the authors of this paper at International Conference on Numerical Analysis and Applied Mathematics (ICNAAM 2017).

UWSNs consist of elements of different types: First, we have *surface stations*, which pass data to a ship or to a data-collecting station located on the sea shore; second, we have *sensor nodes* deployed at various tiers in water or at the sea bed. The sensors, which are deployed in water, can function as sensors measuring the requested data or as transporters of information from seabed sensors. In any case, information collected from all sensors must be passed to surface stations. From these it can be collected either by a ship passing by or, alternatively, transmitted to a data-collecting station located on the sea shore. The ship or the data-collecting stations are *central nodes*.

Denote H the set of all elements of an arbitrary UWSN. Suppose that all elements are capable of handling (i.e., receiving or transmitting) data in the same way. Also suppose that they perform the same set of tasks. Thus they are, from the mathematical point of view, interchangeable and equal (of course, with respect to their functionality as sinks and sensor nodes). The aim of the system is to collect information. Therefore, our elements of H must communicate data. This should be done ideally upwards, towards the surface. As we have mentioned above, there are different ways of passing information. In our model we concentrate on multipath and cluster routing approach (see Figure 1 and Figure 2). For details concerning these see Ayaz et al. and Li et al. [4,29]. Multipath routing protocols (Figure 1), forward the data packets to the sink via other nodes while in cluster based routing protocols (Figure 2), data packets are first aggregated to the respective cluster heads and only then forwarded via other cluster heads to the sink. For our purposes, we denote the i–th cluster by cl_i. Its cluster head will be denoted by CH_i. We call non-CH nodes *ordinary* and sinks will be treated as cluster heads.

Now, suppose that the elements of our system are clustered. In other words, some elements of H function as cluster heads, i.e., masters, while others are ordinary. The data aggregation process goes as follows: Within their cluster, the ordinary elements pass information to their cluster head while between clusters, i.e., supposedly over longer distances, only cluster heads communicate. At a given point in time, each cluster has the unique cluster head, and each element can belong to exactly one cluster. We denote the i–th cluster by cl_i and its cluster head by CH_i.

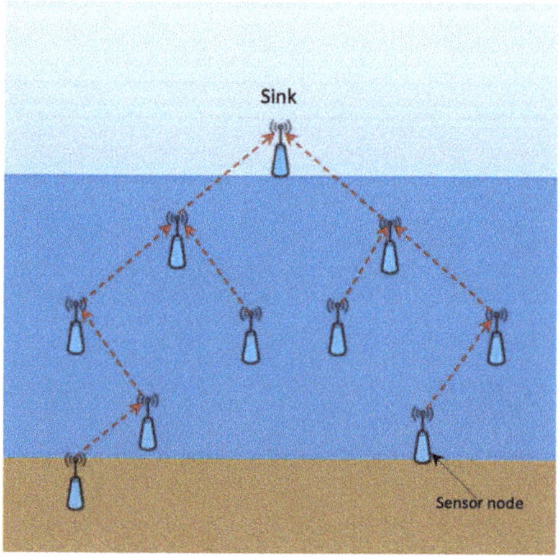

Figure 1. Multipath approach to UWSN data aggregation. Notice the oriented communication between nodes.

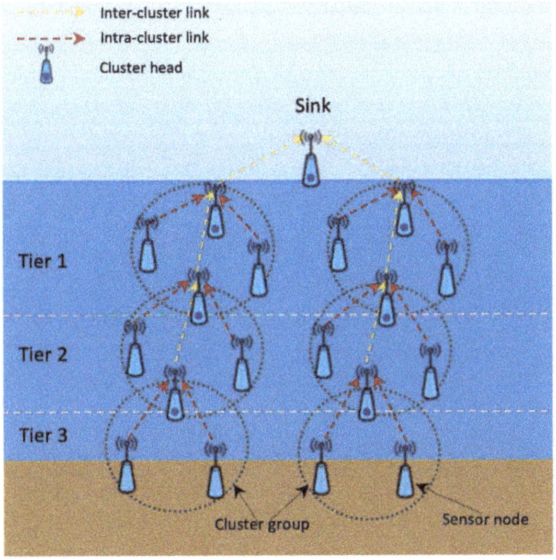

Figure 2. Cluster based approach to UWSN data aggregation—idealized deployment. The tiers need not be horizontal, we usually regard distance towards sink instead of depth.

Now, for a given pair $a, b \in H$, regard a binary hyperoperation, where $a * b$ is, for arbitrary $a, b \in H$, defined by:

$$a * b = \begin{cases} \{a,b\} \cup [a \cdot b)_{\leq} & \text{for } (a = CH_i, b = CH_j) \text{ or } a, b \in cl_i \\ \{a,b\} & \text{for } (a \neq CH_i \text{ or } b \neq CH_j) \text{ and } (a \in cl_i, b \in cl_j, i \neq j) \end{cases} \quad (1)$$

By $[a \cdot b)_{\leq}$ we mean a set $\{x \in H \mid a \cdot b \leq x\}$, where $a \cdot b$ is a result of a single-valued binary operation such that $a \cdot b$ is, for arbitrary $a, b \in H$, defined by:

$$a \cdot b = \begin{cases} CH_i & \text{for } a, b \in cl_i \\ CH_k & \text{for } a = CH_i, b = CH_j, i \neq j \\ s & \text{for } ((a \neq CH_i \text{ or } b \neq CH_j) \text{ and } (a \in cl_i, b \in cl_j, i \neq j)) \text{ or } a = s \text{ or } b = s \end{cases} \quad (2)$$

and CH_k is such a cluster head that $CH_i \leq CH_k$, $CH_j \leq CH_k$, where $a \leq b$ is a relation between elements of H such that: (1) $s \leq s$, $s \leq CH_i$ and $CH_i \leq s$ for all clusters cl_i, (2) within the same cluster cl_i we have $a_j \leq CH_i$ for all $a_j \in cl_i$ while mutually different ordinary elements of the cluster are incomparable, (3) between clusters for $a = CH_i$, $b = CH_j$ the fact that $a \leq b$ means that the tier of b (measured towards the surface) is smaller than or equal to the tier of a, and (4) in all other cases a and b are not related. By CH_k above we mean a cluster head on the closest tier above both CH_i and CH_j. Of course, CH_k always exists yet need not be unique as there may be more cluster heads at this closest tier. In such a case, we choose the most suitable one or regard all cluster heads as equal. Notice that, in our definitions, the fact that $CH_i \leq CH_j$ and simultaneously $CH_j \leq CH_i$ does not mean that $CH_i = CH_j$, rather it only means that CH_i and CH_j are on the same tier. If we are able to chose the most suitable cluster head (further on we remark that we are), the relation "$\leq$" (restricted to $H \setminus \{s\}$) becomes partial ordering and we can write $CH_k = \sup\{CH_i, CH_j\}$ (with respect to the relation "$\leq$"). Finally, the element s is an element of H reserved for situations when a and b fail to communicate. It is artificially added to our set of elements H or we can agree that one (given the actual sensor deployment is of course carefully chosen) of elements of H will be s. In this way, technically speaking, we should in fact write $H_e = H \cup \{s\}$, where H_e could mean "expanded". Of course, if we choose the option of $s \in H$, then $H_e = H$.

Under these definitions, $a \cdot b$ is the element in which the data from a and b meet, and $a * b$ is the path in which the data from both a and b can spread. The facts that $a \cdot b = s$ or $a * b = \{a, b\}$ or $a * b = \{a, b, s\}$ all stand for communication failure.

Lemma 1. [1] *$(H, \leq)$ is a quasi-ordered set.*

Suppose now that we have arbitrary $a, b \in H$. Since the result of $a \cdot b$ is such an element of H in which the data from a and b meet, it is natural to suppose that $a \cdot b = b \cdot a$, i.e., that "$\cdot$" is commutative. However, we can suppose this only on condition that there exists such an algorithms that $a \cdot b = CH_k = CH_l = b \cdot a$ for arbitrary clusters cl_k, cl_l. Further on suppose that such an algorithm exists, i.e., that $(H, \cdot)$ is a commutative groupoid. The following lemma is obvious.

Lemma 2. [1] *If $(H, \cdot)$ is a commutative groupoid, then $(H, *)$ is a commutative hypergroupoid.*

In the following lemma notice that weak associativity of the hyperoperation is defined as $a * (b * c) \cap (a * b) * c \neq \emptyset$ for all $a, b, c \in H$; a quasi-hypergroup is a reproductive hypergroupoid.

Lemma 3. [1] *The hypergroupoid $(H, *)$ is a H_v–group, i.e., a weak associative quasi-hypergroup.*

Lemma 4. [1] *The quasi-ordering "$\leq$" and the operation "$\cdot$" are compatible, i.e., for all $a, b \in H$ such that $a \leq b$ and an arbitrary $c \in H$ there is $a \cdot c \leq b \cdot c$ and $c \cdot a \leq c \cdot b$.*

Now, denote $H_{CH} \subseteq H$ the set of cluster heads. This notation enables us to regard both clustering based systems and multipath systems because the fact that $H_{CH} = H$ means that every element of H is a cluster head, i.e., the system is in fact a multipath one. In such a case the model simplifies substantially. This is because there is no need for the special element s and we do not distinguish between communication within and between clusters. The operation "$\cdot$" defined by Equation (2) reduces to $a \cdot b = c$ (we still suppose that it is commutative) and, consequently, the hyperoperation Equation (1) reduces to $a * b = \{a, b\} \cup [a \cdot b)_{\leq}$, in both cases for all $a, b \in H$.

Lemma 5. [1] *If we are able to uniquely identify CH_k in Equation (2), then $(H_{CH}, \cdot, \leq)$ is a partially ordered semigroup.*

Finally, what is $x \in [a)_{\leq}$? This means that $a \leq x$, i.e., that the data from the element a reach the element x. Thus, if x is a sink, than the fact that $x \in [a)_{\leq}$ means that the data from a can be successfully collected. What we want is that, if we denote S the set of all sinks, for all $a \in H$ there exists at least one $x \in S$ such that $x \in [a)_{\leq}$, which means that data from all elements of our network H can be successfully collected. Of course, in order to achieve this, it is crucial to have an algorithm for unique determination of CH_k in Equation (2). Yet clustering algorithms such as the Distributed Underwater Clustering Scheme (DUCS) [3] or Low Energy Adaptive Clustering Hierarchy (LEACH) protocol can provide this.

3. Use of the Theory of Quasi-Automata

In Definition 2, the concept of quasi-order hypergroup is defined. Chvalina and Chvalinová [25] relate these to the theory of quasi-automata, i.e., automata without output. For an automaton they construct a quasi-order hypergroup of its state set and show that the automaton is connected if and only if the state hypergroup is inner irreducible as well as strogly connected, i.e., we can reach any state from any other state, if and only if the state hypergroup is (in a special way) cyclic. In other words, if we look at the problem of data aggregation from the point of view of the automata theory, where every step is an application of the transition function with the initial state "data aggregation to begin" and the desirable state "data from all elements collected" (or rather "useful data from all elements sent" since every CH not only receives data but also separates useful data from useless ones), we should be interested in constructing such automata or studying their properties.

We call the concept defined below quasi-automaton even though this term is not much frequent (we do this to be consistent with some earlier papers on hyperstructure theory). In fact, we could speak of semiautomata or deterministic finite automata (the below mentioned paper Chvalina and Chvalinová [25] uses a general term automaton; however, notice that [25], p. 107, plain text, defines automaton in the way of Definition 4, which is a definition adopted by the authors of [25] in later years). For an overall discussion of the concepts and the reasons for our choice of the name see Novák et al. [30]. For some further reading and applications see also Hošková et al. [31–33].

Definition 4. *By a quasi–automaton we mean a structure* $\mathbb{A} = (I, S, \delta)$ *such that* $I \neq \emptyset$ *is a monoid,* $S \neq \emptyset$ *and* $\delta : I \times S \to S$ *satisfies the following condition:*

1. *There exists an element* $e \in I$ *such that* $\delta(e, s) = s$ *for any state* $s \in S$*;*
2. $\delta(y, \delta(x, s)) = \delta(xy, s)$ *for any pair* $x, y \in I$ *and any state* $s \in S$.

The set I *is called the input set or input alphabet, the set* S *is called the state set and the mapping* δ *is called next-state or transition function. Condition 2 is called GMAC (Generalized Mixed Associativity Condition).*

In [25], Chvalina and Chvalinová defined what they called a state hypergroup of an automaton. This is in fact a state set with a special hyperoperation, defined by means of the transition function. In this way, the concept of a state hypergroup is fixed to the automata theory. However, the way of defining this concept is a parallel to the concept of quasi-order hypergroups, which means that state hypergroups of quasi-automata are quasi-order hypergroups. The fact that the below defined $(S, \circ)$ is a hypergroup, or rather quasi-order hypergroup, (hence the name state hypergroup) was proved in [25]. (Notice that in [25] I and S are swapped.)

Definition 5. *Let* $\mathbb{A} = (I, S, \delta)$ *be an automaton. We define a binary hyperoperation "*$\circ$*" on the state set* S *by:*

$$s \circ t = \delta(I^*, s) \cup \delta(I^*, t) \tag{3}$$

for any pair of states $s, t \in S$*, where* A^* *is a free monoid of words over the (non-empty) alphabet* A*. The hyperstructure* $(S, \circ)$ *is called state hypergroup of the automaton* $\mathbb{A}$.

Some properties of automata following from properties of its state hypergroup are proved in [25]. This includes the properties of being connected or separated.

Definition 6. *Let* $\mathbb{A} = (I, S, \delta)$ *be a quasi-automaton. A quasi-automaton* $\mathbb{B} = (I, S_1, \delta_1)$ *such that* $S_1 \subseteq S$ *and* δ_1 *is a restriction of* δ *on* $I \times S_1$ *and* $\delta(a, s) \in S_1$ *for any state* $s \in S_1$ *and any word* $a \in I^*$*, is called a sub quasi-automaton of* $\mathbb{A}$*. A sub quasi-automaton* $\mathbb{B} = (I, S_1, \delta_1)$ *of a quasi-automaton* $\mathbb{A} = (I, S, \delta)$ *is called separated if* $\delta(S \setminus S_1, I^*) \cap S_1 = \emptyset$*. A quasi-automaton is called connected if it does not posses any separated proper subautomaton. A quasi-automaton* $\mathbb{A} = (I, S, \delta)$ *is called strongly connected if for any states* $s, t \in S$ *there exists a word* $a \in I^*$ *such that* $\delta(a, s) = t$.

If in quasi–automata we suppose that the input set I is a semihypergroup instead of a free monoid, we arrive at the concept of a quasi–multiautomaton. When defining this concept, caution must be exercised when adjusting the conditions imposed on the transition function δ as on the left-hand side of condition 2 we get a state while on the right-hand side we get *a set of states*. However, in the dichotomy *deterministic — nondeterministic*, quasi–multiautomata still are deterministic because the range of δ is S. The difference between the transition function of a quasi-automaton and the transition function of a quasi-multiautomaton is that in quasi-automata the state achieved by applying y in a state, which is the result of application of x in s, is the same as the state achieved by applying xy in s, while condition (4) says that it is one of the many states achievable by applying any command from $x * y$ in state s.

Definition 7. *A quasi–multiautomaton is a triad $\mathbb{A} = (I, S, \delta)$, where $(I, *)$ is a semihypergroup, S is a non–empty set, and $\delta : I \times S \to S$ is a transition map satisfying the condition:*

$$\delta(b, \delta(a, s)) \in \delta(a * b, s) \text{ for all } a, b \in I, s \in S. \tag{4}$$

The hyperstructure $(I, *)$ is called the input semihypergroup of the quasi–multiautomaton $\mathbb{A}$ (I alone is called the input set or input alphabet), the set S is called the state set of the quasi–multiautomaton $\mathbb{A}$, and δ is called next-state or transition function. Elements of the set S are called states, elements of the set I are called input symbols.

Further on, we will make use of the above mentioned concepts to model the process of data aggregation. Notice that in Novák et al. [30], Cartesian composition of automata resulting in a quasi-multiautomaton is used to describe a task from collective robotics. Moreover, in Chvalina et al. [18], the issue of state sets and input sets having the form of vectors and matrices (of both numbers and special classes of functions) is discussed in the context of quasi-multiautomata.

In Figure 2 we can see that the elements of the UWSN are divided into several tiers. Also, the nodes are grouped into clusters. The process of data aggregation happens as follows: First, data is collected in cluster heads and then transmitted between cluster heads towards the surface, i.e., "upwards". Obviously, we can only transmit the amount of data that the capacity of available memory allows. Suppose that clusters cover areas of more or less the same size, i.e., it does not matter how many nodes there are in respective clusters. Now, regard a set of vectors:

$$S_v = \{\vec{v} = (v_1, v_2, \ldots, v_n) \mid v_i \in \langle 0; 1 \rangle; i \in \{1, \ldots, n\}\} \tag{5}$$

of such a number of components that corresponds to the number of tiers (with index 1 meaning surface and index n meaning seabed or the deepest tier). The components v_i carry information about how much total memory all cluster heads at a given tier has been used. In other words, $v_2 = 0.6$ means that at the second tier 60% memory capacity has been used, regardless of whether this 60% means that every cluster head at this level has 40% free capacity or whether 6 out of 10 cluster heads already have no available memory while 4 are 100% free.

The process of data aggregation starts with $\vec{v} = (0, \ldots, 0)$, i.e., at the moment when all cluster heads have empty memory. Since, at first data are collected within clusters, $\vec{v}$ immediately becomes non-zero. The process of communication between cluster heads is described by the change of $\vec{v}$ by means of multiplying $\vec{v}$ by a square matrix of real numbers:

$$I_M = \left\{ \mathbf{A}_k = \begin{bmatrix} a_{11} & \ldots & a_{1n} \\ \ldots & \ldots & \ldots \\ a_{n1} & \ldots & a_{nn} \end{bmatrix} \mid a_{ij} \in \langle 0; 1 \rangle \text{ for } i \geq j, a_{ij} = 0 \text{ for } i < j; i, j \in \{1, \ldots, n\}; \|\mathbf{A}_k\|_1 \leq 1 \right\},$$

where $\|\mathbf{A}_k\|_1$ is the column norm of $\mathbf{A}_k$. In other words, I_M is a subset of the set of upper triangular matrices, i.e., we can also write:

$$I_M = \left\{ \mathbf{A}_k = \begin{bmatrix} a_{11} & 0 & 0 & \ldots & 0 \\ a_{21} & a_{22} & 0 & \ldots & 0 \\ \vdots & \vdots & \ddots & \vdots & \vdots \\ a_{n1} & a_{n2} & a_{n3} & \ldots & a_{nn} \end{bmatrix} \mid \|\mathbf{A}_k\|_1 \leq 1 \right\} \tag{6}$$

Now, these two sets S_v and I_M will be linked with a transition function δ by:

$$\delta(\mathbf{A}, \vec{v}) = \vec{v} \cdot \mathbf{A} \tag{7}$$

for all $\mathbf{A} \in I_M$ and all $\vec{v} \in S_v$. If we regard the usual matrix multiplication (only swapped), i.e., $\mathbf{A} \odot \mathbf{B} = \mathbf{B} \cdot \mathbf{A}$ for all $\mathbf{A}, \mathbf{B} \in I_M$, then $(I_M, \odot)$ is a monoid such that the free monoid $I_M^* = I_M$. Thus we can regard the triple (I_M, S_v, δ) and study whether it is a quasi-automaton. In our context, the operation of creating words from input symbols of our alphabet I will be matrix multiplication, i.e., a word will be a product of matrices, i.e., again a matrix.

Theorem 1. *The triple (I_M, S_v, δ) is a quasi-automaton.*

Proof. The identity matrix E_n is the neutral element of I_M. Property 1 of Definition 4 holds trivially. Verification of Property 2 is also straightforward:

$$\delta(\mathbf{A}, \delta(\mathbf{B}, \vec{v})) = \delta(\mathbf{A}, \vec{v} \cdot \mathbf{B}) = (\vec{v} \cdot \mathbf{B}) \cdot \mathbf{A} = \vec{v} \cdot \mathbf{B} \cdot \mathbf{A} = \delta(\mathbf{B} \cdot \mathbf{A}, \vec{v}) = \delta(\mathbf{A} \odot \mathbf{B}, \vec{v}).$$

□

The set I_M consists of matrices such that the column norm $||\mathbf{A}||_1$ is at most one. In the following Remark, we show that without this condition the set I_M would not be closed with respect to "$\odot$".

Remark 1. *Suppose two matrices $\mathbf{A}, \mathbf{B} \in I_M$. If we denote:*

$$\mathbf{A} = \begin{bmatrix} a_{11} & 0 & 0 & \dots & 0 \\ a_{21} & a_{22} & 0 & \dots & 0 \\ \vdots & \vdots & \ddots & \vdots & \vdots \\ a_{n1} & a_{n2} & a_{n3} & \dots & a_{nn} \end{bmatrix} \qquad \mathbf{B} = \begin{bmatrix} b_{11} & 0 & 0 & \dots & 0 \\ b_{21} & b_{22} & 0 & \dots & 0 \\ \vdots & \vdots & \ddots & \vdots & \vdots \\ b_{n1} & b_{n2} & b_{n3} & \dots & b_{nn} \end{bmatrix},$$

then,

$$\mathbf{B} \odot \mathbf{A} = \mathbf{A} \cdot \mathbf{B} = \begin{bmatrix} \sum_{i=1}^{1} a_{1i} b_{i1} & 0 & 0 & \dots & 0 \\ \sum_{i=1}^{2} a_{2i} b_{i2} & \sum_{i=2}^{2} a_{2i} b_{i2} & 0 & \dots & 0 \\ \vdots & \vdots & \ddots & \vdots & \vdots \\ \sum_{i=1}^{n} a_{ni} b_{in} & \sum_{i=2}^{n} a_{ni} b_{in} & \sum_{i=3}^{n} a_{ni} b_{in} & \dots & \sum_{i=n}^{n} a_{ni} b_{in} \end{bmatrix}$$

and it is obvious that the column norm $||\mathbf{B} \odot \mathbf{A}||_1$ will not exceed 1, i.e., $\mathbf{B} \odot \mathbf{A} \in I_M$. Indeed, suppose that $\mathbf{A}$ is an all-ones matrix upper triangular matrix (which, of course violates the condition that $||\mathbf{A}||_1 \leq 1$). Then all the sums in $\mathbf{B} \odot \mathbf{A}$ reduce to $\sum_{i=j}^{n} b_{ij}$ which are (due to the fact that $||\mathbf{B}||_1 \leq 1$) smaller than 1. If moreover $||\mathbf{A}||_1 \leq 1$, none of the sums becomes greater. Of course, in I_M we could have used the row norm instead of column one with the same result.

Example 1. *Regard sensor nodes deployed under water, which are divided into four tiers with tier 1 being $0 - 25$ m, tier 2 being $25 - 50$ m, tier 3 being $50 - 75$ m, and tier 4 being $75 - 100$ m under water. Every tier has an arbitrary number of clusters, i.e., an arbitrary number of cluster heads, each with the same memory capacity. Then by vector e.g., $\vec{v} = (0.3; 0.1; 0; 0.5)$, we describe such a state of the system that, at a certain* *point in time, 30% data has been collected (or rather, 30% memory capacity has been used) from tier 1 with the numbers being 10% from tier 2, 0% from tier 3, and 50% from tier 4. Now regard a matrix:*

$$\mathbf{A} = \begin{bmatrix} 0 & 0 & 0 & 0 \\ 1 & 1 & 0 & 0 \\ 0 & 0 & 0 & 0 \\ 0 & 0 & 1 & 0 \end{bmatrix} \in I_M,$$

which we apply on $\vec{v}$. Then $\delta(\mathbf{A}, \vec{v}) = (0.1; 0.1; 0.5; 0)$, which describes the new state of the system. Of course, the first 30% of data (represented by the first component of $\vec{v}$) did not get lost, which we can regard as output, i.e., to each quasi-automaton we can assign output, meaning that data from the upmost tier 1 have been processed, sent out of the system. Furthermore, in our example, data from tier 2 were transferred to tier 1 (and left in tier 2, i.e., in this case we have a backup copy). Also, data from tier 4 were transferred to tier 3, which had been empty, and simultaneously deleted in tier 4. It can be easily observed that the issue of suitable construction of matrices used for quasi-automata inputs is a topic for further research which will be closely linked to various aspects of optimization theory.

When describing the state hypergroup of (I_M, S_v, δ), we must first of all decide whether or not $(S_v, \circ)$ defined by Equation (3) is trivial. However, this task is rather simple.

Lemma 6. *The state hypergroup $(S_v, \circ)$ of (I_M, S_v, δ) is not trivial, i.e., there exist states $s, t \in S_v$ such that $\emptyset \neq \delta(I_M^*, s) \cup \delta(I_M^*, t) \neq S_v$.*

Proof. The transition function δ is defined on (I_M, S_v, δ) by $\delta(\mathbf{A}, \vec{v}) = \vec{v}\mathbf{A}$ for all $\mathbf{A} \in I_M$ and all $\vec{v} \in S_v$. Moreover, by Equation (5), components of vectors from S_v are real numbers from interval $\langle 0; 1\rangle$, and by (6) matrices from I_M are upper triangular with entries taken from the same interval. Finally, in a text preceding Theorem 1 we have already mentioned that $I_M^* = I_M$. Thus, for all $s, t \in I_M$ and all $\mathbf{A} \in I_M$, we have that:

$$\delta(\mathbf{A}, \vec{s}) = \vec{s}\mathbf{A} = (s_1, s_2, \ldots, s_n) \begin{bmatrix} a_{11} & 0 & 0 & \ldots & 0 \\ a_{21} & a_{22} & 0 & \ldots & 0 \\ \vdots & \vdots & \ddots & \vdots & \vdots \\ a_{n1} & a_{n2} & a_{n3} & \ldots & a_{nn} \end{bmatrix} = \left(\sum_{i=1}^{n} s_i a_{i1}, \sum_{i=2}^{n} s_i a_{i2}, \ldots \sum_{i=n-1}^{n} s_i a_{in-1}, s_n a_{nn} \right).$$

Now suppose that $s_n = 0.1$. Obviously in this case there cannot be $\delta(\mathbf{A}, \vec{s}) = S_v$ because we are not able to find such a matrix $\mathbf{A} \in I_M^* = I_M$ that $s_n a_{nn} = 0.9$ as this would require $a_{nn} = 9$ which is out of the interval $\langle 0; 1\rangle$ in which all entries of $\mathbf{A}$ should be. Naturally, there are infinitely many of these choices. The fact that $\delta(I_M^*, s) \cup \delta(I_M^*, t) \neq \emptyset$ is obvious. □

Regard now the set $\hat{I}_M$ which will be different from I_M because the condition of the column norm will be dropped and we will assume that $b_{ii} \geq b_{ij}$ for all $i \in \{1, 2, \ldots, n\}$ and $j \in \{1, \ldots, i\}$, i.e., that each diagonal element will be greater than all other elements in the given row.

On $\hat{I}_M$ we define a hyperoperation "$*$" by:

$$\mathbf{A} * \mathbf{B} = \left\{ \begin{bmatrix} c_{11} & 0 & 0 & \ldots & 0 \\ c_{21} & c_{22} & 0 & \ldots & 0 \\ \vdots & \vdots & \ddots & \vdots & \vdots \\ c_{n1} & c_{n2} & c_{n3} & \ldots & c_{nn} \end{bmatrix} \mid c_{ii} \in \langle a_{ii} \cdot b_{ii}; 1\rangle \text{ and } c_{ij} \in \langle a_{ij} \cdot b_{ij}; a_{ii} \cdot b_{ii}\rangle \text{ for } i \neq j \right\} \quad (8)$$

for all $\mathbf{A}, \mathbf{B} \in \hat{I}_M$, where "$\cdot$" is the usual multiplication of real numbers and c_{ij} belongs to a closed interval bounded by the product $a_{ij} \cdot b_{ij}$ and either 1 or the product of the corresponding diagonal elements.

Remark 2. *Notice that by removing the assumption regarding the column norm we are making our model more real-life because we regard memory losses and overflows. Indeed, suppose we have a vector e.g., $\vec{v} = (0.8; 0.7)$ and a matrix e.g., $\mathbf{A} = \begin{bmatrix} 1 & 0 \\ 1 & 1 \end{bmatrix}$. Then $\delta(\mathbf{A}, \vec{v}) = \vec{v}\mathbf{A} = (1.5; 0.7)$, which (since vector components are taken from the interval $\langle 0; 1 \rangle$ because they describe percentages of memory capacity) must be reduced to $(1; 0.7)$.*

Theorem 2. *The pair $(\hat{I}_M, *)$ is a commutative hypergroup.*

Proof. Commutativity of the hyperoperation "$*$" is obvious because we work with entry-wise multiplication of real numbers.

For the test of associativity, i.e., $\mathbf{A} * (\mathbf{B} * \mathbf{C}) = (\mathbf{A} * \mathbf{B}) * \mathbf{C}$ for all triples $\mathbf{A}, \mathbf{B}, \mathbf{C} \in \hat{I}_M$, recall that we will be comparing sets of matrices of the same dimension, elements of which fall within intervals defined by Equation (8). As far as diagonal elements are concerned, there is obviously no problem because we make use of the multiplication of real numbers. For the non-zero, non-diagonal elements the lower bounds of the intervals are the products of their smallest elements, i.e., $a_{ij}(b_{ij}c_{ij}) = (a_{ij}b_{ij})c_{ij}$, while the upper bounds of the intervals are products of their greatest possible, i.e., diagonal, elements i.e., $a_{ii}(b_{ii}c_{ii}) = (a_{ii}b_{ii})c_{ii}$. Thus, in all cases, associativity of the hyperoperation follows from the associativity of (positive) real numbers (smaller than 1).

Reproductive axiom is also obviously valid because since the null matrix $\mathbf{0}$ is an element of $\hat{I}_M$ and $\mathbf{A} * \mathbf{0} = \hat{I}_M$. □

Notice that in the following theorem, $\delta(\mathbf{A}, \vec{v})$ is the usual vector and matrix multiplication with the additional condition that no diagonal component is allowed to exceed 1. For its motivation see Remark 2.

Theorem 3. *The triple $(\hat{I}_M, S_v, \delta)$ is a quasi-multiautomaton, where, for all $\mathbf{A} \in \hat{I}_M$ and all $\vec{v} \in S_v$,*

$$\delta(\mathbf{A}, \vec{v}) = \left(\sum_{i=1}^{n} v_i a_{i1}, \dots, \sum_{i=1}^{n} v_i a_{in} \right), \tag{9}$$

where in case that $\sum_{i=1}^{n} v_i a_{ik} > 1$ for some $k \in \{1, 2, \dots, n\}$, we set, by default, $\sum_{i=1}^{n} v_i a_{ik} = 1$.

Proof. Thanks to Theorem 2, we shall verify condition Equation (4) only. The left-hand side of the condition is:

$$\delta(\mathbf{A}, \delta(\mathbf{B}, \vec{v})) = \delta(\mathbf{A}, \vec{v} \cdot \mathbf{B}) = \delta\left(\mathbf{A}, \left(\sum_{i=1}^{n} v_i b_{i1}, \dots, \sum_{i=1}^{n} v_i b_{in}\right)\right) = \left(\sum_{i=1}^{n} v_i b_{i1}, \dots, \sum_{i=1}^{n} v_i b_{in}\right) \cdot \begin{bmatrix} a_{11} & \dots & a_{1n} \\ \dots & \dots & \dots \\ a_{n1} & \dots & a_{nn} \end{bmatrix},$$

which, if we regard that **A**, **B** are upper triangular matrices, is:

$$\Big(\sum_{i=1}^{n} v_i b_{i1} a_{11} + \sum_{i=2}^{n} v_i b_{i2} a_{21} + \ldots + \sum_{i=n-1,}^{n} v_i b_{in-1,} a_{n-1,1} + \sum_{i=n}^{n} v_i b_{in} a_{n1},$$
$$\sum_{i=1}^{n} v_i b_{i1} 0 + \sum_{i=2}^{n} v_i b_{i2} a_{22} + \ldots + \sum_{i=n-1,}^{n} v_i b_{in-1,} a_{n-1,2} + \sum_{i=n}^{n} v_i b_{in} a_{n2},$$
$$\sum_{i=1}^{n} v_i b_{i1} 0 + \sum_{i=2}^{n} v_i b_{i2} 0 + \ldots + \sum_{i=n-1,}^{n} v_i b_{in-1,} a_{n-1,n-1} + \sum_{i=n}^{n} v_i b_{in} a_{nn}, \ldots$$
$$\ldots, \sum_{i=1}^{n} v_i b_{i1} 0 + \sum_{i=2}^{n} v_i b_{i2} 0 + \ldots + \sum_{i=n-1,}^{n} v_i b_{in-1,} 0 + \sum_{i=n}^{n} v_i b_{in} a_{nn}\Big)$$

On the right-hand side of the condition we get:

$$\delta(\mathbf{A} * \mathbf{B}, \vec{v}) = \left(\left\langle \sum_{i=1}^{n} v_i a_{i1} b_{i1}; 1\right\rangle, \left\langle \sum_{i=2}^{n} v_i a_{i2} b_{i2}; 1\right\rangle, \ldots, \left\langle \sum_{i=n-1}^{n} v_i a_{in-1} b_{in-1}; 1\right\rangle, \left\langle \sum_{i=n}^{n} v_i a_{in} b_{in}; 1\right\rangle\right),$$

which is a vector of intervals upper-bounded by 1 such that their lower bounds are sums of lower bounds of respective intervals in the definition of the hyperoperation in Equation (8) after multiplication by vector $\vec{v}$ has been performed. (Notice that we work with non-negative matrix and vector entries only.) We have to show that:

$$\begin{aligned}
\sum_{i=1}^{n} v_i a_{i1} b_{i1} &\leq \sum_{i=1}^{n} v_i b_{i1} a_{11} + \sum_{i=2}^{n} v_i b_{i2} a_{21} + \ldots + \sum_{i=n-1,}^{n} v_i b_{in-1,} a_{n-1,1} + \sum_{i=n}^{n} v_i b_{in} a_{n1} \\
\sum_{i=2}^{n} v_i a_{i2} b_{i2} &\leq \sum_{i=1}^{n} v_i b_{i1} 0 + \sum_{i=2}^{n} v_i b_{i2} a_{22} + \ldots + \sum_{i=n-1,}^{n} v_i b_{in-1,} a_{n-1,2} + \sum_{i=n}^{n} v_i b_{in} a_{n2} \\
&\cdots \\
\sum_{i=n-1}^{n} v_i a_{in-1} b_{in-1} &\leq \sum_{i=1}^{n} v_i b_{i1} 0 + \sum_{i=2}^{n} v_i b_{i2} 0 + \ldots + \sum_{i=n-1,}^{n} v_i b_{in-1,} a_{n-1,n-1} + \sum_{i=n}^{n} v_i b_{in} a_{nn} \\
\sum_{i=n}^{n} v_i a_{in} b_{in} &\leq \sum_{i=1}^{n} v_i b_{i1} 0 + \sum_{i=2}^{n} v_i b_{i2} 0 + \ldots + \sum_{i=n-1,}^{n} v_i b_{in-1,} 0 + \sum_{i=n}^{n} v_i b_{in} a_{nn}
\end{aligned}$$

i.e.,

$$\begin{aligned}
\sum_{i=1}^{n} v_i a_{i1} b_{i1} &\leq \sum_{i=1}^{n} v_i b_{i1} a_{11} + \sum_{i=2}^{n} v_i b_{i2} a_{21} + \ldots + \sum_{i=n-1,}^{n} v_i b_{in-1,} a_{n-1,1} + \sum_{i=n}^{n} v_i b_{in} a_{n1} \\
\sum_{i=2}^{n} v_i a_{i2} b_{i2} &\leq \sum_{i=2}^{n} v_i b_{i2} a_{22} + \ldots + \sum_{i=n-1,}^{n} v_i b_{in-1,} a_{n-1,2} + \sum_{i=n}^{n} v_i b_{in} a_{n2} \\
&\cdots \\
\sum_{i=n-1}^{n} v_i a_{in-1} b_{in-1} &\leq \sum_{i=n-1,}^{n} v_i b_{in-1,} a_{n-1,n-1} + \sum_{i=n}^{n} v_i b_{in} a_{nn} \\
\sum_{i=n}^{n} v_i a_{in} b_{in} &\leq \sum_{i=n}^{n} v_i b_{in} a_{nn}
\end{aligned}$$

Now, if we move the sums on the left-hand side of the inequalities to the right-hand side and expand all the sums, we get for the first sum, i.e., for the first component of the vectors:

$$\begin{aligned} v_1a_{11}b_{11} + v_2a_{21}b_{21} + v_3a_{31}b_{31} + \ldots + v_na_{n1}b_{n1} \leq\ & v_1b_{11}a_{11} + v_2b_{21}a_{11} + \ldots + v_nb_{n1}a_{11} + \\ & + v_2b_{22}a_{21} + v_3b_{32}a_{21} + \ldots + v_nb_{n2}a_{21} + \\ & \vdots \\ & + v_{n-1}b_{n-1,n-1}a_{n-1,1} + v_nb_{n,n-1}a_{n-1,1} + \\ & + v_nb_{nn}a_{n1}, \end{aligned}$$

which, after we put elements $v_i, i \in \{1, 2, \ldots, n\}$ in front of the brackets, means that:

$$\begin{aligned} 0 \leq\ & v_2(b_{21}a_{11} + b_{22}a_{21} - a_{21}b_{21}) + \\ & v_3(b_{31}a_{11} + b_{32}a_{21} + b_{33}a_{31} - a_{31}b_{31}) + \\ & \vdots \\ & v_{n-1}(b_{n-1,1}a_{11} + b_{n-1,2}a_{21} + \ldots + b_{n-1,n-1}a_{n-1,1} - a_{n-1,1}b_{n-1,1}) + \\ & v_n(b_{n1}a_{11} + b_{n2}a_{21} + \ldots b_{nn}a_{n1} - a_{n1}b_{n1}) \end{aligned}$$

which, since all numbers are non-negative, will be greater than zero if $b_{ii} \geq b_{i1}$ for all $i \in \{2, 3, \ldots n\}$. In a completely analogous manner we get for the second sum, i.e., for the second component of the vectors:

$$\begin{aligned} v_2a_{22}b_{22} + v_3a_{32}b_{32} + \ldots + v_na_{n2}b_{n2} \leq\ & v_2b_{22}a_{22} + v_3b_{32}a_{22} + \ldots + v_nb_{n2}a_{22} + \\ & + v_3b_{33}a_{32} + v_4b_{43}a_{32} + \ldots + v_nb_{n3}a_{32} \\ & \vdots \\ & + v_{n-1}b_{n-1,n-1}a_{n-1,2} + v_nb_{n,n-1}a_{n-1,2} + \\ & + v_nb_{nn}a_{n2}, \end{aligned}$$

which will give us condition that $b_{ii} \geq b_{i2}$ for all $i \in \{3, 4, \ldots n\}$, etc. until the $(n-1)^{th}$ component will result in condition $b_{ii} \geq b_{i,n-1}$ for $i = n$; and the first and the last condition $b_{ii} \geq b_{ii}$ for all $i \in \{1, n\}$ hold trivially.

Thus, altogether we get that $b_{ii} \geq b_{ij}$ for all $i \in \{1, 2, \ldots, n\}$ and $j \in \{1, \ldots, i\}$. In other words, each diagonal element must be greater than all other elements in the given row, which is exactly what we suppose $\hat{I}_M$ to be. □

Remark 3. *If we want to find out whether we are able to reach any state from any other state with the help of our transition functions (i.e., test strong conectedness), we must take into account that the fact that we regard diagonal matrices means that for known vectors* $\vec{u}, \vec{v}$ *and an unknown matrix* $\mathbf{A}$ *the equation* $\vec{u}\mathbf{A} = \vec{v}$ *results in a linear system in the row echelon form and the task of finding the unknown matrix* $\mathbf{A}$ *is equivalent to finding its solution. Of course, we have to consider the fact that* $\vec{v}$ *might have components being zero, which will affect the solvability of the linear system.*

Example 2. *The hyperoperation on the input set, i.e., the fact that we work with hypergroups, gives us a better tool for controlling the process of data aggregation. Suppose that we have the same vector as in Example 1, i.e.,* $\vec{v} = (0.3; 0.1; 0; 0.5)$. *Furthermore, assume the same context as in Example 1. Let the matrix be:*

$$\mathbf{A} = \begin{bmatrix} 1 & 0 & 0 & 0 \\ 0.9 & 1 & 0 & 0 \\ 0.2 & 1 & 1 & 0 \\ 0.01 & 0.3 & 1 & 0 \end{bmatrix} \in \hat{I}_M.$$

Suppose that in our quasi-multiautomaton this matrix is applied on vector $\vec{v}$, *i.e.*

$$\delta(\mathbf{A}, \vec{v}) = (0.3; 0.1; 0; 0.5) \cdot \begin{bmatrix} 1 & 0 & 0 & 0 \\ 0.9 & 1 & 0 & 0 \\ 0.2 & 1 & 1 & 0 \\ 0.01 & 0.3 & 1 & 0 \end{bmatrix}.$$

Now, focus on the multiplication of $\vec{v}$ *and the first column of* $\mathbf{A}$. *We get that:*

$$(0.3 \cdot 1 + 0.1 \cdot 0.9 + 0 \cdot 0.2 + 0.5 \cdot 0.01; a; b; c) = (0.395; a; b; c),$$

which means that the capacity of tier 1 is filled by 39.5% *in such a way that the original* 30% *was accompanied by further* 9.5% *from tier 2 (which had originally been filled up by* 10%, *i.e., we have a* 1% *transmission loss),* 0% *from tier 3, and* 0.5% *from tier 4. In this case it is obvious that the transmission from tier 4 is extremely difficult with a substantial data loss. Theoretically, we could model the process of data aggregation with matrices with all-one columns. However, this would not be a real-life case. When looking for an optimal transition input (i.e., an optimal matrix), we could regard matrices such as those in Equation (8) with increasing entries. The definition of the hyperoperation (8) means that from a certain moment on, the memory capacity could be filled to maximum and data will start being lost, i.e., not stored in cluster heads on upper tiers anymore. However, this depends also on the initial state of the system, i.e., on the initial vector. In the case of our vector,* $\vec{v}$, *even matrix:*

$$\begin{bmatrix} 1 & 0 & 0 & 0 \\ 1 & 1 & 0 & 0 \\ 1 & 1 & 1 & 0 \\ 1 & 1 & 1 & 1 \end{bmatrix},$$

i.e., a matrix with maximal possible entries will secure successful data aggregation because vector $\vec{v}$ *suggests that the process started when all tiers had enough free memory capacity in all cluster heads without any risk of memory overflow.*

4. Conclusions and Future Work

In Remark 3, we saw in a general case a strong connectedness of our quasi(-multi) automata was not secured. Moreover, in Lemma 6, we observed that the state hypergroup of the quasi-automaton (I_M, S_v, δ) was not trivial. These two facts are motivations for future research into properties of quasi-(multi)automata, especially in their interpretation of our real-life problem, i.e., UWSN design and data aggregation. Furthermore, the range of lemmas included in Section 2 provides a good starting point for linking our context to the already established results of the algebraic hyperstructure theory. Finally, the issue of optimal choice of input matrices is another line of possible research.

Author Contributions: Investigation, M.N., Š.K. and K.O.; Methodology, M.N., Š.K. and K.O.; Supervision, M.N., Š.K. and K.O.; Writing—original draft, M.N., Š.K. and K.O.; Writing—review & editing, M.N., Š.K. and K.O.

Funding: The first author was supported by the FEKT-S-17-4225 grant of Brno University of Technology.

Conflicts of Interest: The authors declare no conflict of interest.

References

1. Novák, M.; Ovaliadis, K.; Křehlík, Š. A hyperstructure model of Underwater Wireless Sensor Network (UWSN) design. In Proceedings of International Conference on Numerical Analysis and Applied Mathematics (ICNAAM 2017), Thessaloniki, Greece, 25–30 September 2017.
2. Wahid, A.; Dongkyun, K. Analyzing Routing Protocols for Underwater Wireless Sensor Networks. *IJCNIS* **2010**, *2*, 253–261.
3. Domingo, M.C.; Prior, R. A distributed clustering scheme for underwater wireless sensor networks. In Proceedings of the 2007 IEEE 18th International Symposium on Personal, Indoor and Mobile Radio Communications, Athens, Greece, 3–7 September 2007.
4. Ayaz M. et al. A survey on routing techniques in underwater wireless sensor networks. *J. Netw. Comput. Appl.* **2011**, *34*, 1908–1927. [CrossRef]
5. Ovaliadis, K.; Savage, N. Cluster protocols in underwater sensor networks: A research review. *J. Eng. Sci. Technol. Rev.* **2014**, *7*, 171–175. [CrossRef]
6. Rault, T.; Abdelmadjid, B.; Yacine, C. Energy efficiency in wireless sensor networks: A top-down survey. *Comput. Netw.* **2014**, *67*, 104–122. [CrossRef]
7. Abougamila, S.; Elmorsy, M.; Elmallah, E.S. A graph theoretic approach to localization under uncertainty. In Proceedings of the 2018 IEEE International Conference on Communications, ICC 2018, Kansas City, MO, USA, 20–24 May 2018.
8. Domingo, M.C. Optimal placement zones of wireless nodes in Underwater Wireless Sensor Networks with shadow zones, In Proceedings of the 2009 2nd IFIP Wireless Days (WD), Paris, France, 15–17 December 2009.
9. Jiang, P.; Wang, X.; Jiang, L. Node deployment algorithm based on connected tree for underwater sensor networks. *Sensors* **2015**, *15*, 16763–16785. [CrossRef]
10. Corsini, P.; Leoreanu, V. *Applications of Hyperstructure Theory*. Kluwer Academic Publishers: Dordrecht, The Netherlands, 2003.
11. Comer, S. D. Combinatorial aspects of relation. *Algebr. Universalis* **1984**, *18*, 77–94. [CrossRef]
12. Corsini, P. Binary relations and hypergroupoids. *Ital. J. Pure Appl. Math.* **2000**, *7*, 11–18.
13. Corsini, P. Hyperstructures associated with ordered sets. *Bull. Greek Math. Soc.* **2003**, *48*, 7–18.
14. Cristea, I.; Ştefănescu, M. Hypergroups and n-ary relations. *European J. Combin.* **2010**, *31*, 780–789. [CrossRef]
15. De Salvo, M.; Lo Faro, G. Hypergroups and binary relations. *Multi. Val. Logic* **2002**, *8*, 645–657.
16. Pickett, H. E. Homomorphisms and subalgebras of multialgebras. *Pac. J. Math.* **1967**, *21*, 327–342. [CrossRef]
17. Chvalina, J. *Functional Graphs, Quasi-ordered Sets and Commutative Hypergroups*; Masaryk University: Brno, Czech Republic, 1995. (in Czech)
18. Chvalina, J.; Křehlík, Š.; Novák, M. Cartesian composition and the problem of generalising the MAC condition to quasi-multiautomata. *An. Şt. Univ. Ovidius Constanţa* **2016**, *24*, 79–100.
19. Křehlík, Š.; Novák, M. From lattices to H_v–matrices. *An. Şt. Univ. Ovidius Constanţa* **2016**, *24*, 209–222. [CrossRef]
20. Novák, M. Some basic properties of EL–hyperstructures. *Eur. J. Combin.* **2013**, *34*, 446–459. [CrossRef]
21. Novák, M.; Cristea, I. Composition in EL–hyperstructures. *Hacet. J. Math. Stat.* **2019**, *48*, 45–58. [CrossRef]
22. Novák, M., Křehlík, Š. EL–hyperstructures revisited. *Soft Comput.* **2018**, *22*, 7269–7280. [CrossRef]
23. Novák, M.; Křehlík, Š.; Cristea, I. Cyclicity in EL–hypergroups. *Symmetry* **2018**, *10*, 611. [CrossRef]
24. Chvalina, J. Commutative hypergroups in the sense of Marty and ordered sets, In Proceedings of the Summer School on General Algebra and Ordered Sets, Olomouc, Czech Republic, 12–15 August 1994; pp. 19–30.
25. Chvalina, J.; Chvalinová, L. State hypergroups of automata. *Acta Math. et Inform. Univ. Ostraviensis* **1996**, *4*, 105–120.
26. Comer, S.D. Some problems on hypergroups. In *Algebraic Hyperstructures and Applications*, Vougiouklis, T., Ed.; World Scientific Publishing: Singapore, 1991; pp. 67–74.
27. Massouros, G.G.; Mittas, J. D. Languages, Automata and Hypercompositional Structures. In *Algebraic Hyperstructures and Applications*, Vougiouklis, T., Ed.; World Scientific Publishing: Singapore, 1991; pp. 137–147.
28. Heidari, D.; Davvaz, B. On ordered hyperstructures. *UPB Sci. Bull. Ser. A* **2011**, *73*, 85–96.
29. Li, N.; Martínez, J.F.; Meneses Chaus, J.; Eckert, M. A survey on underwater acoustic sensor network routing protocols. *Sensors* **2016**, *16*, 414. [CrossRef]

30. Novák, M.; Křehlík, Š.; Staněk, D. n–ary Cartesian composition of automata. *Soft Comput.* **2019**. [CrossRef]
31. Hošková, Š. Discrete transformation hypergroups. In Proceedings of 4th International Conference Aplimat, Bratislava, Slovakia, 1–4 February 2005; pp. 275–279.
32. Hošková, Š; Chvalina, J. A survey of investigations of the Brno research group in the hyperstructure theory since the last AHA Congress. In Proceedings of the AHA 2008: 10th International Congress-Algebraic Hyperstructures And Applications, Brno, Czech Republic, 3–9 September 2008.
33. Hošková, Š.; Chvalina, J.; Račková, P. Hypergroups of integral operators in connections with transformation structures. *AiMT* **2006**, *1*, 105–117.

symmetry

MDPI

Article

An m-Polar Fuzzy Hypergraph Model of Granular Computing

Anam Luqman [1], Muhammad Akram [1,*] and Ali N.A. Koam [2]

1 Department of Mathematics, University of the Punjab, New Campus, P.O. Box 54590, Lahore, Pakistan; anamluqman7@yahoo.com
2 Department of Mathematics, College of Science, Jazan University, New Campus, P.O. Box 2097, Jazan, Saudi Arabia; akoum@jazanu.edu.sa
* Correspondence: m.akram@pucit.edu.pk

Received: 12 March 2019; Accepted: 1 April 2019; Published: 3 April 2019

check for updates

Abstract: An m-polar fuzzy model plays a vital role in modeling of real-world problems that involve multi-attribute, multi-polar information and uncertainty. The m-polar fuzzy models give increasing precision and flexibility to the system as compared to the fuzzy and bipolar fuzzy models. An m-polar fuzzy set assigns the membership degree to an object belonging to $[0,1]^m$ describing the m distinct attributes of that element. Granular computing deals with representing and processing information in the form of information granules. These information granules are collections of elements combined together due to their similarity and functional/physical adjacency. In this paper, we illustrate the formation of granular structures using m-polar fuzzy hypergraphs and level hypergraphs. Further, we define m-polar fuzzy hierarchical quotient space structures. The mappings between the m-polar fuzzy hypergraphs depict the relationships among granules occurring at different levels. The consequences reveal that the representation of the partition of a universal set is more efficient through m-polar fuzzy hypergraphs as compared to crisp hypergraphs. We also present some examples and a real-world problem to signify the validity of our proposed model.

Keywords: m-polar fuzzy hypergraphs; m-polar fuzzy equivalence relation; level hypergraphs; granular computing; application

1. Introduction

Granular computing (GrC) is defined as an identification of techniques, methodologies, tools, and theories that yields the advantages of clusters, groups, or classes, i.e., the granules. The terminology was first introduced by Lin [1]. The fundamental concepts of GrC are utilized in various disciplines, including machine learning, rough set theory, cluster analysis, and artificial intelligence. Different models have been proposed to study the various issues occurring in GrC, including classification of the universe, illustration of granules, and the identification of relations among granules. For example, the procedure of problem solving through GrC can be considered as subdivisions of the problem at multiple levels, and these levels are linked together to construct a hierarchical space structure (HSS). Thus, this is a way of dealing with the formation of granules and the switching between different granularities. Here, the word "hierarchy" implies the methodology of hierarchical analysis in solving a problem and human activities [2]. To understand this methodology, let us consider an example of a national administration in which the complete nation is subdivided into various provinces. Further, we divide every province into various divisions, and so on. The human activities and problem solving involve the simplification of original complicated problem by ignoring some details rather than thinking about all points of the problem. This rationalized model is then further refined till the issue is completely solved. Thus, we resolve and interpret the complex problems from the weaker grain to the stronger one or from highest rank to lowest or from universal to particular, etc. This technique is called hierarchical problem solving. It is further acknowledged that the hierarchical strategy is the only technique that is used by humans to deal with complicated problems, and it enhances competency and efficiency. This strategy is also known as the multi-GrC.

Hypergraphs, as an extension of classical graphs, experience various properties that appear very effective and useful as the basis of different techniques in many fields, including problem solving, declustering, and databases [3]. The real-world problems that are represented and solved using hypergraphs have achieved very good impacts. The formation of hypergraphs is the same as that of granule structures, and the relations between the vertices and hyperedges of hypergraphs can depict the relationships of granules and objects. A hyperedge can contain n vertices representing n-ary relations and hence can provide more effective analysis and description of granules. Many researchers have used hypergraph methods to study the clustering of complex documentation by means of GrC and investigated the database techniques [4,5]. Chen et al. [6] proposed a model of GrC based on the crisp hypergraph. They related a crisp hypergraph to a set of granules and represented the hierarchical structures using series of hypergraphs. They proved a hypergraph model as a visual description of GrC.

Zadeh's [7] fuzzy set (FS) has acquired greater attention by researchers in a wide range of scientific areas, including management sciences, robotics, decision theory, and many other disciplines. Zhang [8] generalized the idea of FSs to the concept of bipolar fuzzy sets (BFSs) whose membership degrees range over the interval $[-1,1]$. An m-polar fuzzy set (m-PFS), as an extension of FS and BFS, was proposed by Chen et al. [9], and it proved that two-PFSs and BFSs are equivalent concepts in mathematics. An m-PFS corresponds to the existence of "multipolar information" because there are many real-world problems that take data or information from n agents ($n \geq 2$). For example, in the case of telecommunication safety, the exact membership degree lies in the interval $[0,1]^n$ $(n \approx 7 \times 10^9)$ as the distinct members are monitored at different times. Similarly, there are many problems that are based on n logic implication operators ($n \geq 2$), including rough measures, ordering results of magazines, fuzziness measures, etc.

To handle uncertainty in the representation of different objects or in the relationships between them, fuzzy graphs (FGs) were defined by Rosenfeld [10]. m-Polar fuzzy graphs and their interesting properties were discussed by Akram et al. [11] to deal with the network models possessing multi-attribute and multipolar data. As an extension of FGs, Kaufmann [12] defined fuzzy hypergraphs. Gong and Wang [13] investigated fuzzy classification by combining fuzzy hypergraphs with the fuzzy formal concept analysis and fuzzy information system. Although, many researchers have explored the construction of granular structures using hypergraphs in various fields, there are many graph theoretic problems that may contain uncertainty and vagueness. To overcome the problems of uncertainty in models of GrC, Wang and Gong [14] studied the construction of granular structures by means of fuzzy hypergraphs. They concluded that the representation of granules and partition is more efficient through the fuzzy hypergraphs. Novel applications and transversals of m-PF hypergraphs were defined by Akram and Sarwar [15,16]. Further, Akram and Shahzadi [17] studied various operations on m-PF hypergraphs. Akram and Luqman [18,19] introduced intuitionistic single-valued and bipolar neutrosophic hypergraphs. The basic purpose of this work is to develop an interpretation of granular structures using m-PF hypergraphs. In the proposed model, the vertex of the m-PF hypergraph denotes an object, and an m-PF hyperedge represents a granule. The "refinement" and "coarsening" operators are defined to switch the different granularities from coarser to finer and vice versa, respectively.

The rest of the paper is arranged as follows: In Section 2, some fundamental concepts of m-PF hypergraphs are reviewed. The uncertainty measures and information entropy of m-polar fuzzy hierarchical quotient space structure are discussed. An m-PFHQSS is developed based on the m-PF equivalence relation. In the next section, we construct an m-PF hypergraph model of GrC. The partition and covering of granules are defined in the same section. The method of bottom-up construction is explained by an algorithm. In Section 4, we construct a model of GrC using level hypergraphs based on the m-PF equivalence relation and explain the procedure of bottom-up construction through an example. Section 5 deals with some concluding remarks and future studies.

2. Fundamental Features of m-Polar Fuzzy Hypergraphs

In this section, we review some basic concepts from [9,11,15,16].

Definition 1. *An m-polar fuzzy set (m-PFS) M on a universal set Z is defined as a mapping* $M{:}Z \to [0,1]^m$. *The membership degree of each element* $z \in Z$ *is represented by* $M(z) = (\mathcal{P}_1 \circ M(z), \mathcal{P}_2 \circ M(z), \mathcal{P}_3 \circ M(z), \ldots, \mathcal{P}_m \circ M(z))$, *where* $\mathcal{P}_j \circ M(z) : [0,1]^m \to [0,1]$ *is defined as the jth projection mapping.*

Note that the mth power of $[0,1]$ (i.e., $[0,1]^m$) is regarded as a partially-ordered set with the point-wise order $\leq$, where m is considered as an ordinal number ($m = n|n < m$ when $m > 0$), $\leq$ is defined as $z_1 \leq z_2$ if and only if $\mathcal{P}_j(z_1) \leq \mathcal{P}_j(z_2)$, for every $1 \leq j \leq m$. $\mathbf{0} = (0,0,\cdots,0)$ and $\mathbf{1} = (1,1,\cdots,1)$ are the smallest and largest values in $[0,1]^m$, respectively.

Definition 2. *Let M be an m-PFS on Z. An m-polar fuzzy relation (m-PFR)* $N = (\mathcal{P}_1 \circ N, \mathcal{P}_2 \circ N, \mathcal{P}_3 \circ N, \ldots, \mathcal{P}_m \circ N)$ *on M is a mapping* $N : M \to M$ *such that* $N(z_1z_2) \leq \inf\{M(z_1), M(z_2)\}$, *for all* $z_1, z_2 \in Z$, *i.e., for each* $1 \leq j \leq m$, $\mathcal{P}_j \circ N(z_1z_2) \leq \inf\{\mathcal{P}_j \circ M(z_1), \mathcal{P}_j \circ M(z_2)\}$, *where* $\mathcal{P}_j \circ M(z)$ *and* $\mathcal{P}_j \circ N(z_1z_2)$ *denote the jth membership degree of an element* $z \in Z$ *and the pair* z_1z_2, *respectively.*

Definition 3. *An m-polar fuzzy graph (m-PFG) on Z is defined as an ordered pair of functions* $G = (C, D)$, *where* $C : Z \to [0,1]^m$ *is an m-polar vertex set and* $D : Z \times Z \to [0,1]^m$ *is an m-polar edge set of G such that* $D(wz) \leq \inf\{C(w), C(z)\}$, *i.e.,* $\mathcal{P}_j \circ D(wz) \leq \inf\{\mathcal{P}_j \circ C(w), \mathcal{P}_j \circ C(z)\}$, *for all* $w, z \in Z$ *and* $1 \leq j \leq m$.

Definition 4. *An m-polar fuzzy hypergraph (m-PFHG) on a non-empty set Z is a pair* $H = (A, B)$, *where* $A = \{M_1, M_2, \ldots, M_r\}$ *is a finite family of m-PFSs on Z and B is an m-PFR on m-PFSs* M_k *such that:*

- $B(E_k) = B(\{z_1, z_2, \cdots, z_l\}) \leq \inf\{M_k(z_1), M_k(z_2), \cdots, M_k(z_l)\}$,
- $\bigcup_{k=1}^{r} supp(M_k) = Z$, *for all* $M_k \in A$ *and for all* $z_1, z_2, \cdots, z_l \in Z$.

Definition 5. *Let* $H = (A, B)$ *be an m-PFHG and* $\tau \in [0,1]^m$. *Then, the* τ*-cut level set of an m-PFS M is defined as,*

$$M_\tau = \{z|\mathcal{P}_j \circ M(z) \geq t_j, 1 \leq j \leq m\}, \tau = (t_1, t_2, \cdots, t_m).$$

$H_\tau = (A_\tau, B_\tau)$ *is called a* τ*-cut level hypergraph of H, where* $A_\tau = \bigcup_{i=1}^{r} M_{i\tau}$.

For further concepts and applications, readers are referred to [20–23].

2.1. Uncertainty Measures of m-Polar Fuzzy Hierarchical Quotient Space Structure

The question of distinct membership degrees of the same object from different scholars has arisen because of various ways of thinking about the interpretation of different functions dealing with the same problem. To resolve this issue, FS was structurally defined by Zhang and Zhang [24], which was based on quotient space (QS) theory and the fuzzy equivalence relation (FER) [25]. This definition provides some new initiatives regarding membership degree, called a hierarchical quotient space structure (HQSS) of an FER. By following the same concept, we develop an HQSS of an m-polar FER.

Definition 6. *An m-polar fuzzy equivalence relation (PFER) on a non-empty finite set Z is called an m-PF similarity relation if it satisfies,*

1. $N(z,z) = (\mathcal{P}_1 \circ N(z,z), \mathcal{P}_2 \circ N(z,z), \ldots, \mathcal{P}_m \circ N(z,z)) = (1,1,\cdots,1)$, *for all* $z \in Z$,
2. $N(u,w) = (\mathcal{P}_1 \circ N(u,w), \mathcal{P}_2 \circ N(u,w), \ldots, \mathcal{P}_m \circ N(u,w)) = (\mathcal{P}_1 \circ N(w,u), \mathcal{P}_2 \circ N(w,u), \ldots, \mathcal{P}_m \circ N(w,u)) = N(w,u)$, *for all* $u, w \in Z$.

Definition 7. *An m-PFER on a non-empty finite set Z is called an m-polar fuzzy equivalence relation (m-PFER) if it satisfies the conditions,*

1. $N(z,z) = (\mathcal{P}_1 \circ N(z,z), \mathcal{P}_2 \circ N(z,z), \ldots, \mathcal{P}_m \circ N(z,z)) = (1,1,\cdots,1)$, *for all* $z \in Z$,
2. $N(u,w) = (\mathcal{P}_1 \circ N(u,w), \mathcal{P}_2 \circ N(u,w), \ldots, \mathcal{P}_m \circ N(u,w)) = (\mathcal{P}_1 \circ N(w,u), \mathcal{P}_2 \circ N(w,u), \ldots, \mathcal{P}_m \circ N(w,u)) = N(w,u)$, *for all* $u, w \in Z$,
3. *for all* $u, v, w \in Z$, $N(u,w) = \sup_{v \in Z}\{\min(N(u,v), N(v,w))\}$, *i.e.*,

$$\mathcal{P}_j \circ N(u,w) = \sup_{v \in Z}\{\min(\mathcal{P}_j \circ N(u,v), \mathcal{P}_j \circ N(v,w))\}, 1 \leq j \leq m.$$

Definition 8. *An m-polar fuzzy quotient space (m-PFQS) is denoted by a triplet* $(Z, \tilde{C}, N)$, *where Z is a finite domain,* $\tilde{C}$ *represents the attributes of Z, and N represents the m-PF relationship between the objects of universe Z, which is called the structure of the domain.*

Definition 9. *Let* z_i *and* z_j *be two objects in the universe Z. The similarity between* z_i, $z_j \in Z$ *having the attribute* $\tilde{c}_k$ *is defined as:*

$$N(z_i, z_j) = \frac{|\tilde{c}_{ik} \cap \tilde{c}_{jk}|}{|\tilde{c}_{ik} \cup \tilde{c}_{jk}|},$$

where $\tilde{c}_{ik}$ *represents that object* z_i *possesses the attribute* $\tilde{c}_k$ *and* $\tilde{c}_{jk}$ *represents that object* z_j *possesses the attribute* $\tilde{c}_k$.

Proposition 1. *Let N be an m-PFR on a finite domain Z and* $N_\tau = \{(x,w) | \mathcal{P}_j \circ N(x,w) \geq t_j, 1 \leq j \leq m\}$, $\tau = (t_1, t_2, \cdots, t_j) \in [0,1]$. *Then,* N_τ *is an ER on Z and is said to be the cut-equivalence relation of N.*

Proposition 1 represents that N_τ is a crisp relation, which is equivalence on Z, and its knowledge space is given as $\xi_{N_\tau}(X) = X/N_\tau$.

The value domain of an ER N on Z is defined as $D = \{N(w,y) | w, y \in Z\}$ such that:

$$\mathcal{P}_j \circ Z(w) \wedge \mathcal{P}_j \circ Z(y) \wedge \mathcal{P}_j \circ N(x,y) > 0, 1 \leq j \leq m.$$

Definition 10. *Let N be an m-PFER on a finite set Z and D be the value domain of N. The set given by* $\xi_Z(N) = \{Z/N_\tau | \tau \in D\}$ *is called the m-polar fuzzy hierarchical quotient space structure (m-PFHQS) of N.*

Example 1. *Let* $Z = \{w_1, w_2, w_3, w_4, w_5, w_6\}$ *be a finite set of elements and* N_1 *be a four-PFER on Z; the relation matrix* $\tilde{M}_{N_1}$ *corresponding to* N_1 *is given as follows:*

$$\tilde{M}_{N_1} = \begin{bmatrix} (1,1,1,1) & (0.4,0.4,0.5,0.5) & (0.5,0.5,0.4,0.4) & (0.5,0.5,0.4,0.4) & (0.5,0.5,0.4,0.4) & (0.5,0.5,0.4,0.4) \\ (0.4,0.4,0.5,0.5) & (1,1,1,1) & (0.8,0.8,0.9,0.9) & (0.8,0.8,0.6,0.6) & (0.8,0.8,0.6,0.6) & (0.6,0.6,0.5,0.5) \\ (0.5,0.5,0.4,0.4) & (0.8,0.8,0.9,0.9) & (1,1,1,1) & (0.6,0.6,0.7,0.7) & (0.6,0.6,0.7,0.7) & (0.6,0.6,0.5,0.5) \\ (0.5,0.5,0.4,0.4) & (0.8,0.8,0.6,0.6) & (0.6,0.6,0.7,0.7) & (1,1,1,1) & (0.7,0.8,0.7,0.8) & (0.6,0.6,0.5,0.5) \\ (0.5,0.5,0.4,0.4) & (0.8,0.8,0.6,0.6) & (0.6,0.6,0.7,0.7) & (0.7,0.8,0.7,0.8) & (1,1,1,1) & (0.6,0.6,0.5,0.5) \\ (0.5,0.5,0.4,0.4) & (0.6,0.6,0.5,0.5) & (0.6,0.6,0.5,0.5) & (0.6,0.6,0.5,0.5) & (0.6,0.6,0.5,0.5) & (1,1,1,1) \end{bmatrix}.$$

Its corresponding m-PFHQSS is given as,

$$\begin{aligned} Z/N_{1\tau_1} = Z/N_{1(t_1,t_2,t_3,t_4)} &= \{\{w_1, w_2, w_3, w_4, w_5, w_6\}\}, \\ Z/N_{1\tau_2} = Z/N_{1(t'_1,t'_2,t'_3,t'_4)} &= \{\{w_1\}, \{w_2, w_3, w_4, w_5, w_6\}\}, \\ Z/N_{1\tau_3} = Z/N_{1(t''_1,t''_2,t''_3,t''_4)} &= \{\{w_1\}, \{w_2, w_3, w_4, w_5\}, \{w_6\}\}, \\ Z/N_{1\tau_4} = Z/N_{1(t'''_1,t'''_2,t'''_3,t'''_4)} &= \{\{w_1\}, \{w_2, w_3\}, \{w_4, w_5\}, \{w_6\}\}, \\ Z/N_{1\tau_5} = Z/N_{1(t''''_1,t''''_2,t''''_3,t''''_4)} &= \{\{w_1\}, \{w_2, w_3\}, \{w_4\}, \{w_5\}, \{w_6\}\}, \\ Z/N_{1\tau_6} = Z/N_{1(t'''''_1,t'''''_2,t'''''_3,t'''''_4)} &= \{\{w_1\}, \{w_2\}, \{w_3\}, \{w_4\}, \{w_5\}, \{w_6\}\}, \end{aligned}$$

where:

$$
\begin{aligned}
0 < \tau_1 &= (t_1, t_2, t_3, t_4) \leq 0.4, \\
0.4 < \tau_2 &= (t_1', t_2', t_3', t_4') \leq 0.5, \\
0.5 < \tau_3 &= (t_1'', t_2'', t_3'', t_4'') \leq 0.6, \\
0.6 < \tau_4 &= (t_1''', t_2''', t_3''', t_4''') \leq 0.7, \\
0.7 < \tau_5 &= (t_1'''', t_2'''', t_3'''', t_4'''') \leq 0.8, \\
0.8 < \tau_6 &= (t_1''''', t_2''''', t_3''''', t_4''''') \leq 1.
\end{aligned}
$$

Hence, a four-PFQSS is given as $\xi_{Z(N_1)} = \{Z/N_{\tau_1}, Z/N_{\tau_2}, Z/N_{\tau_3}, Z/N_{\tau_4}, Z/N_{\tau_5}, Z/N_{\tau_6}\}$ *and is shown in Figure 1.*

It is worth noting that the same HQSS can be formed by different four-PFERs. For instance, the relation matrix $\tilde{M}_{N_2}$ *of four-PFER generates the same HQSS as given by* $\tilde{M}_{N_1}$*. The relation matrix* $\tilde{M}_{N_2}$ *is given as,*

$$
\tilde{M}_{N_1} = \begin{bmatrix}
(1,1,1,1) & (0.2,0.2,0.5,0.5) & (0.6,0.6,0.4,0.4) & (0.6,0.6,0.4,0.4) & (0.6,0.6,0.4,0.4) & (0.6,0.6,0.4,0.4) \\
(0.2,0.2,0.5,0.5) & (1,1,1,1) & (0.2,0.2,0.5,0.5) & (0.2,0.2,0.5,0.5) & (0.2,0.2,0.5,0.5) & (0.2,0.2,0.5,0.5) \\
(0.6,0.6,0.4,0.4) & (0.2,0.2,0.5,0.5) & (1,1,1,1) & (0.7,0.7,0.7,0.7) & (0.7,0.7,0.7,0.7) & (0.7,0.7,0.7,0.7) \\
(0.6,0.6,0.4,0.4) & (0.2,0.2,0.5,0.5) & (0.7,0.7,0.7,0.7) & (1,1,1,1) & (0.8,0.8,0.7,0.8) & (0.6,0.6,0.5,0.5) \\
(0.6,0.6,0.4,0.4) & (0.2,0.2,0.5,0.5) & (0.7,0.7,0.7,0.7) & (0.8,0.8,0.7,0.8) & (1,1,1,1) & (0.6,0.6,0.5,0.5) \\
(0.6,0.6,0.4,0.4) & (0.2,0.2,0.5,0.5) & (0.7,0.7,0.7,0.7) & (0.6,0.6,0.5,0.5) & (0.6,0.6,0.5,0.5) & (1,1,1,1)
\end{bmatrix}.
$$

Furthermore, assuming the number of blocks in every distinct layer of this HQSS, a pyramid model can also be constructed as shown in Figure 2.

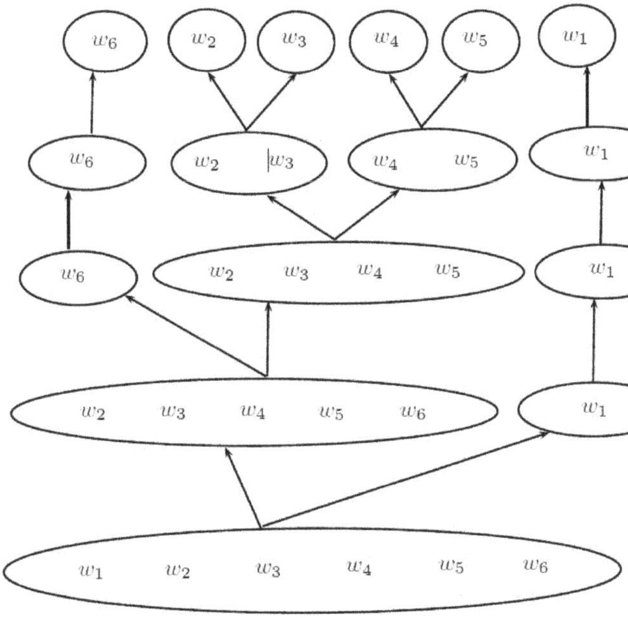

Figure 1. A four-polar fuzzy hierarchical quotient space structure.

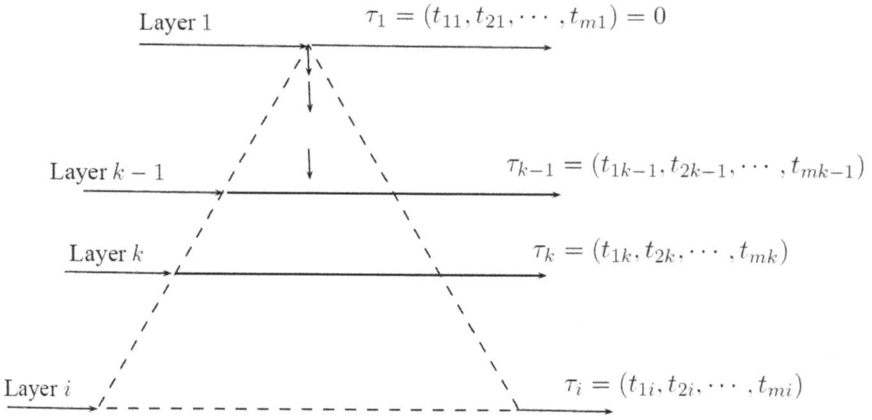

Figure 2. Pyramid model of m-PFHQSS.

2.2. Information Entropy of m-PFHQSS

Definition 11. *Let N be an m-PFER on Z. Let* $\xi_Z(N) = \{Z(\tau_1), Z(\tau_2), Z(\tau_3), \cdots, Z(\tau_j)\}$ *be its corresponding HQSS, where* $\tau_i = (t_{1i}, t_{2i}, \cdots, t_{mi})$, $i = 1, 2, \cdots, j$ *and* $Z(\tau_j) < Z(\tau_{j-1}) < \cdots < Z(\tau_1)$. *Then, the partition sequence of* $\xi_Z(N)$ *is given as* $P(\xi_Z(N)) = \{P_1, P_2, P_3, \cdots, P_j\}$, *where* $P_i = |Z(\tau_i)|$, $i = 1, 2, \cdots, j$, *and* $|.|$ *denotes the number of elements in a set.*

Definition 12. *Let N be an m-PFER on Z. Let* $\xi_Z(N) = \{Z(\tau_1), Z(\tau_2), Z(\tau_3), \cdots, Z(\tau_j)\}$ *be its corresponding HQSS, where* $\tau_i = (t_{1i}, t_{2i}, \cdots, t_{mi})$, $i = 1, 2, \cdots, j$, *and* $Z(\tau_j) < Z(\tau_{j-1}) < \cdots < Z(\tau_1)$, $P(\xi_Z(N)) = \{P_1$, *and* $P_2, \cdots, P_j\}$ *is the partition sequence of* $\xi_Z(N)$. *Assume that* $Z(\tau_i) = \{Z_{i1}, Z_{i2}, \cdots, Z_{iP_i}\}$. *The information entropy* $E_{Z(\tau_i)}$ *is defined as* $E_{Z(\tau_i)} = -\sum_{r=1}^{P_i} \frac{|Z_{ir}|}{|Z|} ln\left(\frac{|Z_{ir}|}{|Z|}\right)$.

Theorem 1. *Let N be an m-PFER on Z. Let* $\xi_Z(N) = \{Z(\tau_1), Z(\tau_2), Z(\tau_3), \cdots, Z(\tau_j)\}$ *be its corresponding HQSS, where* $\tau_i = (t_{1i}, t_{2i}, \cdots, t_{mi})$, $i = 1, 2, \cdots, j$, *then the entropy sequence* $E(\xi_Z(N)) = \{E_{Z(\tau_1)}, E_{Z(\tau_2)}, \cdots, E_{Z(\tau_j)}\}$ *increases monotonically and strictly.*

Proof. The terminology of HQSS implies that $Z(\tau_j) < Z(\tau_{j-1}) < \cdots < Z(\tau_1)$, i.e., $Z(\tau_{j-1})$ is a quotient subspace of $Z(\tau_j)$. Suppose that $Z(\tau_i) = \{Z_{i1}, Z_{i2}, \cdots, Z_{iP_i}\}$ and $Z(\tau_{i-1}) = \{Z_{(i-1)1}, Z_{(i-1)2}, \cdots, Z_{(i-1)P_{(i-1)}}\}$, then every sub-block of $Z(\tau_{i-1})$ is an amalgam of sub-blocks of $Z(\tau_i)$. WLOG, it is assumed that only one sub-block $Z_{i-1,j}$ in $Z(\tau_{i-1})$ is formed by the combination of two sub-blocks Z_{ir}, Z_{is} in $Z(\tau_i)$, and all other remaining blocks are equal in both sequences. Thus,

$$\begin{aligned} E_{Z(\tau_{j-1})} &= -\sum_{r=1}^{P_{i-1}} \frac{|Z_{i-1,r}|}{|Z|} \ln\left(\frac{|Z_{i-1,r}|}{|Z|}\right) \\ &= -\sum_{r=1}^{P_{j-1}} \frac{|Z_{i-1,r}|}{|Z|} \ln\left(\frac{|Z_{i-1,r}|}{|Z|}\right) - \sum_{r=j+1}^{P_{i-1}} \frac{|Z_{i-1,r}|}{|Z|} \ln\left(\frac{|Z_{i-1,r}|}{|Z|}\right) - \frac{|Z_{i-1,j}|}{|Z|} \ln\left(\frac{|Z_{i-1,j}|}{|Z|}\right) \\ &= -\sum_{r=1}^{P_{j-1}} \frac{|Z_{i,r}|}{|Z|} \ln\left(\frac{|Z_{i,r}|}{|Z|}\right) - \sum_{r=j+1}^{P_i} \frac{|Z_{i,r}|}{|Z|} \ln\left(\frac{|Z_{i,r}|}{|Z|}\right) - \frac{|Z_{i,r}| + |Z_{i,s}|}{|Z|} \ln\left(\frac{|Z_{i,r}| + |Z_{i,s}|}{|Z|}\right). \end{aligned}$$

Since:

$$\begin{aligned}\frac{|Z_{i,r}|+|Z_{i,s}|}{|Z|}\ln\left(\frac{|Z_{i,r}|+|Z_{i,s}|}{|Z|}\right) &= \frac{|Z_{i,r}|}{|Z|}\ln\left(\frac{|Z_{i,r}|+|Z_{i,s}|}{|Z|}\right)+\frac{|Z_{i,s}|}{|Z|}\ln\left(\frac{|Z_{i,r}|+|Z_{i,s}|}{|Z|}\right)\\ &> \frac{|Z_{i,r}|}{|Z|}\ln\left(\frac{|Z_{i,r}|}{|Z|}\right)+\frac{|Z_{i,s}|}{|Z|}\ln\left(\frac{|Z_{i,s}|}{|Z|}\right).\end{aligned}$$

Therefore, we have:

$$\begin{aligned}E_{Z(\tau_{j-1})} &< -\sum_{r=1}^{P_{j-1}}\frac{|Z_{i,r}|}{|Z|}\ln\left(\frac{|Z_{i,r}|}{|Z|}\right)-\sum_{r=j+1}^{P_i}\frac{|Z_{i,r}|}{|Z|}\ln\left(\frac{|Z_{i,r}|}{|Z|}\right)-\frac{|Z_{i,r}|}{|Z|}\ln\left(\frac{|Z_{i,r}|}{|Z|}\right)-\frac{|Z_{i,s}|}{|Z|}\ln\left(\frac{|Z_{i,s}|}{|Z|}\right)\\ &= E_{Z(\tau_j)},(2\le j\le n).\end{aligned}$$

Hence, $E_{Z(\tau_1)} < E_{Z(\tau_2)} < E_{Z(\tau_2)} < \cdots < E_{Z(\tau_j)}$. □

Definition 13. *Let $Z=\{s_1,s_2,s_3,\cdots,s_n\}$ be a non-empty set of the universe, and let $P_d(Z)=\{Z_1,Z_2,Z_3,\cdots,Z_d\}$ be a partition space (PS) of Z, where $|P_d(Z)|=d$, then $P_d(Z)$ is called the d-order partition space(d-OPS) on Z.*

Definition 14. *Let Z be a finite non-empty universe, and let $P_d(Z)=\{Z_1,Z_2,Z_3,\cdots,Z_d\}$ be a d-OPS on Z. Let $|Z_1|=l_1,|Z_2|=l_2,\cdots,|Z_d|=l_d$, and the sequence $\{l_1,l_2,\cdots,l_d\}$ is arranged in increasing order, then we got a new sequence $\chi(d)=\{l'_1,l'_2,\cdots,l'_d\}$, which is also increasing and called a sub-block sequence of $P_d(Z)$.*

Note that two different d-OPSs on Z may possess the similar sub-block sequence $\chi(d)$.

Definition 15. *Let Z be a finite non-empty universe, and let $P_d(Z)=\{Z_1,Z_2,Z_3,\cdots,Z_d\}$ be a partition space of Z. Suppose that $\chi_1(d)=\{l'_1,l'_2,\cdots,l'_d\}$ is a sub-block sequence of $P_d(Z)$, then the ω-displacement of $\chi_1(d)$ is defined as an increasing sequence $\chi_2(d)=\{l'_1,l'_2,\cdots,l'_r+1,\cdots,l'_s-1,\cdots,l'_d\}$, where $r<s$, $l'_r+1<l'_s-1$.*

An ω-displacement is obtained by subtracting one from some bigger term and adding one to some smaller element such that the sequence keeps its increasing property.

Theorem 2. *A single time ω-displacement $\chi_2(d)$, which is derived from $\chi_1(d)$, satisfies $E(\chi_1(d))<E(\chi_2(d))$.*

Proof. Let $\chi_1(d)=\{l'_1,l'_2,\cdots,l'_d\}$ and $\chi_2(d)=\{l'_1,l'_2,\cdots,l'_r+1,\cdots,l'_s-1,\cdots,l'_d\}$, $l'_1+l'_2+\cdots+l'_d=k$, then we have:

$$E(\chi_2(t))=-\sum_{j=1}^{d}\frac{l'_l}{k}\ln\frac{l'_l}{k}+\frac{l'_r}{k}\ln\frac{l'_r}{k}+\frac{l'_s}{k}\ln\frac{l'_s}{k}-\frac{l'_r+1}{k}\ln\frac{l'_r+1}{k}-\frac{l'_s-1}{k}\ln\frac{l'_s-1}{k}.$$

Let $g(z)=-\frac{z}{k}\ln\frac{z}{k}-\frac{l-z}{k}\ln\frac{l-z}{k}$, where $l=l'_r+l'_s$ and $g'(z)=\frac{1}{k}\ln\frac{l-z}{z}$. Suppose that $g'(z)=0$, then we obtain a solution, i.e., $z=\frac{l}{2}$. Furthermore, $g''(z)=\frac{-l}{k(l-z)z}<0$, $0\le z\le\frac{l}{2}$, and $g(z)$ is increasing monotonically. Let $z_1=l'_r$ and $z_2=l'_r+1$, $l'_r+1<l'_s-1$, i.e., $z_1<z_2\le\frac{l}{2}=\frac{l'_r+l'_s}{2}$. Since $g(z)$ is monotone, then $g(z_2)-g(z_1)>0$. Thus,

$$\frac{l'_r}{k}\ln\frac{l'_r}{k}+\frac{l'_s}{k}\ln\frac{l'_s}{k}-\frac{l'_r+1}{k}\ln\frac{l'_r+1}{k}-\frac{l'_s-1}{k}\ln\frac{l'_s-1}{k}>0.$$

Hence,

$$
\begin{aligned}
E(\chi_2(d)) &= -\sum_{j=1}^{d} \frac{l'_l}{k}\ln\frac{l'_l}{k} + \frac{l'_r}{k}\ln\frac{l'_r}{k} + \frac{l'_s}{k}\ln\frac{l'_s}{k} - \frac{l'_r+1}{k}\ln\frac{l'_r+1}{k} - \frac{l'_s-1}{k}\ln\frac{l'_s-1}{k} \\
&> -\left(\frac{l'_r+1}{k}\ln\frac{l'_r+1}{k} + \frac{l'_s-1}{k}\ln\frac{l'_s-1}{k}\right) \\
&> -\sum_{j=1}^{t} \frac{l'_l}{k}\ln\frac{l'_l}{k} \\
&= E(\chi_1(d)).
\end{aligned}
$$

This completes the proof. □

3. An *m*-Polar Fuzzy Hypergraph Model of Granular Computing

A granule is defined as a collection of objects or elements having thhe same attributes or characteristics and can be treated as a single unit.

Definition 16. *An object space is defined as a system (Z, N), where Z is a universe of objects or elements and $N = \{n_1, n_2, n_3, \cdots, n_k\}$, $k = |Z|$ is a family of relations between the elements of Z. For $r \leq k$, $n_r \in N$, $n_r \subseteq Z \times Z \times \cdots \times Z$, if $(z_1, z_2, \cdots, z_r) \subseteq n_r$, then there exists an r-array relation n_r on $(z_1, z_2, \cdots, z_n)$.*

A granule is affiliated with a particular level. The whole view of granules at every level can be taken as a complete description of a particular problem at that level of granularity [6]. An *m*-PFHG formed by the set of relations N and membership degrees $Z(w) = \mathcal{P}_j \circ Z(w)$, $1 \leq j \leq m$ of objects in the space is considered as a specific level of the GrC model. All *m*-PF hyperedges in that *m*-PFHG can be regarded as the complete granule at that particular level.

Definition 17. *A partition of a set Z established on the basis of relations between objects is defined as a collection of non-empty subsets, which are pairwise disjoint and whose union is the whole of Z. These subsets that form the partition of Z are called blocks. Every partition of a finite set Z contains the finite number of blocks. Corresponding to the m-PFHG, the constraints of partition $\psi = \{\mathcal{E}_i | 1 \leq i \leq n\}$ can be stated as follows,*

- *each $\mathcal{E}_i$ is non-empty,*
- *for $i \neq j$, $\mathcal{E}_i \cap \mathcal{E}_j = \emptyset$,*
- $\cup\{supp(\mathcal{E}_i) | 1 \leq i \leq n\} = Z$.

Definition 18. *A covering of a set Z is defined as a collection of non-empty subsets whose union is the whole of Z. The conditions for the covering $c = \{\mathcal{E}_i | 1 \leq i \leq n\}$ of Z are stated as,*

- *each $\mathcal{E}_i$ is non-empty,*
- $\cup\{supp(\mathcal{E}_i) | 1 \leq i \leq n\} = Z$.

The corresponding definitions in classical hypergraph theory are completely analogous to the above Definitions 17 and 18. In a crisp hypergraph, if the hyperedges E_i and E_j do not intersect each other, i.e., E_i, $E_j \in E$, and $E_i \cap E_j = \emptyset$, then these hyperedges form a partition of granules at this level. Furthermore, if E_i, $E_j \in E$, and $E_i \cap E_j \neq \emptyset$, i.e., the hyperedges E_i and E_j intersect each other, then these hyperedges form a covering at this level.

Example 2. *Let $Z = \{w_1, w_2, w_3, w_4, w_5, w_6, w_7, w_8, w_9, w_{10}\}$. The partition and covering of Z is given in Figures 3 and 4, respectively.*

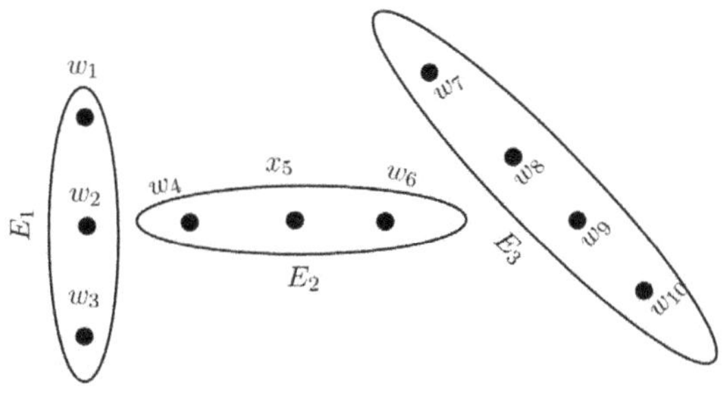

Figure 3. A partition of granules at a level.

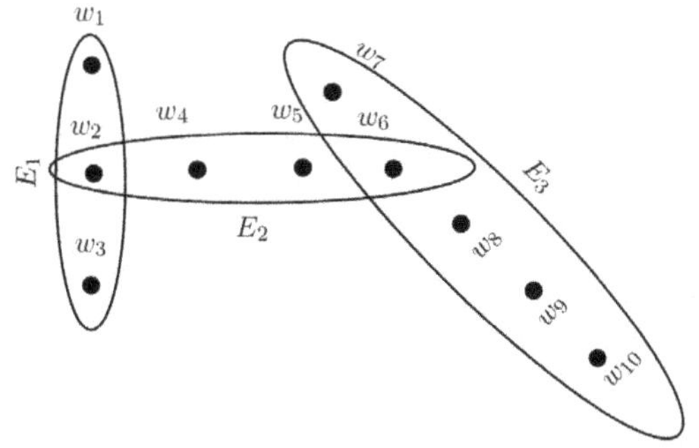

Figure 4. A covering of granules at a level.

A set-theoretic way to study the GrC model uses the following operators in an m-PFHG model.

Definition 19. *Let $\mathcal{G}_1$ and $\mathcal{G}_2$ be two granules in our model and the m-PF hyperedges, and $\mathcal{E}_1$, $\mathcal{E}_2$ represent their external properties. The union of two granules $\mathcal{G}_1 \cup \mathcal{G}_2$ is defined as a larger m-PF hyperedge that contains the vertices of both $\mathcal{E}_1$ and $\mathcal{E}_2$. If $w_i \in \mathcal{G}_1 \cup \mathcal{G}_2$, then the membership degree $\mathcal{G}_1 \cup \mathcal{G}_2(w_i)$ of w_i in larger granule $\mathcal{G}_1 \cup \mathcal{G}_2$ is defined as follows:*

$$\mathcal{P}_j \circ (\mathcal{G}_1 \cup \mathcal{G}_2)(w_i) = \begin{cases} \max\{\mathcal{P}_j \circ (\mathcal{E}_1)(w_i), \mathcal{P}_j \circ (\mathcal{E}_2)(w_i)\}, & \text{if} \quad w_i \in \mathcal{E}_1 \text{ and } w_i \in \mathcal{E}_2, \\ \mathcal{P}_j \circ (\mathcal{E}_1)(w_i), & \text{if} \quad w_i \in \mathcal{E}_1 \text{ and } w_i \notin \mathcal{E}_2, \\ \mathcal{P}_j \circ (\mathcal{E}_2)(w_i), & \text{if} \quad w_i \in \mathcal{E}_2 \text{ and } w_i \notin \mathcal{E}_1, \end{cases}$$

$1 \leq j \leq m$.

Definition 20. *Let $\mathcal{G}_1$ and $\mathcal{G}_2$ be two granules in our model and the m-PF hyperedges, and $\mathcal{E}_1$, $\mathcal{E}_2$ represent their external properties. The union of two granules $\mathcal{G}_1 \cap \mathcal{G}_2$ is defined as a larger m-PF hyperedge that contains the vertices of both $\mathcal{E}_1$ and $\mathcal{E}_2$. If $w_i \in \mathcal{G}_1 \cup \mathcal{G}_2$, then the membership degree $\mathcal{G}_1 \cap \mathcal{G}_2(w_i)$ of w_i in smaller granule $\mathcal{G}_1 \cap \mathcal{G}_2$ is defined as follows:*

$$\mathcal{P}_j \circ (\mathcal{G}_1 \cap \mathcal{G}_2)(w_i) = \begin{cases} \min\{\mathcal{P}_j \circ (\mathcal{E}_1)(w_i), \mathcal{P}_j \circ (\mathcal{E}_2)(w_i)\}, & \text{if} \quad w_i \in \mathcal{E}_1 \text{ and } w_i \in \mathcal{E}_2, \\ \mathcal{P}_j \circ (\mathcal{E}_1)(w_i), & \text{if} \quad w_i \in \mathcal{E}_1 \text{ and } w_i \notin \mathcal{E}_2, \\ \mathcal{P}_j \circ (\mathcal{E}_2)(w_i), & \text{if} \quad w_i \in \mathcal{E}_2 \text{ and } w_i \notin \mathcal{E}_1, \end{cases}$$

$1 \leq j \leq m$.

Definition 21. *Let $\mathcal{G}_1$ and $\mathcal{G}_2$ be two granules in our model and the m-PF hyperedges $\mathcal{E}_1$, $\mathcal{E}_2$ represent their external properties. The difference between two granules $\mathcal{G}_1 - \mathcal{G}_2$ is defined as a smaller m-PF hyperedge that contains those vertices belonging to $\mathcal{E}_1$, but not to $\mathcal{E}_2$.*
Note that, if a vertex $w_i \in \mathcal{E}_1$ and $w_i \notin \mathcal{E}_2$, then:

$$\mathcal{P}_j \circ (\mathcal{E}_1)(w_i) > 0 \text{ and } \mathcal{P}_j \circ (\mathcal{E}_2)(w_i) = 0, 1 \leq j \leq m.$$

Definition 22. *A granule $\mathcal{G}_1$ is said to be the sub-granule of $\mathcal{G}_2$, if each vertex w_i of $\mathcal{E}_1$ also belongs to $\mathcal{E}_2$, i.e., $\mathcal{E}_1 \subseteq \mathcal{E}_2$. In such a case, $\mathcal{G}_2$ is called the super-granule of $\mathcal{G}_1$.*

Note that, if $\mathcal{E}(w_i) = \{0,1\}$, then all the above described operators are reduced to the classical hypergraph theory of GrC.

Formation of Hierarchical Structures

We can interpret a problem at distinct levels of granularities. These granular structures at different levels produce a set of m-PFHGs. The upper set of these hypergraphs constructs a hierarchical structure at distinct levels. The relationships between granules are expressed by the lower level, which represents the problem as a concrete example of granularity. The relationships between granule sets are expressed by the higher level, which represents the problem as an abstract example of granularity. Thus, the single-level structures can be constructed and then can be subdivided into hierarchical structures using the relational mappings between different levels.

Definition 23. *Let $H^1 = (A^1, B^1)$ and $H^2 = (A^2, B^2)$ be two m-PFHGs. In a hierarchy structure, their level cuts are H^1_τ and H^2_τ, respectively, where $\tau = (t_1, t_2, \cdots, t_m)$. Let $\tau \in [0,1]$ and $\mathcal{P}_j \circ \mathcal{E}^1_i \geq t_j, 1 \leq j \leq m$, where $\mathcal{E}^1_i \in B^1$, then a mapping $\phi : H^1_\tau \to H^2_\tau$ from H^1_τ to H^2_τ maps the $\mathcal{E}^1_{\tau_i}$ in H^1_τ to a vertex w^2_i in H^2_τ. Furthermore, the mapping $\phi^{-1} : H^2_\tau \to H^1_\tau$ maps a vertex w^2_i in H^2_τ to the τ-cut of the m-PF hyperedge $\mathcal{E}^1_{\tau_i}$ in $H^1_{\tau_i}$. It can be denoted as $\phi(\mathcal{E}^1_{\tau_i}) = w^2_i$ or $\phi^{-1}(w^2_i) = \mathcal{E}^1_{\tau_i}$, for $1 \leq i \leq n$.*

In an m-PFHG model, the mappings are used to describe the relations among different levels of granularities. At each distinct level, the problem is interpreted w.r.t the m-PF granularity of that level. The mapping associates the different descriptions of the same problem at distinct levels of granularities. There are two fundamental types to construct the method of hierarchical structures, the top-down construction procedure and the bottom-up construction procedure [26].

A formal discussion is provided to interpret an m-PFHG model in GrC, which is more compatible with human thinking. Zhang and Zhang [25] highlighted that one of the most important and acceptable characteristic of human intelligence is that the same problem can be viewed and analyzed at different granularities. Their claim is that the problem can not only be solved using various worlds of granularities, but also can be switched easily and quickly. Hence, the procedure of solving a problem can be considered as the calculations at different hierarchies within that model

A multi-level granularity of the problem is represented by an m-PFHG model, which allows the problem solvers to decompose it into various minor problems and transform it into other granularities. The transformation of the problem in other granularities is performed by using two operators, i.e., zooming-in and zooming-out operators. The transformation from the weaker level to the finer level of granularity is done by the zoom-in operator, and the zoom-out operator deals with the shifting of the problem from coarser to finer granularity.

Definition 24. *Let $H^1 = (A^1, B^1)$ and $H^2 = (A^2, B^2)$ be two m-PFHGs, which are considered as two levels of hierarchical structures, and H^2 owns a coarser granularity than H^1. Suppose $H^1_\tau = (Z^1, E^1_\tau)$ and $H^2_\tau = (Z^2, E^2_\tau)$ are the corresponding τ-level hypergraphs of H^1 and H^2, respectively. Let $e^1_i \in E^1_\tau$, $z^1_j \in Z^1$, $e^2_j \in E^2_\tau$, $z^2_l, z^2_m \in Z^2$, and $z^2_l, z^2_m \in e^2_j$. If $\phi(e^1_i) = z^2_l$, then $n(z^1_j, z^2_m)$ is the relationship between z^1_j, and z^2_m is obtained by the characteristics of granules.*

Definition 25. *Let the hyperedge $\phi^{-1}(z_l)$ be a vertex at a new level and the relation between hyperedges in this level be the same as that of relationship between vertices at the previous level. This is called the zoom-in operator and transforms a weaker level to a stronger level. The function $r(z_j^1, z_m^2)$ defines the relation between vertices of the original level, as well as the new level.*

Let the vertex $\phi(e_i)$ be a hyperedge at a new level, and the relation between vertices at this level is same as that of the relationship between hyperedges at the corresponding level. This is called the zoom-out operator and transforms a finer level to a coarser level.

By using these zoom-in and zoom-out operators, a problem can be viewed at multiple levels of granularities. These operations allow us to solve the problem more appropriately, and granularity can be switched easily at any level of problem solving.

In an m-PFHG model of GrC, the membership degrees of elements reflect the actual situation more efficiently, and a wide variety of complicated problems in uncertain and vague environments can be presented by means of m-PFHGs. The previous analysis concluded that this model of GrC generalizes the classical hypergraph model and fuzzy hypergraph model.

Definition 26 ([6]). *Let H^1 and H^2 be two crisp hypergraphs. Suppose that H^1 owns a finer m-PF granularity than H^2. A mapping from H^1 to H^2 $\psi : H^1 \to H^2$ maps a hyperedge of H^1 to the vertex of H^2, and the mapping $\psi^{-1} : H^2 \to H^1$ maps a vertex of H^2 to the hyperedge of H^1.*

The procedure of bottom-up construction for the level hypergraph model is illustrated in Algorithm 1.

Algorithm 1: The procedure of bottom-up construction for the level hypergraph model.

Step 1. Determine an m-PFER matrix according to the actual circumstances.
Step 2. For fixed $\tau \in [0, 1]$, obtain the corresponding HQSS.
Step 3. Obtain the hyperedges through the HQSS.
Step 4. Granules at the i-level are mapped to the $(i+1)$-level.
Step 5. Calculate the m-polar fuzzy relationships between the vertices of the $(i+1)$-level, and determine the m-PFER matrix.
Step 6. Determine the corresponding HQSS according to τ, which is fixed in Step 2.
Step 7. Get the hyperedges at the $(i+1)$-level, and the $(i+1)$-level of the model is constructed.
Step 8. Step 1–Step 5 are repeated until the whole universe is formulated to a single granule.

Definition 27. *Let N be an m-PFER on Z. A coarse-gained universe Z/N_τ can be obtained by using m-PFER, where $[w_i]_{N_\tau} = \{w_j \in Z | w_i N w_j\}$. This equivalence class $[w_i]_{N_\tau}$ is considered as an hyperedge in the level hypergraph.*

Definition 28. *Let $H_1 = (Z_1, E_1)$ and $H_2 = (Z_2, E_2)$ be level hypergraphs of m-PFHGs, and H_2 has weaker granularity than H_1. Suppose that e_i^1, $e_j^2 \in E_1$ and w_i^2, $w_j^2 \in Z_2$, $i, j = 1, 2, \cdots, n$. The zoom-in operator $\omega : H_2 \to H_1$ is defined as $\omega(w_i^2) = e_i^1$, $e_i^1 \in E_1$. The relations between the vertices of H_2 define the relationships among the hyperedges at the new level. The zoom-in operator of two levels is shown in Figure 5.*

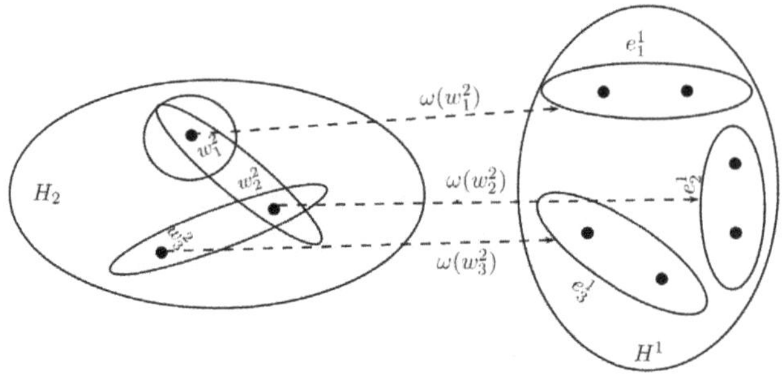

Figure 5. Zoom-in operator.

Remark 1. *For all $Z_2', Z_2'' \subseteq Z_2$, we have $\omega(Z_2') = \bigcup_{w_i^2\in Z_2'} \omega(w_i^2)$ and $\omega(Z_2'') = \bigcup_{w_j^2\in Z_2''} \omega(w_j^2)$.*

Theorem 3. *Let $H_1 = (Z_1, E_1)$ and $H_2 = (Z_2, E_2)$ be two levels and $\omega : H_2 \to H_1$ be the zoom-in operator. Then, for all $Z_2', Z_2'' \subseteq Z_2$, the zoom-in operator satisfies,*

(i) ω maps the empty set to an empty set, i.e., $\omega(\varnothing) = \varnothing$,
(ii) $\omega(Z_2) = E_1$,
(iii) $\omega([Z_2']^c) = [\omega(Z_2')]^c$,
(iv) $\omega(Z_2' \cap Z_2'') = \omega(Z_2') \cap \omega(Z_2'')$,
(v) $\omega(Z_2' \cup Z_2'') = \omega(Z_2') \cup \omega(Z_2'')$,
(vi) $Z_2' \subseteq Z_2''$ if and only if $\omega(Z_2') \subseteq \omega(Z_2'')$.

Proof. (i) It is trivially satisfied that $\omega(\varnothing) = \varnothing$.

(ii) As we know that for all $w_i^2 \in Z_2$, we have $\omega(Z_2') = \bigcup_{w_i^2\in Z_2'} \omega(w_i^2)$, since $\omega(w_i^2) = e_i^1$, we have $\omega(Z_2') = \bigcup_{w_i^2\in Z_2'} \omega(w_i^2) = \bigcup_{e_i^1\in E_1} e_i^1 = E_1$.

(iii) Let $[Z_2']^c = Z_2'$ and $[Z_2'']^c = Z_2''$, then it is obvious that $Z_2' \cap Z_2' = \varnothing$ and $Z_2' \cup Z_2' = Z_2$. It follows from (ii) that $\omega(Z_2) = E_1$, and we denote by W_1' that edge set of H_1 on which the vertex set Z_2' of H_2 is mapped under ω, i.e., $\omega(Z_2') = W_1'$. Then, $\omega([Z_2']^c) = \omega(Z_2') = \bigcup_{w_i^2\in Z_2'} \omega(w_i^2) = \bigcup_{e_i^1\in W_1'} e_i^1 = Z_1'$ and $[\omega(Z_2')]^c = [\bigcup_{w_j^2\in Z_2'} \omega(w_j^2)]^c = [\bigcup_{e_j^1\in E_1'} e_j^1]^c = (E_1')^c$. Since the relationship between hyperedges at the new level is the same as that of relations among vertices at the original level, we have $(E_1')^c = Z_1'$. Hence, we conclude that $\omega([Z_2']^c) = [\omega(Z_2')]^c$.

(iv) Assume that $Z_2' \cap Z_2'' = \tilde{Z}_2$, then for all $w_i^2 \in \tilde{Z}_2$ implies that $w_i^2 \in Z_2'$ and $w_i^2 \in Z_2''$. Further, we have $\omega(Z_2' \cap Z_2'') = \omega(\tilde{Z}_2) = \bigcup_{w_i^2\in \tilde{Z}_2} \omega(w_i^2) = \bigcup_{e_i^1\in \tilde{E}_1} \omega(e_i^1) = \tilde{E}_1$.
$\omega(Z_2') \cap \omega(Z_2'') = \{\bigcup_{w_i^2\in Z_2'} \omega(w_i^2)\} \cap \{\bigcup_{w_j^2\in Z_2''} \omega(w_j^2)\} = \bigcup_{e_i^1\in E_1'} e_i^1 \cap \bigcup_{e_j^1\in E_1''} e_j^1 = E_1' \cap E_1''$. Since the relationship between hyperedges at the new level is the same as that of relations among vertices at the original level, we have $E_1' \cap E_1'' = \tilde{E}_1$. Hence, we conclude that $\omega(Z_2' \cap Z_2'') = \omega(Z_2') \cap \omega(Z_2'')$.

(v) Assume that $Z_2' \cup Z_2'' = \tilde{Z}_2$. Then, we have $\omega(Z_2' \cup Z_2'') = \omega(\tilde{Z}_2) = \bigcup_{w_i^2\in \tilde{Z}_2} \omega(w_i^2) = \bigcup_{e_i^1\in \tilde{E}_1} \omega(e_i^1) = \tilde{E}_1$.
$\omega(Z_2') \cup \omega(Z_2'') = \{\bigcup_{w_i^2\in Z_2'} \omega(w_i^2)\} \cup \{\bigcup_{w_j^2\in Z_2''} \omega(w_j^2)\} = \bigcup_{e_i^1\in E_1'} e_i^1 \cup \bigcup_{e_j^1\in E_1''} e_j^1 = E_1' \cup E_1''$. Since, the relationship between hyperedges at the new level is the same as that of the relations among

the vertices at the original level, we have $E_1' \cup E_1'' = \bar{E}_1$. Hence, we conclude that $\omega(Z_2' \cup Z_2'')= \omega(Z_2') \cup \omega(Z_2'')$.

(vi) First we show that $Z_2' \subseteq Z_2''$ implies that $\omega(Z_2') \subseteq \omega(Z_2'')$. Since $Z_2' \subseteq Z_2''$, which implies that $Z_2' \cap Z_2'' = Z_2'$ and $\omega(Z_2') = \bigcup_{w_i^2 \in Z_2'} \omega(w_i^2) = \bigcup_{e_i^1 \in E_1'} e_i^1 = E_1'$. Furthermore, $\omega(Z_2'') = \bigcup_{w_j^2 \in Z_2''} \omega(w_j^2) = \bigcup_{e_j^1 \in E_1''} e_j^1 = E_1''$. Since the relationship between hyperedges at the new level is the same as that of relations among vertices at the original level, we have $E_1' \subseteq E_1''$, i.e., $\omega(Z_2') \subseteq \omega(Z_2'')$. Hence, $Z_2' \subseteq Z_2''$ implies that $\omega(Z_2') \subseteq \omega(Z_2'')$.
We now prove that $\omega(Z_2') \subseteq \omega(Z_2'')$ implies that $Z_2' \subseteq Z_2''$. Suppose on the contrary that whenever $\omega(Z_2') \subseteq \omega(Z_2'')$, then there is at least one vertex $w_i^2 \in Z_2'$, but $w_i^2 \notin Z_2''$, i.e., $Z_2' \nsubseteq Z_2''$. Since $\omega(w_i^2) = e_i^1$ and the relationship between hyperedges at the new level is the same as that of relations among vertices at the original level, we have $e_i^1 \in E_1'$, but $e_i^1 \notin E_1''$, i.e., $E_1' \nsubseteq E_1''$, which is a contradiction to the supposition. Thus, we have that $\omega(Z_2') \subseteq \omega(Z_2'')$ implies that $Z_2' \subseteq Z_2''$. Hence, $Z_2' \subseteq Z_2''$ if and only if $\omega(Z_2') \subseteq \omega(Z_2'')$.
□

Definition 29. *Let $H_1 = (Z_1, E_1)$ and $H_2 = (Z_2, E_2)$ be level hypergraphs of m-PFHGs, and H_2 has weaker granularity than H_1. Suppose that e_i^1, $e_j^2 \in E_1$ and w_i^2, $w_j^2 \in Z_2$, $i, j = 1, 2, \cdots, n$. The zoom-out operator $\sigma : H_1 \to H_2$ is defined as $\sigma(e_i^1) = w_i^2$, $w_i^2 \in Z_2$. The zoom-out operator of two levels is shown in Figure 6.*

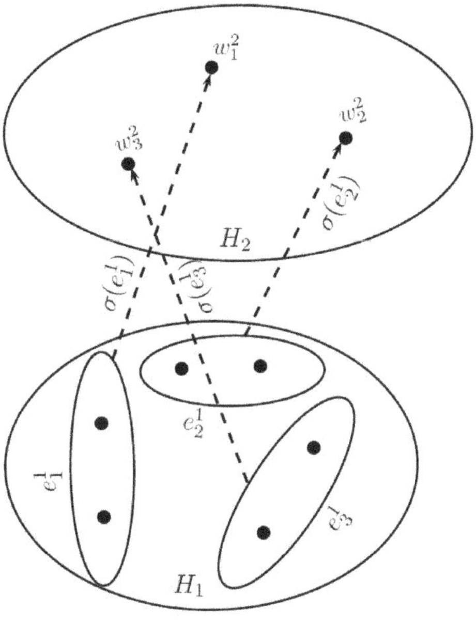

Figure 6. Zoom-out operator.

Theorem 4. *Let $\sigma : H_1 \to H_2$ be the zoom-out operator from $H_1 = (Z_1, E_1)$ to $H_2 = (Z_2, E_2)$, and let $E_1' \subseteq E_1$. Then, the zoom-out operator σ satisfies the following properties,*

(i) $\sigma(\emptyset) = \emptyset$,
(ii) *σ maps the set of hyperedges of H_1 onto the set of vertices of H_2, i.e., $\sigma(E_1) = Z_2$,*
(iii) $\sigma([E_1']^c) = [\sigma(E_1')]^c$.

Proof. (i) This part is trivially satisfied.

(ii) According to the definition of σ, we have $\sigma(e_i^1) = w_i^2$. Since, the hyperedges define a partition of hypergraph, so we have $E_1 = \{e_1^1, e_2^1, e_3^1, \cdots, e_n^1\} = \bigcup_{e_i^1 \in E_1} e_i^1$. Then,

$$\sigma(E_1) = \sigma(\bigcup_{e_i^1 \in E_1} e_i^1) = \bigcup_{e_i^1 \in E_1} \sigma(e_i^1) = \bigcup_{w_i^2 \in Z_2} w_i^2 = Z_2.$$

(iii) Assume that $[E_1']^c = V_1'$, then it is obvious that $E_1' \cap V_1' = \emptyset$ and $E_1' \cup V_1' = E_1$. Suppose on the contrary, there exists at least one vertex $w_i^2 \in \sigma([E_1']^c)$, but $w_i^2 \notin [\sigma(E_1')]^c$. $w_i^2 \in \sigma([E_1']^c)$ implies that $w_i^2 \in \sigma(V_1') \Rightarrow w_i^2 \in \bigcup_{e_i^1 \in V_1'} \sigma(e_i^1) \Rightarrow w_i^2 \in \bigcup_{e_i^1 \in E_1 \setminus E_1'} \sigma(e_i^1)$. Since $w_i^2 \notin [\sigma(E_1')]^c \Rightarrow w_i^2 \in \sigma(E_1') \Rightarrow w_i^2 \in \bigcup_{e_i^1 \in E_1'} \sigma(e_i^1)$, which is a contradiction to our assumption. Hence, $\sigma([E_1']^c) = [\sigma(E_1')]^c$.

□

Definition 30. *Let $H_1 = (Z_1, E_1)$ and $H_2 = (Z_2, E_2)$ be two levels of m-PFHGs, and H_1 possesses a stronger granularity than H_2. Let $E_1' \subseteq E_1$, then $\hat{\sigma}(E_1') = \{e_i^2 | e_i^2 \in E_2, \kappa(e_i^2) \subseteq E_1'\}$ is called the internal zoom-out operator. The operator $\check{\sigma}(E_1') = \{e_i^2 | e_i^2 \in E_2, \kappa(e_i^2) \cap E_1' \neq \emptyset\}$ is called the external zoom-out operator.*

Example 3. *Let $H_1 = (Z_1, E_1)$ and $H_2 = (Z_2, E_2)$ be two levels of m-PFHGs, and H_1 possesses a stronger granularity than H_2, where $E_1 = \{e_1^1, e_2^1, e_3^1, e_4^1, e_5^1, e_6^1\}$, and $E_2 = \{e_1^2, e_2^2, e_3^2\}$. Furthermore, $e_1^2 = \{w_1^2, w_3^2\}$, $e_2^2 = \{w_2^2, w_4^2\}$, $e_3^2 = \{w_5^2, w_6^2\}$, as shown in Figure 7.*

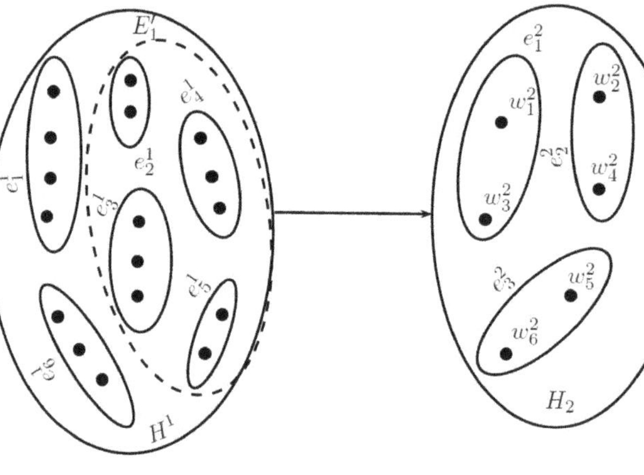

Figure 7. Internal and external zoom-out operators.

Let $E_1' = \{e_2^1, e_3^1, e_4^1, e_5^1\}$ be the subset of hyperedges of H_1, then we cannot zoom-out to H_2 directly; thus, by using the internal and external zoom-out operators, we have the following relations.
$\hat{\sigma}(\{e_2^1, e_3^1, e_4^1, e_5^1\}) = \{e_2^2\}$,
$\check{\sigma}(\{e_2^1, e_3^1, e_4^1, e_5^1\}) = \{e_1^2, e_2^2, e_3^2\}$.

4. A Granular Computing Model of Web Searching Engines

The most fertile way to direct a search on the Internet is through a search engine. A web search engine is defined as a system software that is designed to search for queries on the World Wide Web. A user may utilize a number of search engines to gather information, and similarly, various searchers may make an effective use of the same engine to fulfill their queries. In this section, we construct a GrC model of web searching engines based on four-PFHG. In a web searching hypernetwork, the vertices denote the various search engines. According to the relation set N, the vertices having some relationship are united together as a hyperedge, in which the search engines serve only one user. After assigning the membership degrees to that unit, a four-PF hyperedge is constructed, which is also considered as a granule. A four-PF hyperedge indicates a user that wants to gather some information, and the vertices in that hyperedge represent those search engines that provide relevant data to the user. Let us consider there are ten search engines, and the corresponding four-PFHG $H = (A, B)$ is shown in Figure 8. Note that $A = \{e_1, e_2, e_3, \cdots, e_{10}\}$, and $B = \{U_1, U_2, U_3, U_4, U_5\}$.

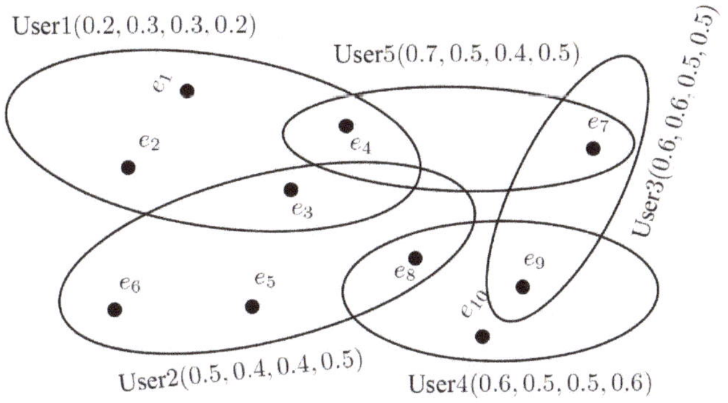

Figure 8. A four-PFHG representation of web searching.

The incidence matrix of four-PFHG is given in Table 1.

Table 1. Incidence matrix.

A	U_1	U_2	U_3	U_4	U_5
e_1	(0.2, 0.3, 0.3, 0.2)	(0, 0, 0, 0)	(0, 0, 0, 0)	(0, 0, 0, 0)	(0, 0, 0, 0)
e_2	(0.2, 0.3, 0.3, 0.2)	(0, 0, 0, 0)	(0, 0, 0, 0)	(0, 0, 0, 0)	(0, 0, 0, 0)
e_3	(0.2, 0.3, 0.3, 0.2)	(0.5, 0.4, 0.4, 0.5)	(0, 0, 0, 0)	(0, 0, 0, 0)	(0, 0, 0, 0)
e_4	(0.2, 0.3, 0.3, 0.2)	(0, 0, 0, 0)	(0, 0, 0, 0)	(0, 0, 0, 0)	(0.7, 0.5, 0.4, 0.5)
e_5	(0, 0, 0, 0)	(0.5, 0.4, 0.4, 0.5)	(0, 0, 0, 0)	(0, 0, 0, 0)	(0, 0, 0, 0)
e_6	(0, 0, 0, 0)	(0.5, 0.4, 0.4, 0.5)	(0, 0, 0, 0)	(0, 0, 0, 0)	(0, 0, 0, 0)
e_7	(0, 0, 0, 0)	(0, 0, 0, 0)	(0.6, 0.6, 0.5, 0.5)	(0, 0, 0, 0)	(0.7, 0.5, 0.4, 0.5)
e_8	(0, 0, 0, 0)	(0, 0, 0, 0)	(0, 0, 0, 0)	(0.6, 0.5, 0.5, 0.6)	(0, 0, 0, 0)
e_9	(0, 0, 0, 0)	(0, 0, 0, 0)	(0.6, 0.6, 0.5, 0.5)	(0.6, 0.5, 0.5, 0.6)	(0, 0, 0, 0)
e_{10}	(0, 0, 0, 0)	(0, 0, 0, 0)	(0, 0, 0, 0)	(0.6, 0.5, 0.5, 0.6)	(0, 0, 0, 0)

An m-PFHG model of GrC illustrates a vague set having some membership degrees. In this model, there are five users that need the search engines to gather information. Note that the membership degrees of these engines are different for the users because whenever a user selects a search engine, he/she considers various factors or attributes. Hence, an m-PFHG in GrC is more meaningful and effective.

Let us suppose that each search engine possesses four attributes, which are core technology, scalability, content processing, and query functionality. The information table for various search engines having these attributes is given in Table 2.

The membership degrees of search engines reveal the percentage of attributes possessed by them, e.g., e_1 owns 70% of core technology, 60% scalability, and 50% provide content processing, and the query functionality of this engine is 70%. The four-PFER matrix describes the similarities between these search engines and is given as follows,

$$\tilde{P}_N = \begin{bmatrix} 1 & 1 & 0.6 & 0.6 & 0.6 & 0.6 & 0.6 & 0.6 & 0.6 & 0 \\ 1 & 1 & 0.7 & 0.7 & 0.7 & 0.7 & 0.7 & 0.7 & 0.7 & 0 \\ 0.6 & 0.7 & 1 & 0.8 & 0.8 & 0.8 & 0.8 & 0.8 & 0.8 & 0 \\ 0.6 & 0.7 & 0.8 & 1 & 0.6 & 0.6 & 0.6 & 0.6 & 0.6 & 0.6 \\ 0.6 & 0.7 & 0.8 & 0.6 & 1 & 0.5 & 0.5 & 0.5 & 0.5 & 0 \\ 0.6 & 0.7 & 0.8 & 0.6 & 0.5 & 1 & 0.6 & 0.6 & 0.6 & 0 \\ 0.6 & 0.7 & 0.8 & 0.6 & 0.5 & 0.6 & 1 & 0.7 & 0.7 & 0.7 \\ 0.6 & 0.7 & 0.8 & 0.6 & 0.5 & 0.6 & 0.7 & 1 & 0.8 & 0.8 \\ 0.6 & 0.7 & 0.8 & 0.6 & 0.5 & 0.6 & 0.7 & 0.8 & 1 & 0.8 \\ 0 & 0 & 0 & 0.6 & 0 & 0 & 0.7 & 0.8 & 0.8 & 1 \end{bmatrix},$$

where $1 = (1,1,1,1)$, $0 = (0,0,0,0)$, $0.5 = (0.5,0.5,0.5,0.5)$, $0.6 = (0.6,0.6,0.6,0.6)$, $0.7 = (0.7,0.7,0.7,0.7)$, and $0.8 = (0.8,0.8,0.8,0.8)$. Let $\tau = (t_1,t_2,t_3,t_4) = (0.7,0.7,0.7,0.7)$, then its corresponding HQSS is given as follows,

$$\begin{aligned} Z/N_\tau = Z/N_{(0.7,0.7,0.7,0.7)} = \; & \{\{e_1,e_2\},\{e_1,e_2,e_3,e_4,e_5,e_6,e_7,e_8,e_9\},\{e_2,e_3,e_5\}, \\ & \{e_2,e_3,e_4,e_5,e_6,e_7,e_8,e_9\},\{e_2,e_3,e_4\},\{e_2,e_3,e_6\}, \\ & \{e_2,e_3,e_7,e_8,e_9,e_{10}\},\{e_7,e_8,e_9,e_{10}\}\}. \end{aligned}$$

Note that $n_1 = n_5 = n_7 = n_{10} = \{\emptyset\}$, $n_2 = \{(e_1,e_2)\}$, $n_3 = \{(e_2,e_3,e_4),(e_2,e_3,e_5),(e_2,e_3,e_6)\}$, $n_4 = \{(e_7,e_8,e_9,e_{10})\}$, $n_6 = \{e_2,e_3,e_7,e_8,e_9,e_{10}\}$, $n_8 = \{e_2,e_3,e_4,e_5,e_6,e_7,e_8,e_9\}$, $n_9 = \{e_1,e_2,e_3,e_4,e_5,e_6,e_7,e_8,e_9\}$. Hence, a single level of the four-PFHG model is constructed and is shown in Figure 9.

Table 2. The information table.

A	Core Technology	Scalability	Content Processing	Query Functionality
e_1	0.7	0.6	0.5	0.7
e_2	0.6	0.5	0.5	0.6
e_3	0.7	0.8	0.8	0.7
e_4	0.8	0.6	0.6	0.8
e_5	0.7	0.5	0.5	0.7
e_6	0.7	0.6	0.5	0.7
e_7	0.6	0.5	0.5	0.6
e_8	0.7	0.8	0.8	0.7
e_9	0.8	0.6	0.6	0.8
e_{10}	0.7	0.8	0.8	0.7

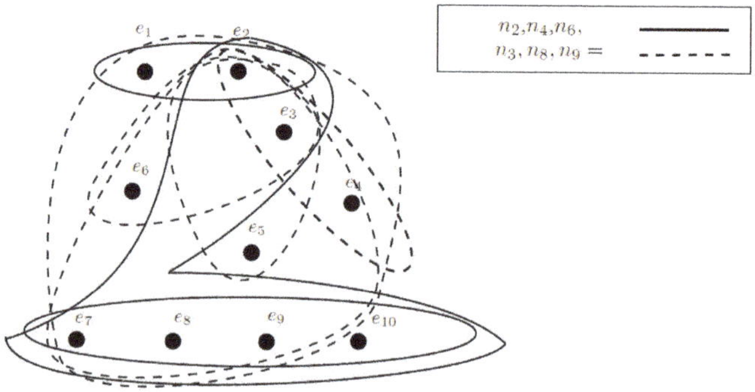

Figure 9. A single-level model of four-PFHG.

Thus, we can obtain eight hyperedges $E_1 = \{e_1, e_2\}$, $E_2 = \{e_2, e_3, e_4\}$, $E_3 = \{e_2, e_3, e_5\}$, $E_4 = \{e_2, e_3, e_6\}$, $E_5 = \{e_7, e_8, e_9, e_{10}\}$, $E_6 = \{e_2, e_3, e_7, e_8, e_9, e_{10}\}$, $E_7 = \{e_2, e_3, e_4, e_5, e_6, e_7, e_8, e_9\}$, $E_8 = \{e_1, e_2, e_3, e_4, e_5, e_6, e_7, e_8, e_9\}$. The procedure of constructing this single-level model is explained in the following flowchart Figure 10.

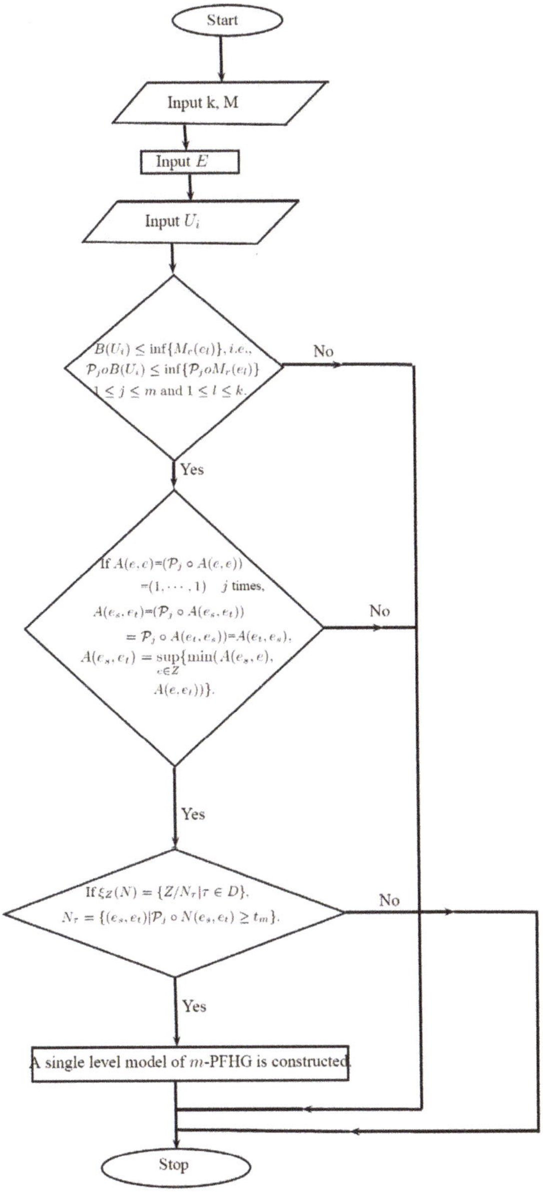

Figure 10. Flowchart of the single-level model of m-PFHG.

We now illustrate the procedure of bottom-up construction through a concrete example and also explain our method through Algorithm 2.

Example 4. *Let $H = (A, B)$ be a three-PFHG as shown in Figure 11. Let $Z = \{z_1, z_2, z_3, z_4, z_5, z_6, z_7, z_8, z_9, z_{10}, z_{11}, z_{12}\}$, and $B = \{\mathcal{E}_1, \mathcal{E}_2, \mathcal{E}_3, \mathcal{E}_4, \mathcal{E}_5\}$.*

For $t_1 = 0.5$, $t_2 = 0.5$, and $t_3 = 0.6$, the $(0.5, 0.5, 0.6)$-level hypergraph of H is given in Figure 12.

By considering the fixed t_1, t_2, t_3 and following Algorithm 2, the bottom-up construction of this model is given in Figure 13.

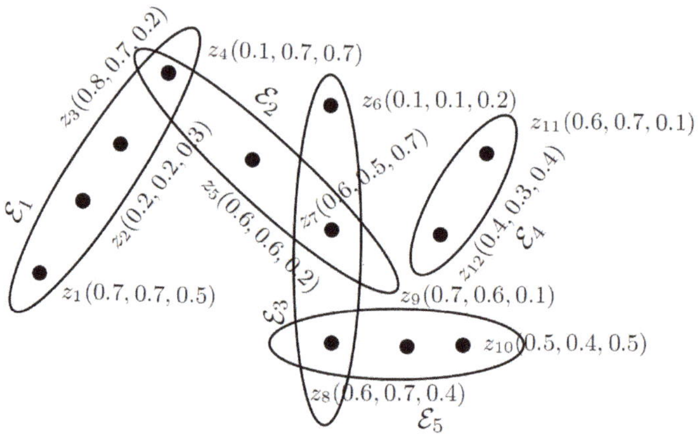

Figure 11. A three-polar fuzzy hypergraph.

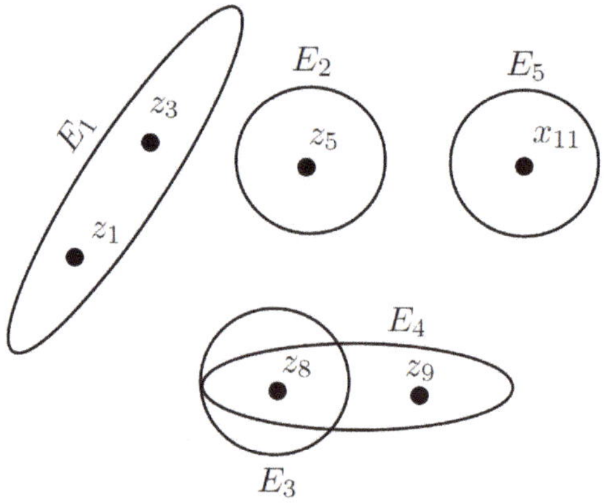

Figure 12. $(0.5, 0.5, 0.6)$-level hypergraph of H.

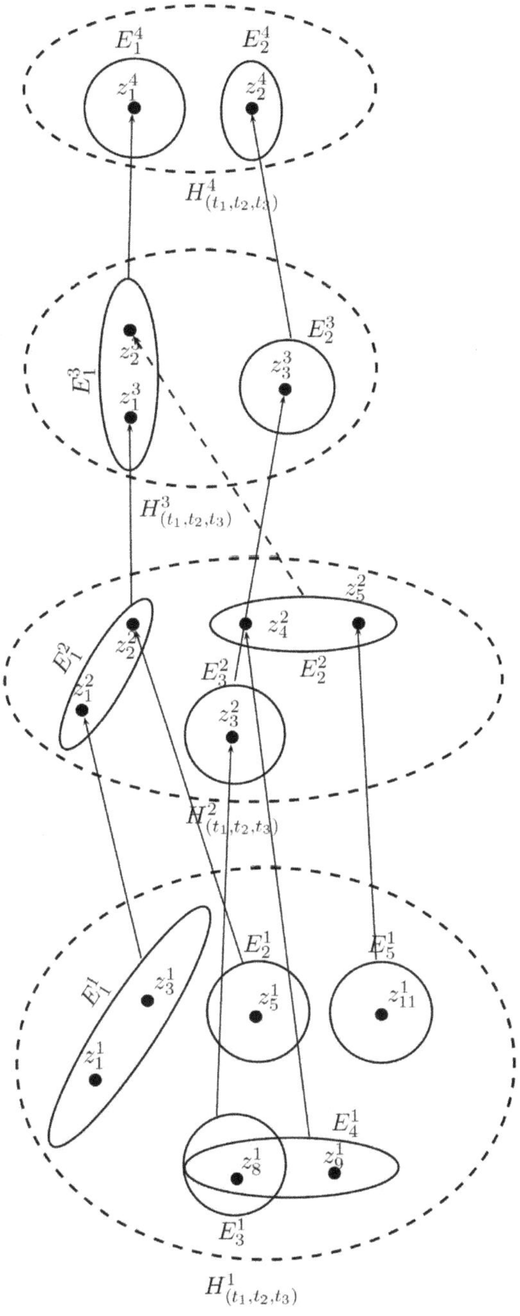

Figure 13. Bottom-up construction procedure.

The possible method for the bottom-up construction is described in Algorithm 2.

Algorithm 2: Algorithm for the method of the bottom-up construction.

1. clc
2. $\mathcal{P}_j \circ z_i$=input('$\mathcal{P}_j \circ z_i$='); T=input('τ='); q=1;
3.   while q==1
4.   [r, m]=size($\mathcal{P}_j \circ z_i$);N=zeros(r, r);N=input('N='); [r_1, r]=size(N); D=ones(r_1, m)+1;
5.   for l=1:r_1
6.   if N(l,:)==zeros(1, r)
7.   D(l,:)=zeros(1, m);
8.   else
9.   for k=1:r
10.   if N(l, k)==1
11.   for j=1:m
12.   D(l, j)=min(D(l, j),$\mathcal{P}_j \circ z_i$(k, j));
13.   end
14.   else
15.   s=0;
16.   end
17.   end
18.   end
19.   end
20.   D
21. $\mathcal{P}_j \circ \mathcal{E}_i$=input('$\mathcal{P}_j \circ \mathcal{E}_i$=');
22.   if size($\mathcal{P}_j \circ \mathcal{E}_i$)==[$r_1$, m]
23.   if $\mathcal{P}_j \circ \mathcal{E}_i$ <=D
24.   if size(T)==[1, m]
25.   S=zeros(r_1, r);s=zeros(r_1, 1);
26.   for l=1:r_1
27.   for k=1:r
28.   if N(l, k)==1
29.   if $\mathcal{P}_j \circ z_i$(k,:)>=T(1,:)
30.   S(l, k)=1;
31.   s(l, 1)=s(l, 1)+1;
32.   else
33.   S(l, k)=0;
34.   end
35.   end
36.   end
37.   end
38.   S
39.   if s==ones(r_1, 1)
40.   q=2;
41.   else
42. $\mathcal{P}_j \circ z_i = \mathcal{P}_j \circ \mathcal{E}_i$;
43.   end
44.   else
45.   fprintf('error')
46.   end
47.   else
48.   fprintf('error')
49.   end
50.   else
51.   fprintf('error')
52.   end
53.   end

5. Conclusions

Granular computing (GrC) is a general framework to deal with granules, and it captures our ability to explore the real-world problems at different levels of granularity. It enables us to change these granularities during problem solving. The fundamental concepts of GrC are discussed in various environments, including cluster analysis, rough set theory, and artificial intelligence. The main objective of research in the field of GrC is to construct an effective computational model to deal with large amounts of information and knowledge. An m-PFS, as an extension of FS and BFS, is used to deal with multipolar information and multi-object network models. In this paper, we have used an m-PFHG model such that the benefits of m-PFSs and hypergraphs theory are combined to represent GrC. We have given a visual description of GrC and considered this model to examine the multi-polarity in GrC. After concise review of m-PFSs and m-PFHGs, we have applied these concepts in the formation of HQSSs. In this hypergraph model, we have obtained the granules as hyperedges from m-PFER, and the partition of the universe has been defined using these granules. Further, we have defined the zoom-in and zoom-out operators to give a description of the mappings between two hypergraphs. We have constructed a GrC model for web searching engines, which is based on multi-polar information. Moreover, we have discussed a concrete example to reveal the validity of our proposed model, and the procedure is represented through an algorithm. We aim to broaden our study to: (1) m-polar fuzzy directed hypergraphs, (2) fuzzy rough soft directed hypergraphs, (3) granular computing based on fuzzy rough and rough fuzzy hypergraphs, (4) hesitant m-polar fuzzy hypergraphs, and (5) the q-Rung picture fuzzy concept lattice.

Author Contributions: A.L., M.A. and A.N.A.K. conceived of the presented concept. A.L. and M.A. developed the theory and performed the computations. A.N.A.K. verified the analytical methods.

Acknowledgments: The authors are grateful to the Editor of the Journal and anonymous referees for their detailed and valuable comments.

Conflicts of Interest: The authors declare no conflict of interest.

References

1. Lin, T.Y. Granular computing: From rough sets and neighborhood systems to information granulation and computing with words. In Proceedings of the European Congress on Intelligent Techniques and Soft Computing, Aachen, Germany, 8–12 September 1997.
2. Zhang, L.; Zhang, B. *Hierarchy and Multi-Granular Computing, Quotient Space Based Problem Solving*; Tsinghua University Press: Beijing, China, 2014; pp. 45–103.
3. Berge, C. *Graphs and Hypergraphs*; North-Holland Publishing Company: Amsterdam, The Netherlands, 1973.
4. Liu, Q.; Jin, W.B.; Wu, S.Y.; Zhou, Y.H. Clustering research using dynamic modeling based on granular computing. In Proceedings of the IEEE International Conference on Granular Computing, Beijing, China, 25–27 July 2005; pp. 539–543.
5. Wong, S.K.M.; Wu, D. Automated mining of granular database scheme. In Proceeding of the IEEE International Conference on Fuzzy Systems, St. Louis, MO, USA, 25–28 May 2002; pp. 690–694.
6. Chen, G.; Zhong, N.; Yao, Y. A hypergraph model of granular computing. In Proceedings of the IEEE International Conference on Granular Computing, Hangzhou, China, 26–28 August 2008; pp. 130–135.
7. Zadeh, L.A. Fuzzy sets. *Inf. Control* **1965**, *8*, 338–353. [CrossRef]
8. Zhang, W.-R. Bipolar fuzzy sets and relations: A computational framework forcognitive modeling and multiagent decision analysis. In Proceedings of the First International Joint Conference of the North American Fuzzy Information Processing Society Biannual Conference, the Industrial Fuzzy Control and Intellige, San Antonio, TX, USA, 18–21 December 1994; pp. 305–309.
9. Chen, J.; Li, S.; Ma, S.; Wang, Z. m-Polar fuzzy sets: An extension of bipolar fuzzy sets. *Sci. World J.* **2014**, *2014*, 416530. [CrossRef] [PubMed]
10. Rosenfeld, A. Fuzzy graphs. In *Fuzzy Sets and Their Applications to Cognitive and Decision Processes*; Academic Press: New York, NY, USA, 1975; pp. 77–95.

11. Akram, M. *m*-Polar fuzzy graphs. In *Studies in Fuzziness and Soft Computing*; Springer: Berlin, Germany, 2019; Volume 371, pp. 1–284.
12. Kaufmann, A. *Introduction a la Thiorie des Sous-Ensemble Flous*; Masson: Paris, France, 1977.
13. Gong, Z.; Wang, Q. On the connection of fuzzy hypergraph with fuzzy information system. *J. Intell. Fuzzy Syst.* **2017**, *44*, 1665–1676. [CrossRef]
14. Wang, Q.; Gong, Z. An application of fuzzy hypergraphs and hypergraphs in granular computing. *Inf. Sci.* **2018**, *429*, 296–314. [CrossRef]
15. Akram, M.; Sarwar, M. Novel applications of *m*-polar fuzzy hypergraphs. *J. Intell. Fuzzy Syst.* **2017**, *32*, 2747–2762. [CrossRef]
16. Akram, M.; Sarwar, M. Transversals of *m*-polar fuzzy hypergraphs with applications. *J. Intell. Fuzzy Syst.* **2017**, *33*, 351–364. [CrossRef]
17. Akram, M.; Shahzadi, G. Hypergraphs in *m*-polar fuzzy environment. *Mathematics* **2018**, *6*, 28. [CrossRef]
18. Akram, M.; Luqman, A. Intuitionistic single-valued neutrosophic hypergraphs. *Opsearch* **2017**, *54*, 799–815. [CrossRef]
19. Akram, M.; Luqman, A. Bipolar neutrosophic hypergraphs with applications. *J. Intell. Fuzzy Syst.* **2017**, *33*, 1699–1713. [CrossRef]
20. Mordeson, J.N.; Nair, P.S. *Fuzzy Graphs and Fuzzy Hypergraphs*, 2nd ed.; Physica Verlag: Heidelberg, Germany, 2001.
21. Parvathi, R.; Akram, M.; Thilagavathi, S. Intuitionistic fuzzy shortest hyperpath in a network. *Inf. Process. Lett.* **2013**, *113*, 599–603.
22. Yang, J.; Wang, G.; Zhang, Q. Knowledge distance measure in multigranulation spaces of fuzzy equivalence relation. *Inf. Sci.* **2018**, *448*, 18–35. [CrossRef]
23. Zhang, H.; Zhang, R.; Huang, H.; Wang, J. Some picture fuzzy Dombi Heronian mean operators with their application to multi-attribute decision-making. *Symmetry* **2018**, *10*, 593. [CrossRef]
24. Zhang, L.; Zhang, B. *The Theory and Applications of Problem Solving-Quotient Space Based Granular Computing*; Tsinghua University Press: Beijing, China, 2007
25. Zhang, L.; Zhang, B. The structural analysis of fuzzy sets. *J. Approx. Reason.* **2005**, *40*, 92–108. [CrossRef]
26. Yao, Y.Y. A partition model of granular computing. In *Transactions on Rough Sets I. Lecture Notes in Computer Science*; Peters, J.F., Skowron, A., Grzymala-Busse, J.W., Kostek, B., Swiniarski, R.W., Szczuka, M.S., Eds.; Springer: Berlin/Heidelberg, Germany, 2004.

symmetry

MDPI

Article

A Novel Description on Edge-Regular q-Rung Picture Fuzzy Graphs with Application

Muhammad Akram [1,*], Amna Habib [1] and Ali N. A. Koam [2]

[1] Department of Mathematics, University of the Punjab, New Campus, Lahore 54590, Pakistan; amnahabib96@gmail.com
[2] Department of Mathematics, College of Science, Jazan University, New Campus, P.O. Box 2097, Jazan 45142, Saudi Arabia; akoum@jazanu.edu.sa
* Correspondence: m.akram@pucit.edu.pk

Received: 5 March 2019; Accepted: 1 April 2019; Published: 4 April 2019

Abstract: Picture fuzzy model is a generalized structure of intuitionistic fuzzy model in the sense that it not only assigns the membership and nonmembership values in the form of orthopair (μ, ν) to an element, but it assigns a triplet (μ, η, ν), where η denotes the neutral degree and the difference $\pi = 1 - (\mu + \eta + \nu)$ indicates the degree of refusal. The q-rung picture fuzzy set(q-RPFS) provides a wide formal mathematical sketch in which uncertain and vague conceptual phenomenon can be precisely and rigorously studied because of its distinctive quality of vast representation space of acceptable triplets. This paper discusses some properties including edge regularity, total edge regularity and perfect edge regularity of q-rung picture fuzzy graphs (q-RPFGs). The work introduces and investigates these properties for square q-RPFGs and q-RPF line graphs. Furthermore, this study characterizes how regularity and edge regularity of q-RPFGs structurally relate. In addition, it presents the concept of ego-networks to extract knowledge from large social networks under q-rung picture fuzzy environment with algorithm.

Keywords: q-rung picture fuzzy graphs; edge regular; perfect edge regular; square q-rung picture fuzzy graphs; q-rung picture fuzzy line graphs; ego networks

1. Introduction

Fuzzy sets (FSs), coined by Zadeh [1], have become one of the emerging areas in contemporary technologies of information processing. Recent studies spread across various areas from control, pattern recognition, and knowledge-based systems to computer vision and artificial life. Fuzzy sets implicate capturing, portraying and dealing with linguistic notions-items with indefinite boundaries and thus it came forth as a new approach to incorporate uncertainty. Picture fuzzy sets (PFSs), created by Cuong [2,3], a direct extension of Atanassov's [4] intuitionistic fuzzy sets (IFSs), which itself extends the Zadeh's notion of fuzzy sets, are the sets characterized by not only membership $\mu : X \to [0,1]$, and nonmembership $\nu : X \to [0,1]$ functions but also neutral function $\eta : X \to [0,1]$, where the difference $\pi = 1 - (\mu + \eta + \nu)$ is called refusal function. Thus PFSs may be satisfactory in cases whenever expert's evaluations are of types: yes, abstain, no, refusal. Cuong suggested 'voting' as a paradigm of his proposed concept. Since picture fuzzy sets are suitable for capturing imprecise, uncertain and inconsistent information, therefore they can be applied to many decision-making processes such as: solution choice, financial forecasting, estimating risks in business etc. Son [5] introduced some clustering algorithms based on picture fuzzy sets with applications to time series and weather forecasting. Thong [6] developed a hybrid model relating picture fuzzy clustering for medical diagnosis. There have been significant attempts which explored this concept, one can see [7–13]. It is noteworthy that the constraint $\mu + \nu \leq 1$ in IFS limits the selection of orthopairs from a triangular region.

In order to increase the adeptness of IFS, Yager and Abbasov [14] proposed the notion of Pythagorean fuzzy sets (PyFSs) in 2013, which replace the constraint of IFS with $\mu^2 + \nu^2 \leq 1$. This notion also limits the selection of orthopairs from unit circular region in the first quadrant. In 2017, Yager [15] introduced q-rung orthopair fuzzy sets (q-ROFSs) as a new generalization of orthopair fuzzy sets (i.e., IFS and PyFS), which further relax the constraint of orthopair membership grades with $\mu(x)^q + \nu(x)^q \leq 1$ $(q \geq 1)$. Analogously, since picture fuzzy sets confine the selection of triplets only from a tetrahedron, as shown in Figure 1, the spherical fuzzy sets (SFSs), proposed by Gündoğdu and Kahraman [16], have given strength to the idea of picture fuzzy sets. Although both PFS and SFS can easily reflect the ambiguous character of subjective assessments, they still have apparent variations. The membership functions μ, η, and ν of PFSs are required to meet the condition $\mu + \eta + \nu \leq 1$. However, these functions in SFSs satisfy the constraint $\mu^2 + \eta^2 + \nu^2 \leq 1$. Which indicates that SFSs somehow expanded the space of admissible triplets. In 2018, Li et al. [17] proposed the q-RPFS model which inherits the virtues of both q-ROFS and PFS. Being a suitable model for capturing imprecise and inconsistent data, it not only assigns three membership degrees to an element, but also alleviate the constraint of picture and spherical fuzzy sets to a great extent with $\mu(u)^q + \eta(u)^q + \nu(u)^q \leq 1$ $(q \geq 1)$. Figure 1 shows that the space of admissible triplets expands with increasing q. For example, if an expert provides the positive, neutral and negative membership values to an object as 0.9, 0.7 and 0.5, respectively. It is immediately seen that $0.9 + 0.7 + 0.5 \geq 1$, and $0.9^2 + 0.7^2 + 0.5^2 \geq 1$. Such a case can neither be explained by PFS nor by SFS. However, it is appropriate to use q-RPFS because $0.9^q + 0.7^q + 0.5^q \leq 1$ for sufficiently large value of q. Thus immense number of triplets are qualified for q-RPF model due to its elastic bounding constraint. It can be observed that the surface $\mu(u)^q + \eta(u)^q + \nu(u)^q \leq 1$ bounds a portion of first octant, whose volume approaches to the unit cube's volume as the parameter q approaches to infinity. Of course the constraint condition of q-RPF model provides a sense of interdependence of membership functions μ, η and ν. This fact makes this notion considerably more close to natural world than that of prior notions. Some remarkable attempts can be seen in [18–21].

Graphs theory, a dynamic field in both theory and applications, allows graphs as the most important abstract data structures in many fields. Graphs can not translate all the phenomenons of real world scenarios adequately due to the uncertainty and vagueness of parameters. These positions direct to define fuzziness in graph theoretic concepts. Zadeh [22] originated the idea of fuzzy relation. Kaufmann [23] introduced fuzzy graph to state uncertainty in networks. Rosenfeld [24] presented more concepts relating fuzzy graphs. Parvathi and Karunambigai introduced intuitionistic fuzzy graphs(IFGs) [25]. Afterward, IFGs were examined by Akram and Davvaz [26]. Naz et al. [27] discussed the notion of Pythagorean fuzzy graphs. Habib et al. [28] investigated q-ROFGs. The flexible nature of these notions make them vast research area, so far in [29–34]. To manage the cases requiring opinions of types: yes, abstain, no and refusal in graph-theocratic concepts in a broad manner, recently, Akram and Habib [35] introduced the concept of q-RPFGs and defined their regularity. The concept of regularity of fuzzy graphs has led to many developments in their structural theory as they play important role in combinatorics and theoretical computer science. First, Gani and Radha [36] defined degree, total degree and regularity of fuzzy graphs. Degree and total degree of an edge is introduced by Radha and Kumaravel [37]. Cary [38] initiated the idea of perfectly regular and perfectly edge regular fuzzy graphs. Several related works on regularity and edge regularity can be viewed in [35,39–46]. In this paper, we discuss some properties of q-RPFGs, namely edge regularity, total edge regularity and perfect edge regularity. We introduce and investigate these properties for square q-RPFGs and q-RPF line graphs. Furthermore, we provide a brief characterization on structural relationships between regularity and edge regularity of several q-RPFGs. In addition, we present the idea of ego-networks to extract knowledge from large social networks under q-rung picture fuzzy environment with algorithm, as an application of the proposed concept.

We now review the concept of q-RPFS [17] which is necessary to proceed further.

Definition 1. *Let X be a universe of discourse. A q-rung picture fuzzy set(q-RPFS) $\mathscr{P}$ on X given by*

$$\mathscr{P} = \{\langle u, \mu_{\mathscr{P}}(u), \eta_{\mathscr{P}}(u), \nu_{\mathscr{P}}(u)\rangle \mid u \in X\}$$

is characterized by a positive membership function $\mu_{\mathscr{P}} : X \to [0,1]$, a neutral/abstinence membership function $\eta_{\mathscr{P}} : X \to [0,1]$, and a negative membership function $\nu_{\mathscr{P}} : X \to [0,1]$ such that $0 \leq \mu^q_{\mathscr{P}}(u) + \eta^q_{\mathscr{P}}(u) + \nu^q_{\mathscr{P}}(u) \leq 1$ for all $u \in X$. Moreover, $\pi_{\mathscr{P}}(u) = \sqrt[q]{1 - \mu^q_{\mathscr{P}}(u) - \eta^q_{\mathscr{P}}(u) - \nu^q_{\mathscr{P}}(u)}$ is called a q-rung picture fuzzy index or degree of refusal membership of u to the set $\mathscr{P}$, where $q \geq 1$.

The q-rung picture fuzzy set based models may be satisfactory in conjunctures when human opinions concerning responses: yes, abstain, no, and refusal are encountered. Voting can be considered as a paradigm of such environments as human voters can be split into four classes of those who: vote for, abstain, vote against, refusal of voting. The graphical structure of q-RPFS in Figure 1 provides a gradual increase in spaces bounded by surfaces $\mu(x)^q + \eta(x)^q + \nu(x)^q = 1$, obtained by varying q in the first octant. For $q = 1$, the space bounded by surface $x + y + z = 1$ in the first octant is equivalent to the volume occupied by a tetrahedron ABCD. For $q = 2$, the q-RPFS reduces to spherical fuzzy set which covers more space of acceptable triplets than PFS as it is bounded by surface $x^2 + y^2 + z^2 = 1$ in the first octant, which is equivalent to the volume occupied by unit sphere in the first octant. It can be seen that for $q = 3$, the q-RPFS can accept more triplets than picture and spherical fuzzy sets as it covers up more space and for $q = 4$, the space of q-RPFS is vast than all preceding spaces. Since the volume occupied by a surface covers the volume occupied by all prior surfaces; therefore, any element belongs to a particular q-RPFS, must qualify for all picture fuzzy sets of higher rungs (i.e., greater than q).

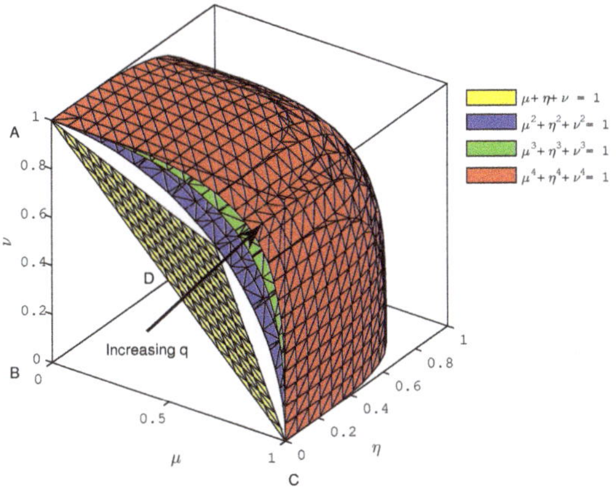

Figure 1. Comparison of Spaces of q-RPFS.

2. Edge Regular q-Rung Picture Fuzzy Graphs

In this section, we first provide some basic definitions relating regularity of q-RPFGs, defined in [35], which will be used for further developments.

Definition 2. *A q-rung picture fuzzy graph on a non-empty set X is a pair $\mathscr{G} = (\mathscr{P}, \mathscr{Q})$ with $\mathscr{P}$ a q-rung picture fuzzy set on X, and $\mathscr{Q}$ a q-rung picture fuzzy relation on X such that*

$$\begin{cases} \mu_{\mathscr{Q}}(uv) \leq \mu_{\mathscr{P}}(u) \wedge \mu_{\mathscr{P}}(v), \\ \eta_{\mathscr{Q}}(uv) \leq \eta_{\mathscr{P}}(u) \wedge \eta_{\mathscr{P}}(v), \\ \nu_{\mathscr{Q}}(uv) \leq \nu_{\mathscr{P}}(u) \vee \nu_{\mathscr{P}}(v), \end{cases}$$

and $0 \leq \mu^q_{\mathscr{Q}}(uv) + \eta^q_{\mathscr{Q}}(uv) + \nu^q_{\mathscr{Q}}(uv) \leq 1$ *for all* $u, v \in X$, *where* $\mu_{\mathscr{Q}} : X \times X \longrightarrow [0,1]$, $\eta_{\mathscr{Q}} : X \times X \longrightarrow [0,1]$ *and* $\nu_{\mathscr{Q}} : X \times X \longrightarrow [0,1]$ *represents the positive membership, neutral membership, and negative membership functions of* $\mathscr{Q}$, *respectively.*

Example 1. *Consider a graph* $G = (P, Q)$ *such that* $P = \{a, b, c, d, e\}$ *and* $Q = \{ac, ad, bc, be, cd, ce\} \subseteq P \times P$. *Let* $\mathscr{P}$ *be a 4-RPFS on* P *and* $\mathscr{Q}$ *be a 4-RPFR on* P, *defined by*

$\mathscr{P}$	a	b	c	d	e
$\mu_{\mathscr{P}}$	0.8	0.7	0.55	0.7	0.9
$\eta_{\mathscr{P}}$	0.7	0.85	0.8	0.9	0.4
$\nu_{\mathscr{P}}$	0.6	0.67	0.77	0.5	0.55

$\mathscr{Q}$	ac	ad	bc	be	cd	ce
$\mu_{\mathscr{Q}}$	0.5	0.6	0.55	0.7	0.5	0.5
$\eta_{\mathscr{Q}}$	0.7	0.7	0.8	0.3	0.8	0.25
$\nu_{\mathscr{Q}}$	0.65	0.6	0.56	0.65	0.66	0.75

Routine computations show that $\mathscr{G} = (\mathscr{P}, \mathscr{Q})$, *displayed in Figure 2, is a 4-RPFG.*

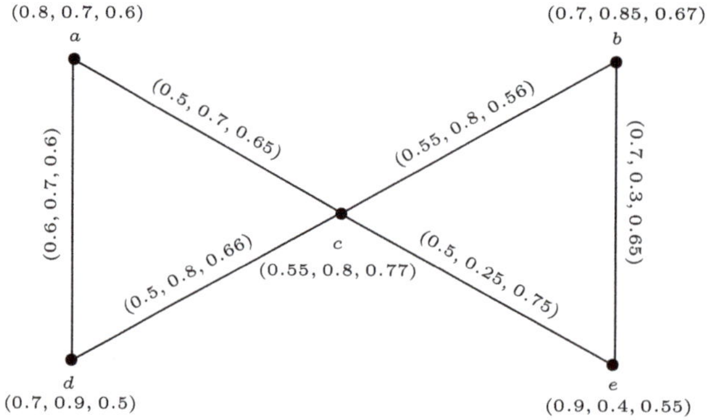

Figure 2. A 4-RPFG $\mathscr{G}$.

Definition 3. *Let* $\mathscr{G} = (\mathscr{P}, \mathscr{Q})$ *be a q-RPFG on underlying crisp graph* $G = (P, Q)$. *If* $\mu_{\mathscr{Q}}(uv) = \mu_{\mathscr{P}}(u) \wedge \mu_{\mathscr{P}}(v)$, $\eta_{\mathscr{Q}}(uv) = \eta_{\mathscr{P}}(u) \wedge \eta_{\mathscr{P}}(v)$ *and* $\nu_{\mathscr{Q}}(uv) = \nu_{\mathscr{P}}(u) \vee \nu_{\mathscr{P}}(v)$ *for all* $uv \in Q$, *then* $\mathscr{G} = (\mathscr{P}, \mathscr{Q})$ *is called a strong q-rung picture fuzzy graph.*

Definition 4. *Let* $\mathscr{G} = (\mathscr{P}, \mathscr{Q})$ *be a q-RPFG on underlying crisp graph* $G = (P, Q)$. *If* $\mu_{\mathscr{Q}}(uv) = \mu_{\mathscr{P}}(u) \wedge \mu_{\mathscr{P}}(v)$, $\eta_{\mathscr{Q}}(uv) = \eta_{\mathscr{P}}(u) \wedge \eta_{\mathscr{P}}(v)$ *and* $\nu_{\mathscr{Q}}(uv) = \nu_{\mathscr{P}}(u) \vee \nu_{\mathscr{P}}(v)$ *for all* $u, v \in P$, *then* $\mathscr{G} = (\mathscr{P}, \mathscr{Q})$ *is called a complete q-rung picture fuzzy graph.*

Definition 5. *Let* $\mathscr{G} = (\mathscr{P}, \mathscr{Q})$ *be a q-RPFG defined on* $G = (P, Q)$. *The order of* $\mathscr{G}$ *is denoted by* $O(\mathscr{G})$, *and defined as*

$$O(\mathscr{G}) = \left(\sum_{u \in P} \mu_{\mathscr{P}}(u), \sum_{u \in P} \eta_{\mathscr{P}}(u), \sum_{u \in P} \nu_{\mathscr{P}}(u) \right).$$

Definition 6. *Let* $\mathscr{G} = (\mathscr{P}, \mathscr{Q})$ *be a q-RPFG defined on* $G = (P, Q)$. *The size of* $\mathscr{G}$ *is denoted by* $S(\mathscr{G})$, *and defined as*

$$S(\mathscr{G}) = \left(\sum_{uv \in P} \mu_{\mathscr{Q}}(uv), \sum_{uv \in P} \eta_{\mathscr{Q}}(uv), \sum_{uv \in P} \nu_{\mathscr{Q}}(uv) \right).$$

Example 2. *The order and size of q-rung picture fuzzy graph displayed in Figure 2 are* $O(\mathscr{G}) = (3.65, 3.65, 3.09)$ *and* $S(\mathscr{G}) = (3.35, 3.55, 3.12)$, *respectively.*

Definition 7. *Let $\mathscr{G} = (\mathscr{P}, \mathscr{Q})$ be a q-RPFG defined on $G = (P, Q)$. The degree of a vertex u of $\mathscr{G}$ is denoted by $d_{\mathscr{G}}(u) = (d_\mu(u), d_\eta(u), d_\nu(u))$, and defined as*

$$d_{\mathscr{G}}(u) = \left(\sum_{uv \in Q} \mu_{\mathscr{Q}}(uv), \sum_{uv \in Q} \eta_{\mathscr{Q}}(uv), \sum_{uv \in Q} \nu_{\mathscr{Q}}(uv) \right).$$

Definition 8. *Let $\mathscr{G} = (\mathscr{P}, \mathscr{Q})$ be a q-RPFG defined on $G = (P, Q)$. The total degree of a vertex u of $\mathscr{G}$ is denoted by $td_{\mathscr{G}}(u) = (td_\mu(u), td_\eta(u), td_\nu(u))$, and defined as*

$$td_{\mathscr{G}}(u) = \left(\sum_{uv \in Q} \mu_{\mathscr{Q}}(uv) + \mu_{\mathscr{P}}(u), \sum_{uv \in Q} \eta_{\mathscr{Q}}(uv) + \eta_{\mathscr{P}}(u), \sum_{uv \in Q} \nu_{\mathscr{Q}}(uv) + \nu_{\mathscr{P}}(u) \right).$$

Example 3. *Consider a 4-RPFG $\mathscr{G}$ displayed in Figure 2. The degree, and total degree of vertex a in $\mathscr{G}$ is given as $d_{\mathscr{G}}(a) = (\mu_{\mathscr{Q}}(ac) + \mu_{\mathscr{Q}}(ad),\ \eta_{\mathscr{Q}}(ac) + \eta_{\mathscr{Q}}(ad),\ \nu_{\mathscr{Q}}(ac) + \nu_{\mathscr{Q}}(ad)) = (0.5 + 0.6,\ 0.7 + 0.7,\ 0.65 + 0.6) = (1.1, 1.4, 1.25)$, and $td_{\mathscr{G}}(a) = (d_\mu(a) + \mu_{\mathscr{P}}(a),\ d_\eta(a) + \eta_{\mathscr{P}}(a),\ d_\nu(a) + \nu_{\mathscr{P}}(a)) = (1.1 + 0.8,\ 1.4 + 0.7,\ 1.25 + 0.6) = (1.9, 2.1, 1.85)$, respectively.*

Definition 9. *Let $\mathscr{G} = (\mathscr{P}, \mathscr{Q})$ be a q-RPFG defined on $G = (P, Q)$. If each vertex of $\mathscr{G}$ has same degree, that is, $d_{\mathscr{G}}(u) = (k_1, k_2, k_3)$ for all $u \in P$, then $\mathscr{G}$ is called (k_1, k_2, k_3)-regular q-RPFG.*

Example 4. *Consider a 3-RPFG $\mathscr{G} = (\mathscr{P}, \mathscr{Q})$ as displayed in Figure 3.*

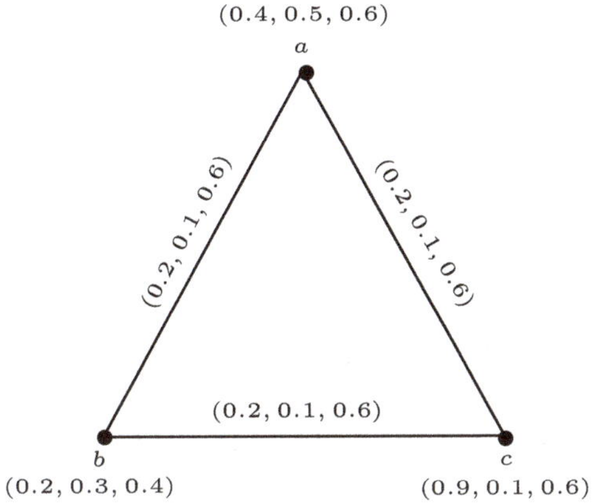

Figure 3. A regular 3-RPFG $\mathscr{G}$.

We see that the degree of each vertex in $\mathscr{G}$ is $d_{\mathscr{G}}(a) = d_{\mathscr{G}}(b) = d_{\mathscr{G}}(c) = (0.4, 0.2, 1.2)$. Hence, $\mathscr{G}$ is $(0.2, 0.1, 0.6)$-regular.

Definition 10. *Let $\mathscr{G} = (\mathscr{P}, \mathscr{Q})$ be a q-RPFG defined on $G = (P, Q)$. If each vertex of $\mathscr{G}$ has same total degree, that is, $td_{\mathscr{G}}(u) = (l_1, l_2, l_3)$ for all $u \in P$, then $\mathscr{G}$ is called (l_1, l_2, l_3)-totally regular q-RPFG.*

Example 5. *Consider a 2-RPFG $\mathscr{G} = (\mathscr{P}, \mathscr{Q})$ as displayed in Figure 4.*

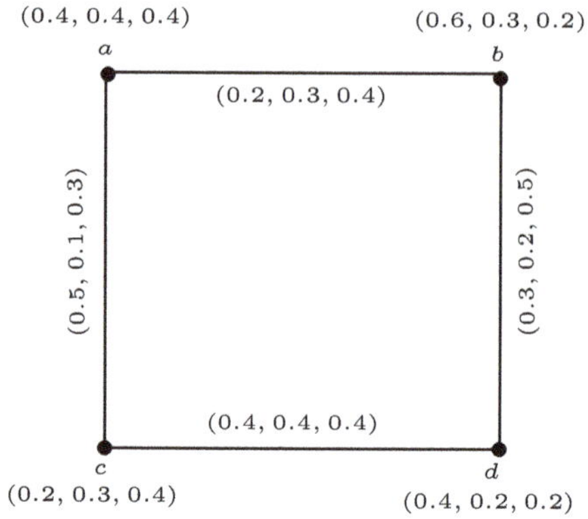

Figure 4. A totally regular 2-RPFG $\mathscr{G}$.

$\mathscr{G}$ *is* $(1.1, 0.8.1.1)$*-totally regular* 2*-RPFG since* $td_{\mathscr{G}}(a) = td_{\mathscr{G}}(b) = td_{\mathscr{G}}(c) = td_{\mathscr{G}}(d) = (1.1, 0.8, 1.1)$.

Definition 11. *A perfectly regular q-RPFG is a q-RPFG that is both regular and totally regular.*

Example 6. *Consider a* 3*-rung picture fuzzy graph* $\mathscr{G} = (\mathscr{P}, \mathscr{Q})$ *as shown in Figure 5.*

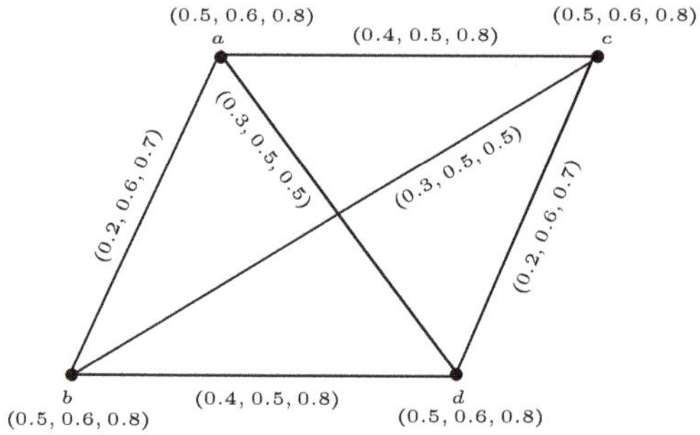

Figure 5. Perfectly Regular 3-RPFG.

Clearly, 3*-RPFG* $\mathscr{G}$ *is regular since* $d_{\mathscr{G}}(a) = d_{\mathscr{G}}(b) = d_{\mathscr{G}}(c) = d_{\mathscr{G}}(d) = (0.9, 1.6, 2.0)$*, and is totally regular since* $td_{\mathscr{G}}(a) = td_{\mathscr{G}}(b) = td_{\mathscr{G}}(c) = td_{\mathscr{G}}(d) = (1.4, 2.2, 2.8)$*. Hence,* $\mathscr{G}$ *is perfectly regular* 3*-RPFG.*

Definition 12. *A q-RPFG* $\mathscr{G} = (\mathscr{P}, \mathscr{Q})$ *defined on* $G = (P, Q)$ *is said to be partially regular if the underlying graph* $G = (P, Q)$ *is regular.*

Definition 13. *A q-RPFG* $\mathscr{G} = (\mathscr{P}, \mathscr{Q})$ *defined on* $G = (P, Q)$*, is said to be full regular if* $\mathscr{G}$ *is both regular, and partially regular.*

The concept of edge regularity has been explored by many researchers on fuzzy graphs, and several of its generalizations. We now give a description on edge regular q-RPFGs. First, we state some definitions in this context.

Definition 14. *Let* $\mathscr{G} = (\mathscr{P}, \mathscr{Q})$ *be a q-RPFG defined on* $G = (P, Q)$*. The edge degree of uv in* $\mathscr{G}$ *is denoted by* $d_{\mathscr{G}}(uv) = (d_{\mu}(uv), d_{\eta}(uv), d_{\nu}(uv))$*, and defined as*

$$d_{\mathscr{G}}(uv) = d_{\mathscr{G}}(u) + d_{\mathscr{G}}(v) - 2(\mu_{\mathscr{Q}}(uv), \eta_{\mathscr{Q}}(uv), \nu_{\mathscr{Q}}(uv)).$$

This is equivalent to

$$d_{\mathscr{G}}(uv) = \Big(\sum_{uw\in Q}\mu_{\mathscr{Q}}(uw), \sum_{uw\in Q}\eta_{\mathscr{Q}}(uw), \sum_{uw\in Q}\nu_{\mathscr{Q}}(uw)\Big) + \Big(\sum_{vw\in Q}\mu_{\mathscr{Q}}(vw), \sum_{vw\in Q}\eta_{\mathscr{Q}}(vw), \sum_{vw\in Q}\nu_{\mathscr{Q}}(vw)\Big)$$
$$- 2(\mu_{\mathscr{Q}}(uv), \eta_{\mathscr{Q}}(uv), \nu_{\mathscr{Q}}(uv))$$
$$= \Big(\sum_{w\neq v}\mu_{\mathscr{Q}}(uw), \sum_{w\neq v}\eta_{\mathscr{Q}}(uw), \sum_{w\neq v}\nu_{\mathscr{Q}}(uw)\Big) + \Big(\sum_{w\neq u}\mu_{\mathscr{Q}}(vw), \sum_{w\neq u}\eta_{\mathscr{Q}}(vw), \sum_{w\neq u}\nu_{\mathscr{Q}}(vw)\Big).$$

Example 7. *Consider a 4-RPFG $\mathscr{G}$ displayed in Figure 6.*

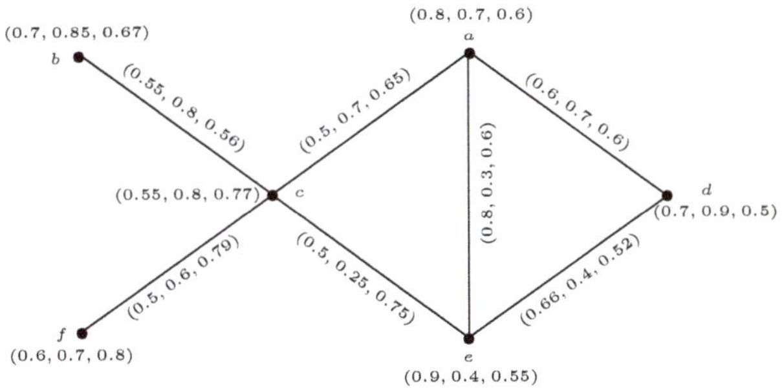

Figure 6. A 4-RPFG $\mathscr{G}$.

The degree of edge ad in $\mathscr{G}$ can be computed as

$$\begin{aligned} d_{\mathscr{G}}(ad) &= d_{\mathscr{G}}(a) + d_{\mathscr{G}}(d) - 2(\mu_{\mathscr{Q}}(ad), \eta_{\mathscr{Q}}(ad), \nu_{\mathscr{Q}}(ad)) \\ &= (1.9, 1.7, 1.85) + (1.26, 1.1, 1.12) - 2(0.6, 0.7, 0.6) = (1.96, 1.4, 1.77). \end{aligned}$$

Theorem 1. *[35] For any q-RPFG $\mathscr{G} = (\mathscr{P}, \mathscr{Q})$ defined on $P = \{v_1, v_2, \ldots, v_n\}$, the following relation for degrees of vertices of $\mathscr{G}$ must hold:*

$$\sum_{j=1}^{n} d_{\mathscr{G}}(v_j) = 2\Big(\sum_{\substack{j=1\\ i>j}}^{n-1}\mu_{\mathscr{Q}}(v_jv_i), \sum_{\substack{j=1\\ i>j}}^{n-1}\eta_{\mathscr{Q}}(v_jv_i), \sum_{\substack{j=1\\ i>j}}^{n-1}\nu_{\mathscr{Q}}(v_jv_i)\Big)$$

for all $1 \leq i \leq n$.

The following theorem is developed to define a comprehensive relationship between the degrees of edges, and the degrees of vertices of q-RPFGs.

Theorem 2. *For any q-RPFG $\mathscr{G} = (\mathscr{P}, \mathscr{Q})$ defined on $P = \{v_1, v_2, \ldots, v_n\}$, if $(\mu_{\mathscr{Q}}(v_iv_j), \eta_{\mathscr{Q}}(v_iv_j), \nu_{\mathscr{Q}}(v_iv_j)) \neq (0,0,0)$ for all $v_i, v_j \in P$ $(i \neq j)$, then the following relation for degrees of edges of $\mathscr{G}$ must hold:*

$$\sum_{\substack{i=1\\ j>i}}^{n-1} d_{\mathscr{G}}(v_iv_j) = (n-2)\sum_{i=1}^{n} d_{\mathscr{G}}(v_i)$$

for all $1 \leq j \leq n$. That is, the sum of degrees of all edges is equal to $(n-2)$ times the sum of degrees of all vertices of $\mathscr{G}$.

Proof. Let $P = \{v_1, v_2, \ldots, v_n\}$, and $\mathscr{G} = (\mathscr{P}, \mathscr{Q})$ be a q-RPFG defined on $G = (P, Q)$. The degree of an edge v_iv_j of q-RPFG $\mathscr{G}$ can be defined as

$$\begin{aligned} d_{\mathscr{G}}(v_iv_j) &= d_{\mathscr{G}}(v_i) + d_{\mathscr{G}}(v_j) - 2(\mu_{\mathscr{Q}}(v_iv_j), \eta_{\mathscr{Q}}(v_iv_j), \nu_{\mathscr{Q}}(v_iv_j)), \\ \sum_{\substack{i,j=1 \\ i\neq j}}^{n} d_{\mathscr{G}}(v_iv_j) &= \sum_{\substack{i,j=1 \\ i\neq j}}^{n} d_{\mathscr{G}}(v_i) + \sum_{\substack{i,j=1 \\ i\neq j}}^{n} d_{\mathscr{G}}(v_j) - 2\sum_{\substack{i,j=1 \\ i\neq j}}^{n} (\mu_{\mathscr{Q}}(v_iv_j), \eta_{\mathscr{Q}}(v_iv_j), \nu_{\mathscr{Q}}(v_iv_j)). \end{aligned}$$

For all $1 \leq j \leq n$, we have

$$\begin{aligned} 2\sum_{\substack{i=1 \\ j>i}}^{n-1} d_{\mathscr{G}}(v_iv_j) &= (n-1)\sum_{i=1}^{n} d_{\mathscr{G}}(v_i) + (n-1)\sum_{j=1}^{n} d_{\mathscr{G}}(v_j) - 2\left(2\sum_{\substack{i=1 \\ j>i}}^{n-1} (\mu_{\mathscr{Q}}(v_iv_j), \eta_{\mathscr{Q}}(v_iv_j), \nu_{\mathscr{Q}}(v_iv_j))\right), \\ 2\sum_{\substack{i=1 \\ j>i}}^{n-1} d_{\mathscr{G}}(v_iv_j) &= 2(n-1)\sum_{i=1}^{n} d_{\mathscr{G}}(v_i) - 2\left[2\left(\sum_{\substack{i=1 \\ j>i}}^{n-1} \mu_{\mathscr{Q}}(v_iv_j), \sum_{\substack{i=1 \\ j>i}}^{n-1} \eta_{\mathscr{Q}}(v_iv_j), \sum_{\substack{i=1 \\ j>i}}^{n-1} \nu_{\mathscr{Q}}(v_iv_j)\right)\right], \\ 2\sum_{\substack{i=1 \\ j>i}}^{n-1} d_{\mathscr{G}}(v_iv_j) &= 2(n-1)\sum_{i=1}^{n} d_{\mathscr{G}}(v_i) - 2\sum_{i=1}^{n} d_{\mathscr{G}}(v_i), \text{ using Theorem 1}, \\ \sum_{\substack{i=1 \\ j>i}}^{n-1} d_{\mathscr{G}}(v_iv_j) &= (n-2)\sum_{i=1}^{n} d_{\mathscr{G}}(v_i). \end{aligned}$$

This completes the proof. □

Next, we state some well known results regarding degrees of edges in q-RPFGs.

Theorem 3. *Let $\mathscr{G} = (\mathscr{P}, \mathscr{Q})$ be a q-RPFG on a cycle $G = (P, Q)$. Then,*

$$\sum_{v_i \in P} d_{\mathscr{G}}(v_i) = \sum_{v_iv_j \in Q} d_{\mathscr{G}}(v_iv_j).$$

Theorem 4. *Let $\mathscr{G} = (\mathscr{P}, \mathscr{Q})$ be a q-RPFG on $G = (P, Q)$. Then,*

$$\sum_{v_iv_j \in Q} d_{\mathscr{G}}(v_iv_j) = \sum_{v_iv_j \in Q} d_G(v_iv_j)\,(\mu_{\mathscr{Q}}(v_iv_j), \eta_{\mathscr{Q}}(v_iv_j), \nu_{\mathscr{Q}}(v_iv_j)),$$

where $d_G(v_iv_j) = d_G(v_i) + d_G(v_j) - 2$ for all $v_i, v_j \in P$.

Theorem 5. *Let $\mathscr{G} = (\mathscr{P}, \mathscr{Q})$ be a q-RPFG on a k-regular graph $G = (P, Q)$. Then,*

$$\sum_{v_iv_j \in Q} d_{\mathscr{G}}(v_iv_j) = 2(k-1)\, S(\mathscr{G}).$$

For proofs of above theorems, readers are referred to [37,41,44].

Definition 15. *The minimum edge degree of q-RPFG $\mathscr{G}$ is defined as $\delta_Q(\mathscr{G}) = (\delta_{Q_\mu}(\mathscr{G}), \delta_{Q_\eta}(\mathscr{G}), \delta_{Q_\nu}(\mathscr{G}))$, where $\delta_{Q_\mu}(\mathscr{G}) = \min\{d_\mu(uv) : uv \in Q\}$ is minimum μ-edge degree of $\mathscr{G}$, $\delta_{Q_\eta}(\mathscr{G}) = \min\{d_\eta(uv) : uv \in Q\}$ is minimum η-edge degree of $\mathscr{G}$, and $\delta_{Q_\nu}(\mathscr{G}) = \min\{d_\nu(uv) : uv \in Q\}$ is minimum ν-edge degree of $\mathscr{G}$.*

Definition 16. *The maximum edge degree of q-RPFG $\mathscr{G}$ is defined as $\Delta_Q(\mathscr{G}) = (\Delta_{Q_\mu}(\mathscr{G}), \Delta_{Q_\eta}(\mathscr{G}), \Delta_{Q_\nu}(\mathscr{G}))$, where $\Delta_{Q_\mu}(\mathscr{G}) = \max\{d_\mu(uv) : uv \in Q\}$ is maximum μ-edge degree of $\mathscr{G}$, $\Delta_{Q_\eta}(\mathscr{G}) = \max\{d_\eta(uv) : uv \in Q\}$ is maximum η-edge degree of $\mathscr{G}$, and $\Delta_{Q_\nu}(\mathscr{G}) = \max\{d_\nu(uv) : uv \in Q\}$ is maximum ν-edge degree of $\mathscr{G}$.*

Example 8. *Consider a q-rung picture fuzzy graph* $\mathscr{G} = (\mathscr{P}, \mathscr{Q})$ *as shown in Figure 6. By routine computations, it is easy to see that the minimum, and maximum edge degree of* $\mathscr{G}$ *are* $\delta_Q(\mathscr{G}) = (1.55, 1.25, 1.87)$ and $\Delta_Q(\mathscr{G}) = (3.01, 3.1, 3.95)$.

Definition 17. *Let* $\mathscr{G} = (\mathscr{P}, \mathscr{Q})$ *be a q-RPFG defined on* $G = (P, Q)$. *The total edge degree of uv in* $\mathscr{G}$ *is denoted by* $td_{\mathscr{G}}(uv) = (td_\mu(uv), td_\eta(uv), td_\nu(uv))$, *and defined as*

$$td_{\mathscr{G}}(uv) = d_{\mathscr{G}}(uv) + (\mu_{\mathscr{Q}}(uv), \eta_{\mathscr{Q}}(uv), \nu_{\mathscr{Q}}(uv)) = d_{\mathscr{G}}(u) + d_{\mathscr{G}}(v) - (\mu_{\mathscr{Q}}(uv), \eta_{\mathscr{Q}}(uv), \nu_{\mathscr{Q}}(uv)).$$

This is equivalent to

$$td_{\mathscr{G}}(uv) = \Big(\sum_{w \neq v} \mu_{\mathscr{Q}}(uw), \sum_{w \neq v} \eta_{\mathscr{Q}}(uw), \sum_{w \neq v} \nu_{\mathscr{Q}}(uw)\Big) + \Big(\sum_{w \neq u} \mu_{\mathscr{Q}}(vw), \sum_{w \neq u} \eta_{\mathscr{Q}}(vw), \sum_{w \neq u} \nu_{\mathscr{Q}}(vw)\Big)$$
$$+ (\mu_{\mathscr{Q}}(uv), \eta_{\mathscr{Q}}(uv), \nu_{\mathscr{Q}}(uv)).$$

Example 9. *Consider a 4-RPFG* $\mathscr{G}$ *displayed in Figure 6. The total degree of edge ad in* $\mathscr{G}$ *can be computed as*

$$\begin{aligned} td_{\mathscr{G}}(ad) &= d_{\mathscr{G}}(ad) + (\mu_{\mathscr{Q}}(ad), \eta_{\mathscr{Q}}(ad), \nu_{\mathscr{Q}}(ad)) \\ &= (1.96, 1.4, 1.77) + (0.6, 0.7, 0.6) = (2.56, 2.1, 2.37). \end{aligned}$$

Theorem 6. *For any q-RPFG* $\mathscr{G} = (\mathscr{P}, \mathscr{Q})$ *defined on* $P = \{v_1, v_2, \ldots, v_n\}$, *if* $(\mu_{\mathscr{Q}}(v_iv_j), \eta_{\mathscr{Q}}(v_iv_j), \nu_{\mathscr{Q}}(v_iv_j)) \neq (0,0,0)$ *for all* $v_i, v_j \in P$ $(i \neq j)$, *then the following relation for total degrees of edges of* $\mathscr{G}$ *must hold:*

$$\sum_{\substack{i=1 \\ j>i}}^{n-1} td_{\mathscr{G}}(v_iv_j) = \Big(n - \frac{3}{2}\Big) \sum_{i=1}^{n} td_{\mathscr{G}}(v_i)$$

for all $1 \leq j \leq n$.

Proof. The proof directly follows from Theorem 2 and Definition 17. □

Theorem 7. *Let* $\mathscr{G} = (\mathscr{P}, \mathscr{Q})$ *be a q-RPFG on* $G = (P, Q)$. *Then,*

$$\sum_{v_iv_j \in Q} td_{\mathscr{G}}(v_iv_j) = \sum_{v_iv_j \in Q} d_G(v_iv_j)\,(\mu_{\mathscr{Q}}(v_iv_j), \eta_{\mathscr{Q}}(v_iv_j), \nu_{\mathscr{Q}}(v_iv_j)) + S(\mathscr{G}),$$

where $d_G(v_iv_j) = d_G(v_i) + d_G(v_j) - 2$ *for all* $v_i, v_j \in P$.

Proof. The total degree of an edge v_iv_j in a q-RPFG is

$$td_{\mathscr{G}}(v_iv_j) = d_{\mathscr{G}}(v_iv_j) + (\mu_{\mathscr{Q}}(v_iv_j), \eta_{\mathscr{Q}}(v_iv_j), \nu_{\mathscr{Q}}(v_iv_j)).$$

Therefore,

$$\begin{aligned} \sum_{v_iv_j \in Q} td_{\mathscr{G}}(v_iv_j) &= \sum_{v_iv_j \in Q} (d_{\mathscr{G}}(v_iv_j) + (\mu_{\mathscr{Q}}(v_iv_j), \eta_{\mathscr{Q}}(v_iv_j), \nu_{\mathscr{Q}}(v_iv_j))) \\ &= \sum_{v_iv_j \in Q} d_{\mathscr{G}}(v_iv_j) + \sum_{v_iv_j \in Q} (\mu_{\mathscr{Q}}(v_iv_j), \eta_{\mathscr{Q}}(v_iv_j), \nu_{\mathscr{Q}}(v_iv_j)) \\ &= \sum_{v_iv_j \in Q} d_G(v_iv_j)\,(\mu_{\mathscr{Q}}(v_iv_j), \eta_{\mathscr{Q}}(v_iv_j), \nu_{\mathscr{Q}}(v_iv_j)) + S(\mathscr{G}), \end{aligned}$$

where $d_G(v_iv_j) = d_G(v_i) + d_G(v_j) - 2$ for all $v_i, v_j \in P$. This completes the proof. □

The concept of edge degree leads to defining edge regularity of q-RPFGs. Formally, we have the following definition:

Definition 18. *Let $\mathscr{G} = (\mathscr{P}, \mathscr{Q})$ be a q-RPFG defined on G = (P, Q). If each edge of $\mathscr{G}$ has same degree, which is*

$$d_{\mathscr{G}}(uv) = (p_1, p_2, p_3) \text{ for all } uv \in Q,$$

then $\mathscr{G}$ is called (p_1, p_2, p_3)-edge regular q-RPFG.

Example 10. *Consider a 3-RPFG $\mathscr{G} = (\mathscr{P}, \mathscr{Q})$ defined on G = (P, Q), where $\mathscr{P}$ be a 3-rung picture fuzzy set on P, and $\mathscr{Q}$ be a 3-rung picture fuzzy relation on P, defined by*

$\mathscr{P}$	a	b	c
$\mu_{\mathscr{P}}$	0.4	0.2	0.9
$\eta_{\mathscr{P}}$	0.5	0.3	0.1
$\nu_{\mathscr{P}}$	0.6	0.4	0.6

$\mathscr{Q}$	ab	bc	ac
$\mu_{\mathscr{Q}}$	0.2	0.2	0.2
$\eta_{\mathscr{Q}}$	0.1	0.1	0.1
$\nu_{\mathscr{Q}}$	0.6	0.6	0.6

We see that $d_{\mathscr{G}}(ab) = d_{\mathscr{G}}(bc) = d_{\mathscr{G}}(ac) = (0.4, 0.2, 1.2)$. Hence, the 3-RPFG $\mathscr{G}$, displayed in Figure 7, is $(0.4, 0.2, 1.2)$-edge regular.

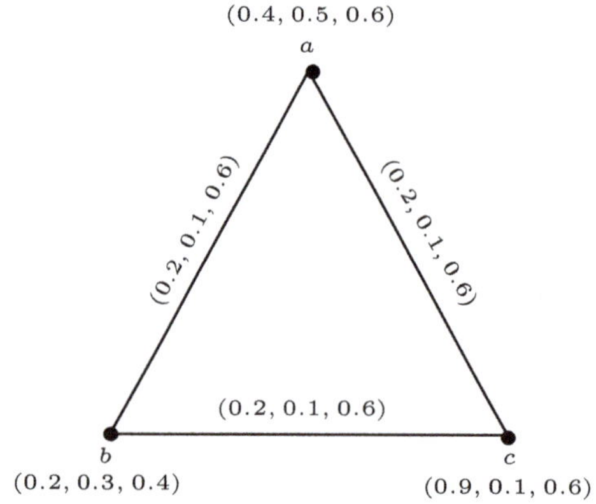

Figure 7. An edge regular 3-RPFG $\mathscr{G}$.

Definition 19. *Let $\mathscr{G} = (\mathscr{P}, \mathscr{Q})$ be a q-RPFG defined on G = (P, Q). If each edge of $\mathscr{G}$ has same total edge degree, that is*

$$td_{\mathscr{G}}(uv) = (q_1, q_2, q_3) \text{ for all } uv \in Q,$$

then $\mathscr{G}$ is called (q_1, q_2, q_3)-total edge regular q-RPFG.

Example 11. *Consider a 3-RPFG $\mathscr{G} = (\mathscr{P}, \mathscr{Q})$ as displayed in Figure 8.*

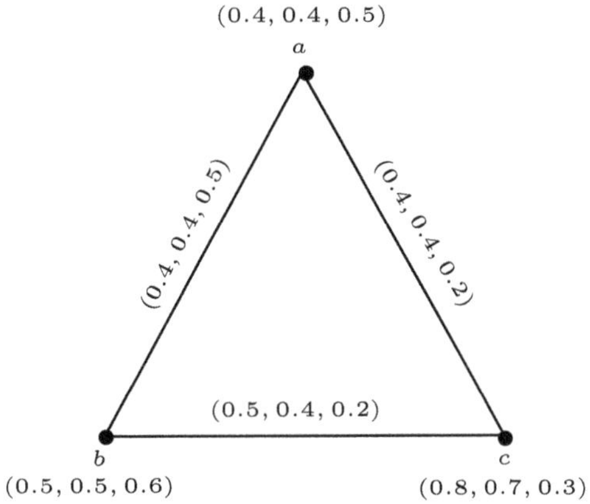

Figure 8. A total edge regular 3-RPFG $\mathscr{G}$.

We see that $\mathscr{G}$ *is* $(1.3, 1.2.0.9)$*-total edge regular 3-RPFG since* $td_{\mathscr{G}}(ab) = td_{\mathscr{G}}(bc) = td_{\mathscr{G}}(ac) = (1.3, 1.2, 0.9)$.

Remark 1.

1. *Any connected q-RPFG with two vertices is edge regular.*
2. *A q-RPFG is edge regular if and only if* $\delta_Q(\mathscr{G}) = \Delta_Q(\mathscr{G})$.

Remark 2. *A complete q-rung picture fuzzy graph need not be edge regular.*
For example, consider a 4*-RPFG* $\mathscr{G}$ *as displayed in Figure 9.*

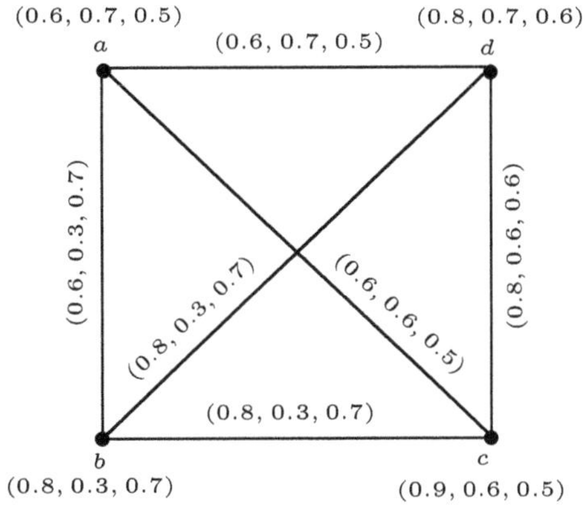

Figure 9. 4-RPFG $\mathscr{G}$.

It is clear that $\mathscr{G}$ *is a complete* 4*-RPFG. However,* $\mathscr{G}$ *is not edge regular since* $d_{\mathscr{G}}(ab) = (2.8, 1.9, 2.4) \neq (2.8, 1.8, 2.5) = d_{\mathscr{G}}(ad)$.

Next, we present a necessary condition for a complete q-RPFG to be edge regular.

Theorem 8. *Let* $\mathscr{G} = (\mathscr{P}, \mathscr{Q})$ *be a complete q-rung picture fuzzy graph on* $G = (P, Q)$*, and* $\mu_{\mathscr{Q}}, \eta_{\mathscr{Q}}$*, and* $\nu_{\mathscr{Q}}$ *are constant functions; then,* $\mathscr{G}$ *is an edge regular q-RPFG.*

Proof. Let $\mathscr{G} = (\mathscr{P}, \mathscr{Q})$ be a complete q-rung picture fuzzy graph defined on $G = (P, Q)$, where $P = \{u_1, u_2, \ldots, u_n\}$. Then, for all $u, v \in P$, $\mu_{\mathscr{Q}}(uv) = \mu_{\mathscr{P}}(u) \wedge \mu_{\mathscr{P}}(v)$, $\eta_{\mathscr{Q}}(uv) = \eta_{\mathscr{P}}(u) \wedge \eta_{\mathscr{P}}(v)$, and $\nu_{\mathscr{Q}}(uv) = \nu_{\mathscr{P}}(u) \vee \nu_{\mathscr{P}}(v)$. Let $\mu_{\mathscr{Q}}(uv) = c_1$, $\eta_{\mathscr{Q}}(uv) = c_2$ and $\nu_{\mathscr{Q}}(uv) = c_3$, for all $uv \in Q$. The completeness of $\mathscr{G}$ implies that each vertex u of $\mathscr{G}$ is connected with $n-1$ vertices by edges, with membership values $(\mu_{\mathscr{Q}}(uv), \eta_{\mathscr{Q}}(uv), \nu_{\mathscr{Q}}(uv)) = (c_1, c_2, c_3)$. Thus, degree of each vertex $u \in P$ can be written as $d_{\mathscr{G}}(u) = (n-1)(c_1, c_2, c_3)$. By definition of edge degree, we have

$$\begin{aligned} d_{\mathscr{G}}(uv) &= d_{\mathscr{G}}(u) + d_{\mathscr{G}}(v) - 2(\mu_{\mathscr{Q}}(uv), \eta_{\mathscr{Q}}(uv), \nu_{\mathscr{Q}}(uv)) \\ &= (n-1)(c_1, c_2, c_3) + (n-1)(c_1, c_2, c_3) - 2(c_1, c_2, c_3) \\ &= 2(n-1)(c_1, c_2, c_3) - 2(c_1, c_2, c_3) \\ &= 2((n-2)c_1, (n-2)c_2, (n-2)c_3) \end{aligned}$$

for all $uv \in Q$. Hence, $\mathscr{G}$ is an edge regular q-RPFG. This completes the proof. □

Remark 3. *An edge regular q-rung picture fuzzy graph may not be total edge regular.*
For example, consider a 6-RPFG $\mathscr{G} = (\mathscr{P}, \mathscr{Q})$ *as displayed in Figure 10.*

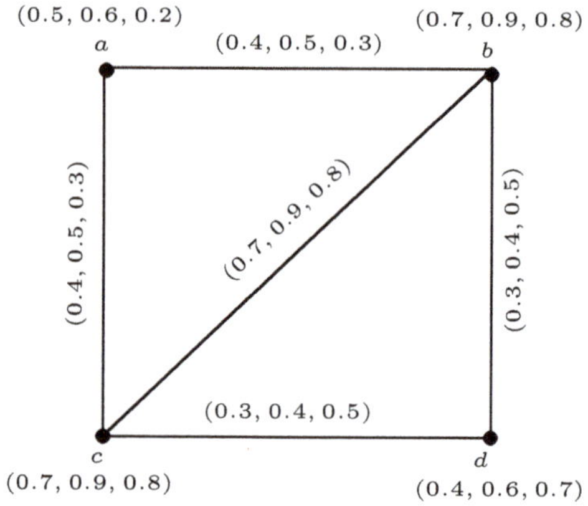

Figure 10. 6-RPFG $\mathscr{G}$.

We see that $\mathscr{G}$ *is edge regular as* $d_{\mathscr{G}}(ab) = d_{\mathscr{G}}(ac) = d_{\mathscr{G}}(bc) = d_{\mathscr{G}}(bd) = d_{\mathscr{G}}(cd) = (1.4, 1.8, 1.6)$. *However,* $\mathscr{G}$ *is not total edge regular 6-RPFG since* $td_{\mathscr{G}}(ab) = (1.8, 2.3, 1.9) \neq (1.7, 2.2, 2.1) = td_{\mathscr{G}}(cd)$.

Remark 4. *A total edge regular q-rung picture fuzzy graph may not be edge regular.*
For example, consider a 3-rung picture fuzzy graph $\mathscr{G} = (\mathscr{P}, \mathscr{Q})$ *as displayed in Figure 8. We see that* $\mathscr{G}$ *is* $(1.3, 1.2, 0.9)$*-total edge regular. However,* $\mathscr{G}$ *is not an edge regular 3-RPFG since* $d_{\mathscr{G}}(ab) = (0.9, 0.8, 0.4) \neq (0.9, 0.8, 0.7) = d_{\mathscr{G}}(ac)$.

Moreover, it is easy to observe that there may exist such q-RPFGs which are both edge regular, and total edge regular—neither edge regular nor total edge regular. Thus, there does not exist any relationship between edge regular q-RPFGs, and total edge regular q-RPFGs. Next, we illustrate a theorem providing characterization for edge regularity of q-rung picture fuzzy graphs.

Theorem 9. *Let* $\mathscr{G} = (\mathscr{P}, \mathscr{Q})$ *be a q-rung picture fuzzy graph on* $G = (P, Q)$, *and* $\mu_{\mathscr{Q}}, \eta_{\mathscr{Q}}$, *and* $\nu_{\mathscr{Q}}$ *are constant functions; then, the following are equivalent:*

(a) $\mathscr{G}$ *is an edge regular q-RPFG.*
(b) $\mathscr{G}$ *is a total edge regular q-RPFG.*

Proof. Let $\mathscr{G} = (\mathscr{P}, \mathscr{Q})$ be a q-RPFG on $G = (P, Q)$. Suppose that, for all $uv \in Q$, $\mu_{\mathscr{Q}}(uv) = c_1$, $\eta_{\mathscr{Q}}(uv) = c_2$ and $\nu_{\mathscr{Q}}(uv) = c_3$. Assume that $\mathscr{G}$ is a (p_1, p_2, p_3)-edge regular q-RPFG. Then, for all $uv \in Q$, $d_{\mathscr{G}}(uv) = (p_1, p_2, p_3)$. By definition of total edge degree,

$$\begin{aligned} td_{\mathscr{G}}(uv) &= d_{\mathscr{G}}(uv) + (\mu_{\mathscr{Q}}(uv),\ \eta_{\mathscr{Q}}(uv),\ \nu_{\mathscr{Q}}(uv)) \\ &= (p_1, p_2, p_3) + (c_1, c_2, c_3) \\ &= (p_1 + c_1, p_2 + c_2, p_3 + c_3) \end{aligned}$$

for all $uv \in Q$. Hence, $\mathscr{G}$ is total edge regular q-RPFG.

Conversely, suppose that $\mathscr{G}$ is a (q_1, q_2, q_3)-total edge regular q-RPFG. Then, by definition of total edge regular

$$\begin{aligned} td_{\mathscr{G}}(uv) &= d_{\mathscr{G}}(uv) + (\mu_{\mathscr{Q}}(uv),\ \eta_{\mathscr{Q}}(uv),\ \nu_{\mathscr{Q}}(uv)) \\ (q_1, q_2, q_3) &= d_{\mathscr{G}}(uv) + (c_1, c_2, c_3) \\ d_{\mathscr{G}}(uv) &= (q_1, q_2, q_3) - (c_1, c_2, c_3) \\ &= (q_1 - c_1, q_2 - c_2, q_3 - c_3) \end{aligned}$$

for all $uv \in Q$. Hence, $\mathscr{G}$ is edge regular q-RPFG. This proves that (a) and (b) are equivalent.

On the other hand, assume that (a), and (b) are equivalent. Suppose, on the contrary, that $\mu_{\mathscr{Q}}$, $\eta_{\mathscr{Q}}$, and $\nu_{\mathscr{Q}}$ are not constant functions. Then, there exist at least one edge xy in Q such that $(\mu_{\mathscr{Q}}(uv),\ \eta_{\mathscr{Q}}(uv),\ \nu_{\mathscr{Q}}(uv)) \neq (\mu_{\mathscr{Q}}(xy),\ \eta_{\mathscr{Q}}(xy),\ and \nu_{\mathscr{Q}}(xy))$. Let $\mathscr{G}$ be a (p_1, p_2, p_3)-edge regular q-RPFG. Then, $d_{\mathscr{G}}(uv) = d_{\mathscr{G}}(xy) = (p_1, p_2, p_3)$. Therefore,

$$\begin{aligned} td_{\mathscr{G}}(uv) &= d_{\mathscr{G}}(uv) + (\mu_{\mathscr{Q}}(uv),\ \eta_{\mathscr{Q}}(uv),\ \nu_{\mathscr{Q}}(uv)) \\ &= (p_1, p_2, p_3) + (\mu_{\mathscr{Q}}(uv),\ \eta_{\mathscr{Q}}(uv),\ \nu_{\mathscr{Q}}(uv)), \end{aligned}$$

$$\begin{aligned} td_{\mathscr{G}}(xy) &= d_{\mathscr{G}}(xy) + (\mu_{\mathscr{Q}}(xy),\ \eta_{\mathscr{Q}}(xy),\ \nu_{\mathscr{Q}}(xy)) \\ &= (p_1, p_2, p_3) + (\mu_{\mathscr{Q}}(xy),\ \eta_{\mathscr{Q}}(xy),\ \nu_{\mathscr{Q}}(xy)). \end{aligned}$$

Since $(\mu_{\mathscr{Q}}(uv),\ \eta_{\mathscr{Q}}(uv),\ \nu_{\mathscr{Q}}(uv)) \neq (\mu_{\mathscr{Q}}(xy),\ \eta_{\mathscr{Q}}(xy),\ \nu_{\mathscr{Q}}(xy))$, therefore, $td_{\mathscr{G}}(uv) \neq td_{\mathscr{G}}(xy)$. Hence, $\mathscr{G}$ is not total edge regular, which gives a contradiction to our assumption.

Now, let $\mathscr{G}$ be a total edge regular q-RPFG. Then, by definition of total edge regular,

$$\begin{aligned} td_{\mathscr{G}}(uv) &= td_{\mathscr{G}}(xy), \\ d_{\mathscr{G}}(uv) + (\mu_{\mathscr{Q}}(uv), \eta_{\mathscr{Q}}(uv), \nu_{\mathscr{Q}}(uv)) &= d_{\mathscr{G}}(xy) + (\mu_{\mathscr{Q}}(xy), \eta_{\mathscr{Q}}(xy), \nu_{\mathscr{Q}}(xy)), \\ d_{\mathscr{G}}(uv) - d_{\mathscr{G}}(xy) &= (\mu_{\mathscr{Q}}(xy), \eta_{\mathscr{Q}}(xy), \nu_{\mathscr{Q}}(xy)) - (\mu_{\mathscr{Q}}(uv), \eta_{\mathscr{Q}}(uv), \nu_{\mathscr{Q}}(uv)) \neq 0, \\ d_{\mathscr{G}}(uv) &\neq d_{\mathscr{G}}(xy). \end{aligned}$$

The fact that $\mathscr{G}$ is not edge regular leads a contradiction to our assumption. Hence, $\mu_{\mathscr{Q}}$, $\eta_{\mathscr{Q}}$, and $\nu_{\mathscr{Q}}$ are constant functions. This completes the proof. □

Definition 20. *A q-RPFG $\mathscr{G} = (\mathscr{P}, \mathscr{Q})$ defined on $G = (P, Q)$ is said to be partially edge regular if the underlying graph $G = (P, Q)$ is edge regular.*

Example 12. *Consider a 5-RPFG $\mathscr{G} = (\mathscr{P}, \mathscr{Q})$ displayed in Figure 11a, and its underlying crisp graph $G = (P, Q)$ displayed in Figure 11b.*

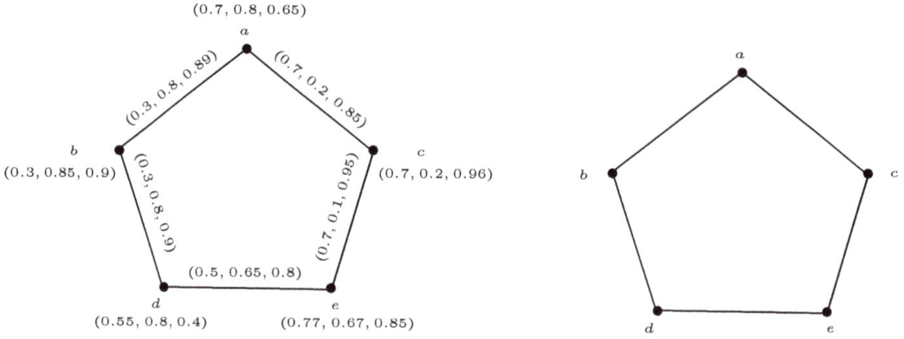

(a) $\mathscr{G}$ (b) G

Figure 11. A partially edge regular 5-RPFG $\mathscr{G}$.

We see that G is 2-edge regular graph. Thus, $\mathscr{G}$ is partially edge regular 5-RPFG.

Definition 21. *A q-RPFG $\mathscr{G} = (\mathscr{P}, \mathscr{Q})$ defined on $G = (P, Q)$ is said to be full edge regular if $\mathscr{G}$ is both edge regular, and partially edge regular.*

Example 13. *Figure 7 can be considered as an example of full edge regular 3-RPFG. Since $\mathscr{G}$ as well as its underlying crisp graph G are edge regular.*

Remark 5. *An edge regular q-rung picture fuzzy graph may not be partially edge regular (or full edge regular). For example, consider the 6-RPFG $\mathscr{G} = (\mathscr{P}, \mathscr{Q})$, displayed in Figure 12a, and its underlying crisp graph $G = (P, Q)$ displayed in Figure 12b.*

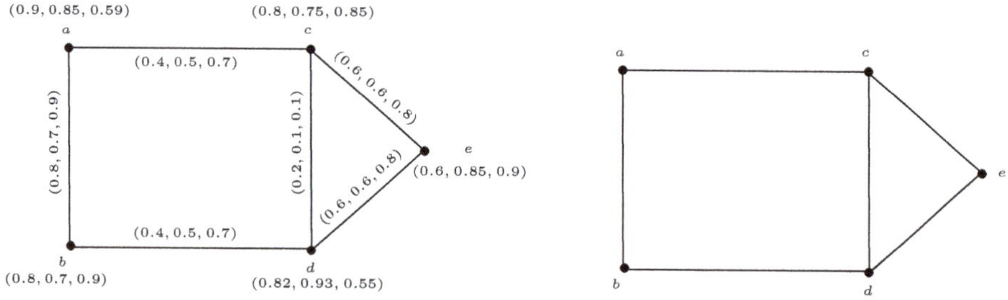

(a) $\mathscr{G}$ (b) G

Figure 12. An edge regular but not partially edge regular 6-RPFG $\mathscr{G}$.

We see that $\mathscr{G}$ is edge regular as $d_{\mathscr{G}}(a) = d_{\mathscr{G}}(b) = d_{\mathscr{G}}(c) = d_{\mathscr{G}}(d) = d_{\mathscr{G}}(e) = (1.2, 1.2, 1.6)$. However, $\mathscr{G}$ is not partially edge regular 6-RPFG since its underlying graph $G = (P, Q)$ is not edge regular.

Remark 6. *A partially edge regular q-rung picture fuzzy graph may not be edge regular (or full edge regular). For example, consider a 5-RPFG $\mathscr{G} = (\mathscr{P}, \mathscr{Q})$ displayed in Figure 11. We see that $\mathscr{G}$ is partially edge regular as its underlying graph G is edge regular. However, $\mathscr{G}$ is not edge regular 5-RPFG since $d_{\mathscr{G}}(ab) = (1.0, 1.0, 1.75) \neq (1.0, 0.9, 1.84) = d_{\mathscr{G}}(ac)$.*

The above remarks show that there does not exist any relationship between edge regular, and partially edge regular (or full edge regular) q-RPFGs. In the following theorem, we develop a relationship between edge regular, and partially edge regular q-RPFGs.

Theorem 10. *Let $\mathscr{G} = (\mathscr{P}, \mathscr{Q})$ be a q-rung picture fuzzy graph on $G = (P, Q)$. If $\mu_{\mathscr{Q}}, \eta_{\mathscr{Q}}$, and $\nu_{\mathscr{Q}}$ are constant functions, then $\mathscr{G}$ is edge regular if and only if $\mathscr{G}$ is partially edge regular.*

Proof. Let $\mathscr{G} = (\mathscr{P}, \mathscr{Q})$ be a q-RPFG such that for all $uv \in Q$, $\mu_{\mathscr{Q}}(uv) = c_1$, $\eta_{\mathscr{Q}}(uv) = c_2$ and $\nu_{\mathscr{Q}}(uv) = c_3$. Then, by definition of edge degree

$$\begin{aligned}
d_{\mathscr{G}}(uv) &= \Big(\sum_{\substack{uw\in Q\\ w\neq v}} \mu_{\mathscr{Q}}(uw), \sum_{\substack{uw\in Q\\ w\neq v}} \eta_{\mathscr{Q}}(uw), \sum_{\substack{uw\in Q\\ w\neq v}} \nu_{\mathscr{Q}}(uw) \Big) + \Big(\sum_{\substack{vw\in Q\\ w\neq u}} \mu_{\mathscr{Q}}(vw), \sum_{\substack{vw\in Q\\ w\neq u}} \eta_{\mathscr{Q}}(vw), \sum_{\substack{vw\in Q\\ w\neq u}} \nu_{\mathscr{Q}}(vw) \Big) \\
&= \sum_{\substack{uw\in Q\\ w\neq v}} (c_1, c_2, c_3) + \sum_{\substack{vw\in Q\\ w\neq u}} (c_1, c_2, c_3) \\
&= (c_1, c_2, c_3)(d_G(u) - 1) + (c_1, c_2, c_3)(d_G(u) - 1) \\
&= (c_1, c_2, c_3)(d_G(u) + d_G(v) - 2) \\
&= (c_1, c_2, c_3)\, d_G(uv)
\end{aligned}$$

for all $uv \in Q$. Hence, $\mathscr{G}$ is edge regular. Assume that $\mathscr{G} = (\mathscr{P}, \mathscr{Q})$ is a (p_1, p_2, p_3)-edge regular q-RPFG. Then, for all $uv \in Q$,

$$\begin{aligned}
(p_1, p_2, p_3) &= (c_1, c_2, c_3)\, d_G(uv), \\
d_G(uv) &= \frac{(p_1, p_2, p_3)}{(c_1, c_2, c_3)}.
\end{aligned}$$

This shows that G is an edge regular q-RPFG. Hence, $\mathscr{G}$ is partially edge regular q-RPFG.

Conversely, assume that $\mathscr{G}$ is partially edge regular q-RPFG. Let G be a p-edge regular graph. Then, for all $uv \in Q$, $d_{\mathscr{G}}(uv) = p\,(c_1, c_2, c_3)$. Consequently, $\mathscr{G}$ is edge regular q-RPFG. This completes the proof. □

Corollary 1. *Let $\mathscr{G} = (\mathscr{P}, \mathscr{Q})$ be a q-RPFG such that $\mu_{\mathscr{Q}}, \eta_{\mathscr{Q}}$, and $\nu_{\mathscr{Q}}$ are constant functions. If $\mathscr{G}$ is an edge regular q-RPFG or a partially edge regular q-RPFG, then $\mathscr{G}$ is a full edge regular q-RPFG.*

Theorem 11. *Let $\mathscr{G} = (\mathscr{P}, \mathscr{Q})$ be a strong q-RPFG such that $\mu_{\mathscr{P}}, \eta_{\mathscr{P}}$, and $\nu_{\mathscr{P}}$ are constant functions. Then, $\mathscr{G}$ is an edge regular q-RPFG if and only if $\mathscr{G}$ is partially edge regular q-RPFG.*

Proof. Let $\mathscr{G} = (\mathscr{P}, \mathscr{Q})$ be a strong q-RPFG defined on $G = (P, Q)$. Then, for all $uv \in Q$, $\mu_{\mathscr{Q}}(uv) = \mu_{\mathscr{P}}(u) \wedge \mu_{\mathscr{P}}(v)$, $\eta_{\mathscr{Q}}(uv) = \eta_{\mathscr{P}}(u) \wedge \eta_{\mathscr{P}}(v)$, and $\nu_{\mathscr{Q}}(uv) = \nu_{\mathscr{P}}(u) \vee \nu_{\mathscr{P}}(v)$. Let $\mu_{\mathscr{P}}, \eta_{\mathscr{P}}$, and $\nu_{\mathscr{P}}$ be constant functions. Then, for all $u \in P$, $\mu_{\mathscr{P}}(u) = c_1$, $\eta_{\mathscr{P}}(u) = c_2$, and $\nu_{\mathscr{P}}(u) = c_3$. Combining both facts, we have

$$\begin{aligned}
\mu_{\mathscr{Q}}(uv) &= \mu_{\mathscr{P}}(u) \wedge \mu_{\mathscr{P}}(v) = c_1 \wedge c_1 = c_1, \\
\eta_{\mathscr{Q}}(uv) &= \eta_{\mathscr{P}}(u) \wedge \eta_{\mathscr{P}}(v) = c_2 \wedge c_2 = c_2, \\
\nu_{\mathscr{Q}}(uv) &= \nu_{\mathscr{P}}(u) \vee \nu_{\mathscr{P}}(v) = c_3 \vee c_3 = c_3.
\end{aligned}$$

Thus, $\mu_{\mathscr{Q}}, \eta_{\mathscr{Q}}$, and $\nu_{\mathscr{Q}}$ are constant functions. Consequently, the result follows from Theorem 10. □

Corollary 2. *Let $\mathscr{G} = (\mathscr{P}, \mathscr{Q})$ be a strong q-RPFG such that $\mu_{\mathscr{P}}, \eta_{\mathscr{P}}$, and $\nu_{\mathscr{P}}$ are constant functions. If $\mathscr{G}$ is an edge regular q-RPFG or a partially edge regular q-RPFG, then $\mathscr{G}$ is a full edge regular q-RPFG.*

Next, we state some results regarding order, and size of edge regular q-RPFG.

Theorem 12. *The size of (p_1, p_2, p_3)-edge regular q-RPFG $\mathscr{G} = (\mathscr{P}, \mathscr{Q})$ on a p-edge regular graph $G = (P, Q)$ is $\frac{q(p_1, p_2, p_3)}{p}$, where $q = |Q|$.*

Theorem 13. *If $\mathscr{G} = (\mathscr{P}, \mathscr{Q})$ is (q_1, q_2, q_3)-total edge regular, and (r_1, r_2, r_3)-partially edge regular q-RPFG, then $S(\mathscr{G}) = \frac{q(q_1,q_2,q_3)}{(r_1,r_2,r_3)+1}$, where $q = |Q|$.*

Theorem 14. *If $\mathscr{G} = (\mathscr{P}, \mathscr{Q})$ is (p_1, p_2, p_3)-edge regular, and (q_1, q_2, q_3)-total edge regular q-RPFG, then $S(\mathscr{G}) = q\,(q_1 - p_1, q_2 - p_2, q_3 - p_3)$, where $q = |Q|$.*

For proofs of above results, readers are referred to [37,41].

2.1. Perfectly Edge Regular q-RPFGs

In this section, we introduce perfect edge regular q-RPFGs taking as a point of departure the respective definition of perfect edge regular fuzzy graphs by Cary [38].

Definition 22. *A perfect edge regular q-RPFG is a q-RPFG that is both edge regular, and total edge regular.*

Example 14. *Consider a 5-rung picture fuzzy graph $\mathscr{G} = (\mathscr{P}, \mathscr{Q})$ as shown in Figure 13.*

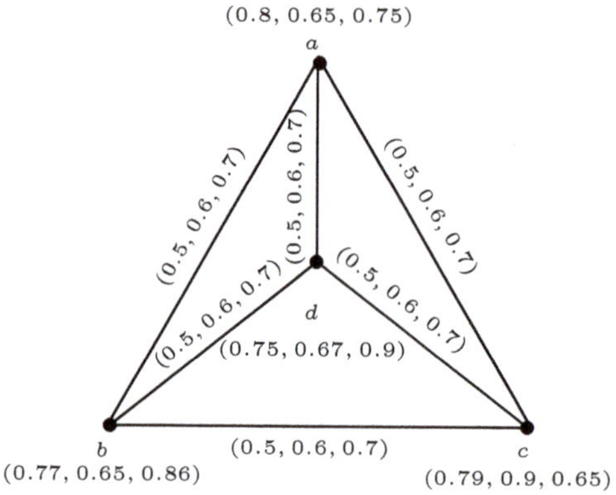

Figure 13. Perfectly edge regular 5-RPFG.

Clearly, $\mathscr{G}$ is $(2.0, 2.4, 2.8)$-edge regular, and $(2.5, 3.0, 3.5)$-total edge regular. Hence, $\mathscr{G}$ is perfect edge regular 5-RPFG.

Theorem 15. *If $\mathscr{G} = (\mathscr{P}, \mathscr{Q})$ is perfect edge regular q-RPFG, then $\mu_{\mathscr{Q}}, \eta_{\mathscr{Q}}$, and $\nu_{\mathscr{Q}}$ are constant functions.*

Proof. Let $\mathscr{G} = (\mathscr{P}, \mathscr{Q})$ be a perfect edge regular q-RPFG defined on $G = (P, Q)$. Then, $\mathscr{G}$ must be edge regular and total edge regular, i.e., for all $uv \in Q$, $d_{\mathscr{G}}(uv) = (p_1, p_2, p_3)$, and $td_{\mathscr{G}}(uv) = (q_1, q_2, q_3)$. By definition of total edge degree,

$$\begin{aligned} d_{\mathscr{G}}(uv) + (\mu_{\mathscr{Q}}(uv), \eta_{\mathscr{Q}}(uv), \nu_{\mathscr{Q}}(uv)) &= (q_1, q_2, q_3), \\ (p_1, p_2, p_3) + (\mu_{\mathscr{Q}}(uv), \eta_{\mathscr{Q}}(uv), \nu_{\mathscr{Q}}(uv)) &= (q_1, q_2, q_3), \\ (\mu_{\mathscr{Q}}(uv), \eta_{\mathscr{Q}}(uv), \nu_{\mathscr{Q}}(uv)) &= (q_1 - p_1, q_2 - p_2, q_3 - p_3). \end{aligned}$$

Since $\mu_{\mathscr{Q}}(uv) = q_1 - p_1$, $\eta_{\mathscr{Q}}(uv) = q_2 - p_2$, and $\nu_{\mathscr{Q}}(uv) = q_3 - p_3$, for all $uv \in Q$, therefore, $\mu_{\mathscr{Q}}, \eta_{\mathscr{Q}}$, and $\nu_{\mathscr{Q}}$ are constant functions. This completes the proof. □

Perfectly edge regular q-RPFGs can be classified by the following theorem.

Theorem 16. *A q-RPFG $\mathscr{G} = (\mathscr{P}, \mathscr{Q})$ is perfect edge regular if and only if it satisfies the following conditions:*

1. $\sum_{uw\in Q}\left(\mu_{\mathscr{Q}}(uw),\eta_{\mathscr{Q}}(uw),\nu_{\mathscr{Q}}(uw)\right)+\sum_{vw\in Q}\left(\mu_{\mathscr{Q}}(vw),\eta_{\mathscr{Q}}(vw),\nu_{\mathscr{Q}}(vw)\right)-2(\mu_{\mathscr{Q}}(uv),\eta_{\mathscr{Q}}(uv),\nu_{\mathscr{Q}}(uv))=\sum_{xz\in Q}\left(\mu_{\mathscr{Q}}(xz),\eta_{\mathscr{Q}}(xz),\nu_{\mathscr{Q}}(xz)\right)+\sum_{yz\in Q}\left(\mu_{\mathscr{Q}}(yz),\eta_{\mathscr{Q}}(yz),\nu_{\mathscr{Q}}(yz)\right)-2(\mu_{\mathscr{Q}}(xy),\eta_{\mathscr{Q}}(xy),\nu_{\mathscr{Q}}(xy))$,
2. $(\mu_{\mathscr{Q}}(uv),\eta_{\mathscr{Q}}(uv),\nu_{\mathscr{Q}}(uv))=(\mu_{\mathscr{Q}}(xy),\eta_{\mathscr{Q}}(xy),\nu_{\mathscr{Q}}(xy))$,

for all $uv,\ xy\in Q$.

Proof. Suppose that $\mathscr{G}=(\mathscr{P},\mathscr{Q})$ is perfect edge regular q-RPFG. Then, it must be regular as well as total edge regular. The fact that $\mathscr{G}$ is edge regular implies Condition 1 so that each edge has the same degree. Condition 2 is followed directly by Theorem 15.

Conversely, suppose that Condition 1 and 2 hold. It is straightforward to see that Condition 1 guarantees the edge regularity of $\mathscr{G}$. Let for any two edges uv and xy, $d_{\mathscr{G}}(uv)=d_{\mathscr{G}}(xy)$. By definition of total edge degree,

$$\begin{aligned} td_{\mathscr{G}}(uv) &= d_{\mathscr{G}}(uv)+(\mu_{\mathscr{Q}}(uv),\eta_{\mathscr{Q}}(uv),\nu_{\mathscr{Q}}(uv)) \\ &= d_{\mathscr{G}}(xy)+(\mu_{\mathscr{Q}}(xy),\eta_{\mathscr{Q}}(xy),\nu_{\mathscr{Q}}(xy)) = td_{\mathscr{G}}(xy). \end{aligned}$$

Thus, $\mathscr{G}$ is total edge regular, as uv and xy are arbitrary edges. Consequently, $\mathscr{G}$ is perfect edge regular. This completes the proof. □

Remark 7. *The above theorem does not constitute a sufficient condition for a q-RPFG to be perfect edge regular. For example, consider a 4-RPFG $\mathscr{G}$ as displayed in Figure 14.*

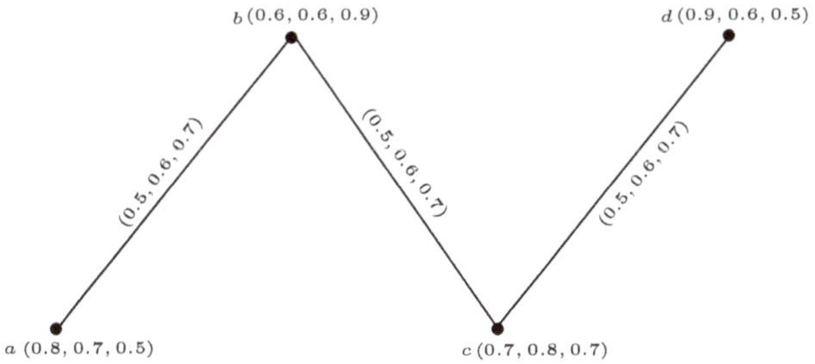

Figure 14. A 4-RPFG $\mathscr{G}$.

We see that $(\mu_{\mathscr{Q}}(uv),\eta_{\mathscr{Q}}(uv),\nu_{\mathscr{Q}}(uv))=(0.5,0.6,0.7)$, for all $uv\in Q$. However, since $d_{\mathscr{G}}(ab)=(0.5,0.6,0.7)\neq(1.0,1.2,1.4)=d_{\mathscr{G}}(bc)$, and $td_{\mathscr{G}}(ab)=(1.0,1.2,1.4)\neq(1.5,1.8,2.1)=td_{\mathscr{G}}(bc)$, therefore, $\mathscr{G}$ is neither edge regular nor totally egde regular. Hence, $\mathscr{G}$ is not a perfect edge regular 4-RPFG.

The following theorem provides a necessary condition for the converse of the above theorem to hold.

Theorem 17. *If $\mathscr{G}=(\mathscr{P},\mathscr{Q})$ is edge regular q-RPFG, and $\mu_{\mathscr{Q}},\eta_{\mathscr{Q}}$, and $\nu_{\mathscr{Q}}$ are constant functions, then $\mathscr{G}$ is perfect edge regular q-RPFG.*

Proof. The proof is directly followed from Theorem 9, which proves that $\mathscr{G}$ is total edge regular q-RPFG. Moreover, by our assumption, $\mathscr{G}$ is edge regular. Hence, $\mathscr{G}$ is perfect edge regular. This completes the proof. □

We next illustrate some results relating order and size of perfect edge regular q-RPFG.

Theorem 18. *If $\mathscr{G} = (\mathscr{P}, \mathscr{Q})$ is perfect edge regular q-RPFG, and $\mu_{\mathscr{Q}}, \eta_{\mathscr{Q}}$, and $\nu_{\mathscr{Q}}$ are constant functions, then the size of $\mathscr{G}$ is $S(\mathscr{G}) = (c_1, c_2, c_3)|Q|$.*

Proof. Obvious. □

Unlike the order of perfectly regular q-RPFG, we can no longer give an explicit formula for the order of perfect edge regular q-RPFG. However, we can bound its order.

Theorem 19. *If $\mathscr{G} = (\mathscr{P}, \mathscr{Q})$ is perfect edge regular q-RPFG, then the order of $\mathscr{G}$ is bounded as:*

$$\sum_{u\in P} \max_{u\neq v}\{\mu_{\mathscr{Q}}(uv), \eta_{\mathscr{Q}}(uv), \nu_{\mathscr{Q}}(uv)\} \leq O(\mathscr{G}) \leq |P|.$$

Proof. Upper bound of $O(\mathscr{G})$ is obvious. We now prove its lower bound. By definition of q-RPFG, $\mu_{\mathscr{Q}}(uv) \leq \mu_{\mathscr{P}}(u) \wedge \mu_{\mathscr{P}}(v)$, $\eta_{\mathscr{Q}}(uv) \leq \eta_{\mathscr{P}}(u) \wedge \eta_{\mathscr{P}}(v)$, and $\nu_{\mathscr{Q}}(uv) \leq \nu_{\mathscr{P}}(u) \vee \nu_{\mathscr{P}}(v)$. Thus, we have $\mu_{\mathscr{P}}(u) \geq \max_{u\neq v} \mu_{\mathscr{Q}}(uv)$, $\eta_{\mathscr{P}}(u) \geq \max_{u\neq v} \eta_{\mathscr{Q}}(uv)$, and $\nu_{\mathscr{P}}(u) \geq \max_{u\neq v} \nu_{\mathscr{Q}}(uv)$. For each $u \in P$, we obtain

$$\sum_{u\in P} \max_{u\neq v}\{\mu_{\mathscr{Q}}(uv), \eta_{\mathscr{Q}}(uv), \nu_{\mathscr{Q}}(uv)\} \leq \sum_{u\in P} (\mu_{\mathscr{P}}(u), \eta_{\mathscr{P}}(u), \mu_{\mathscr{P}}(u)) = O(\mathscr{G}).$$

This completes the proof. □

2.2. Edge Regular Square q-RPFGs

Following [47], we define square q-RPFGs. In this section, we concentrate on edge regularity of square q-RPFGs.

Definition 23. *Let $\mathscr{G} = (\mathscr{P}, \mathscr{Q})$ be a q-RPFG defined on $G = (P, Q)$. The square q-RPFG of $\mathscr{G}$ is denoted by $S^2(\mathscr{G}) = (\mathscr{P}, \mathscr{Q}_{sq})$, and defined as:*

1. *If $uv \in Q$, then $\mu_{\mathscr{Q}_{sq}}(uv) = \mu_{\mathscr{Q}}(uv)$, $\eta_{\mathscr{Q}_{sq}}(uv) = \eta_{\mathscr{Q}}(uv)$, and $\nu_{\mathscr{Q}_{sq}}(uv) = \nu_{\mathscr{Q}}(uv)$.*
2. *If $uv \in Q$, and u, v are joined by a path of length less than or equal to 2 in G, then*

$$\begin{aligned}
\mu_{\mathscr{Q}_{sq}}(uv) &\leq \mu_{\mathscr{P}}(u) \wedge \mu_{\mathscr{P}}(v),\\
\eta_{\mathscr{Q}_{sq}}(uv) &\leq \eta_{\mathscr{P}}(u) \wedge \eta_{\mathscr{P}}(v),\\
\nu_{\mathscr{Q}_{sq}}(uv) &\leq \nu_{\mathscr{P}}(u) \vee \nu_{\mathscr{P}}(v).
\end{aligned}$$

Example 15. *Consider a 6-RPFG $\mathscr{G} = (\mathscr{P}, \mathscr{Q})$ defined on $G = (P, Q)$, where $\mathscr{P}$ is a 6-rung picture fuzzy set on P, and $\mathscr{Q}$ is a 6-rung picture fuzzy relation on P, defined by*

$\mathscr{P}$	a	b	c	d	e
$\mu_{\mathscr{P}}$	0.8	0.82	0.9	0.8	0.6
$\eta_{\mathscr{P}}$	0.75	0.93	0.85	0.7	0.85
$\nu_{\mathscr{P}}$	0.85	0.55	0.59	0.9	0.9

$\mathscr{Q}$	ab	ac	ae	cd	be
$\mu_{\mathscr{Q}}$	0.8	0.6	0.5	0.8	0.6
$\eta_{\mathscr{Q}}$	0.7	0.6	0.7	0.7	0.8
$\nu_{\mathscr{Q}}$	0.7	0.6	0.9	0.5	0.6

By routine computations, it is easy to see that $S^2(\mathscr{G}) = (\mathscr{P}, \mathscr{Q}_{sq})$ in Figure 15b is a square 6-rung picture fuzzy graph of $\mathscr{G}$, where the 6-rung picture fuzzy relation $\mathscr{Q}_{sq}$ can be defined as follows:

$\mathscr{Q}_{sq}$	ab	ac	ae	cd	be	ce	ad	bc
$\mu_{\mathscr{Q}}$	0.8	0.6	0.5	0.8	0.6	0.6	0.8	0.82
$\eta_{\mathscr{Q}}$	0.7	0.6	0.7	0.7	0.8	0.5	0.7	0.85
$\nu_{\mathscr{Q}}$	0.7	0.6	0.9	0.5	0.6	0.9	0.9	0.59

The following theorem defines the degree of an edge in square q-RPFG.

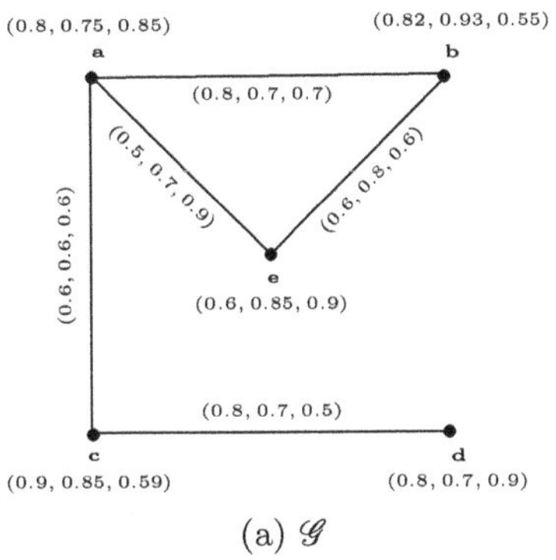

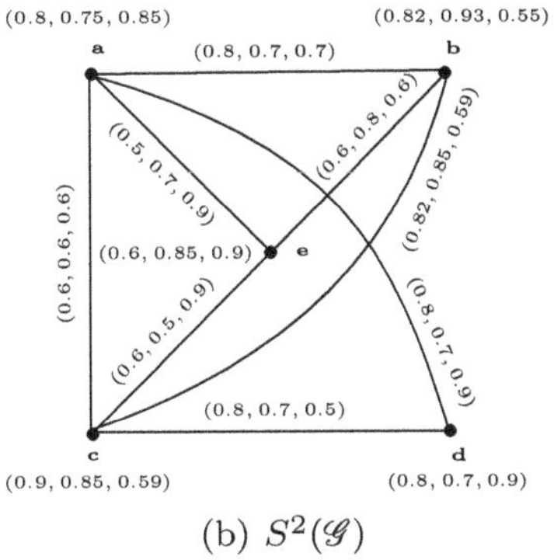

Figure 15. A 6-RPFG $\mathscr{G}$ and its square graph $S^2(\mathscr{G})$.

Theorem 20. *For any q-RPFG $\mathscr{G} = (\mathscr{P}, \mathscr{Q})$, the degree of an edge in its square q-RPFG $S^2(\mathscr{G}) = (\mathscr{P}, \mathscr{Q}_{sq})$ is given by*

1. *For $uv \in Q$,*

$$d_{S^2(\mathscr{G})}(uv) = d_{\mathscr{G}}(uv) + \sum_{\substack{ux \notin Q \\ ux \in Q_{sq}}} (\mu_{\mathscr{P}}(u) \wedge \mu_{\mathscr{P}}(x), \eta_{\mathscr{P}}(u) \wedge \eta_{\mathscr{P}}(x), \nu_{\mathscr{P}}(u) \vee \nu_{\mathscr{P}}(x)) +$$
$$\sum_{\substack{vy \notin Q \\ vy \in Q_{sq}}} (\mu_{\mathscr{P}}(v) \wedge \mu_{\mathscr{P}}(y), \eta_{\mathscr{P}}(v) \wedge \eta_{\mathscr{P}}(y), \nu_{\mathscr{P}}(v) \vee \nu_{\mathscr{P}}(y)).$$

2. *For $uv \in Q_{sq}$ such that $uv \notin Q$,*

$$d_{S^2(\mathscr{G})}(uv) = d_{\mathscr{G}}(u) + d_{\mathscr{G}}(v) + \sum_{\substack{ux \notin Q \\ ux \in Q_{sq}}} (\mu_{\mathscr{P}}(u) \wedge \mu_{\mathscr{P}}(x), \eta_{\mathscr{P}}(u) \wedge \eta_{\mathscr{P}}(x), \nu_{\mathscr{P}}(u) \vee \nu_{\mathscr{P}}(x)) +$$
$$\sum_{\substack{vy \notin Q \\ vy \in Q_{sq}}} (\mu_{\mathscr{P}}(v) \wedge \mu_{\mathscr{P}}(y), \eta_{\mathscr{P}}(v) \wedge \eta_{\mathscr{P}}(y), \nu_{\mathscr{P}}(v) \vee \nu_{\mathscr{P}}(y)) - 2(\mu_{\mathscr{Q}_{sq}}(uv), \eta_{\mathscr{Q}_{sq}}(uv), \nu_{\mathscr{Q}_{sq}}(uv)).$$

Proof. Assume that $S^2(\mathscr{G}) = (\mathscr{P}, \mathscr{Q}_{sq})$ is a square q-RPFG of $\mathscr{G} = (\mathscr{P}, \mathscr{Q})$.

1. If $uv \in Q$, then by definition of square q-RPFG $uv \in Q_{sq}$. The degree of edge uv in $S^2(\mathscr{G})$ is

$$
\begin{aligned}
d_{S^2(\mathscr{G})}(uv) &= d_{S^2(\mathscr{G})}(u) + d_{S^2(\mathscr{G})}(v) - 2(\mu_{\mathscr{Q}_{sq}}(uv), \eta_{\mathscr{Q}_{sq}}(uv), \nu_{\mathscr{Q}_{sq}}(uv)) \\
&= d_{\mathscr{G}}(u) + \sum_{\substack{ux \notin Q \\ ux \in Q_{sq}}} (\mu_{\mathscr{Q}_{sq}}(ux), \eta_{\mathscr{Q}_{sq}}(ux), \nu_{\mathscr{Q}_{sq}}(ux)) + d_{\mathscr{G}}(v) + \sum_{\substack{vy \notin Q \\ vy \in Q_{sq}}} (\mu_{\mathscr{Q}_{sq}}(vy), \eta_{\mathscr{Q}_{sq}}(vy), \\
&\quad \nu_{\mathscr{Q}_{sq}}(vy)) - 2(\mu_{\mathscr{Q}_{sq}}(uv), \eta_{\mathscr{Q}_{sq}}(uv), \nu_{\mathscr{Q}_{sq}}(uv)) \\
&= d_{\mathscr{G}}(u) + d_{\mathscr{G}}(v) - 2(\mu_{\mathscr{Q}}(uv), \eta_{\mathscr{Q}}(uv), \nu_{\mathscr{Q}}(uv))) + \sum_{\substack{ux \notin Q \\ ux \in Q_{sq}}} (\mu_{\mathscr{P}}(u) \wedge \mu_{\mathscr{P}}(x), \eta_{\mathscr{P}}(u) \wedge \eta_{\mathscr{P}}(x), \\
&\quad \nu_{\mathscr{P}}(u) \vee \nu_{\mathscr{P}}(x)) + \sum_{\substack{vy \notin Q \\ vy \in Q_{sq}}} (\mu_{\mathscr{P}}(v) \wedge \mu_{\mathscr{P}}(y), \eta_{\mathscr{P}}(v) \wedge \eta_{\mathscr{P}}(y), \nu_{\mathscr{P}}(v) \vee \nu_{\mathscr{P}}(y)) \\
&= d_{\mathscr{G}}(uv) + \sum_{\substack{ux \notin Q \\ ux \in Q_{sq}}} (\mu_{\mathscr{P}}(u) \wedge \mu_{\mathscr{P}}(x), \eta_{\mathscr{P}}(u) \wedge \eta_{\mathscr{P}}(x), \nu_{\mathscr{P}}(u) \vee \nu_{\mathscr{P}}(x)) + \\
&\quad \sum_{\substack{vy \notin Q \\ vy \in Q_{sq}}} (\mu_{\mathscr{P}}(v) \wedge \mu_{\mathscr{P}}(y), \eta_{\mathscr{P}}(v) \wedge \eta_{\mathscr{P}}(y), \nu_{\mathscr{P}}(v) \vee \nu_{\mathscr{P}}(y)).
\end{aligned}
$$

2. If $uv \in Q_{sq}$ such that $uv \notin Q$, then degree of edge uv in $S^2(\mathscr{G})$ is

$$
\begin{aligned}
d_{S^2(\mathscr{G})}(uv) &= d_{S^2(\mathscr{G})}(u) + d_{S^2(\mathscr{G})}(v) - 2(\mu_{\mathscr{Q}_{sq}}(uv), \eta_{\mathscr{Q}_{sq}}(uv), \nu_{\mathscr{Q}_{sq}}(uv)) \\
&= d_{\mathscr{G}}(u) + \sum_{\substack{ux \notin Q \\ ux \in Q_{sq}}} (\mu_{\mathscr{Q}_{sq}}(ux), \eta_{\mathscr{Q}_{sq}}(ux), \nu_{\mathscr{Q}_{sq}}(ux)) + d_{\mathscr{G}}(v) + \sum_{\substack{vy \notin Q \\ vy \in Q_{sq}}} (\mu_{\mathscr{Q}_{sq}}(vy), \eta_{\mathscr{Q}_{sq}}(vy), \\
&\quad \nu_{\mathscr{Q}_{sq}}(vy)) - 2(\mu_{\mathscr{Q}_{sq}}(uv), \eta_{\mathscr{Q}_{sq}}(uv), \nu_{\mathscr{Q}_{sq}}(uv)) \\
&= d_{\mathscr{G}}(u) + d_{\mathscr{G}}(v) + \sum_{\substack{ux \notin Q \\ ux \in Q_{sq}}} (\mu_{\mathscr{P}}(u) \wedge \mu_{\mathscr{P}}(x), \eta_{\mathscr{P}}(u) \wedge \eta_{\mathscr{P}}(x), \nu_{\mathscr{P}}(u) \vee \nu_{\mathscr{P}}(x)) + \\
&\quad \sum_{\substack{vy \notin Q \\ vy \in Q_{sq}}} (\mu_{\mathscr{P}}(v) \wedge \mu_{\mathscr{P}}(y), \eta_{\mathscr{P}}(v) \wedge \eta_{\mathscr{P}}(y), \nu_{\mathscr{P}}(v) \vee \nu_{\mathscr{P}}(y)) - 2(\mu_{\mathscr{Q}_{sq}}(uv), \eta_{\mathscr{Q}_{sq}}(uv), \nu_{\mathscr{Q}_{sq}}(uv)).
\end{aligned}
$$

This completes the proof. □

Remark 8. *If $\mathscr{G} = (\mathscr{P}, \mathscr{Q})$ is edge regular q-RPFG, then $S^2(\mathscr{G}) = (\mathscr{P}, \mathscr{Q}_{sq})$ may not be edge regular. For example, consider a 3-RPFG $\mathscr{G}$, and its square graph $S^2(\mathscr{G})$ as shown in Figure 16.*

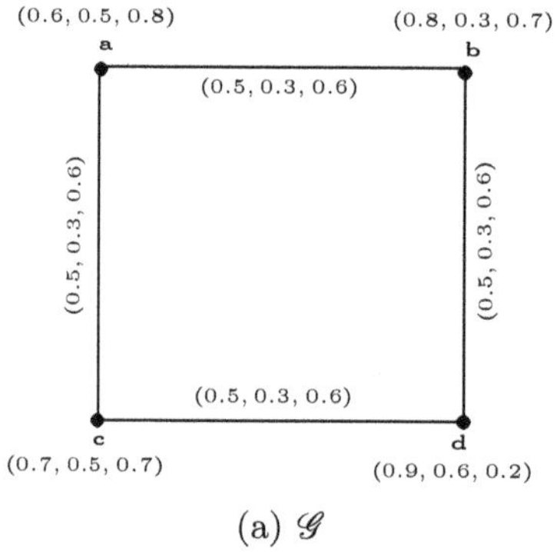

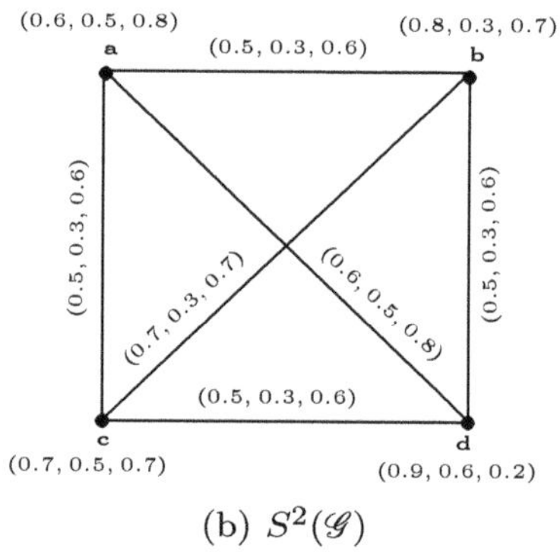

Figure 16. A 3-RPFG $\mathscr{G}$ and its square graph $S^2(\mathscr{G})$.

We see that $\mathscr{G}$ is (1.0, 0.6, 1.2)-edge regular 3-RPFG. However, since $d_{S^2(\mathscr{G})}(ab) = (2.3, 1.4, 2.7) \neq (2.0, 1.2, 2.4) = d_{S^2(\mathscr{G})}(ad)$. Hence, $S^2(\mathscr{G})$ is not edge regular.

Remark 9. *If $S^2(\mathscr{G}) = (\mathscr{P}, \mathscr{Q}_{sq})$ is edge regular q-RPFG, then $\mathscr{G} = (\mathscr{P}, \mathscr{Q})$ may not be edge regular. For example, consider a 5-RPFG $\mathscr{G}$, and its square graph $S^2(\mathscr{G})$ as shown in Figure 17.*

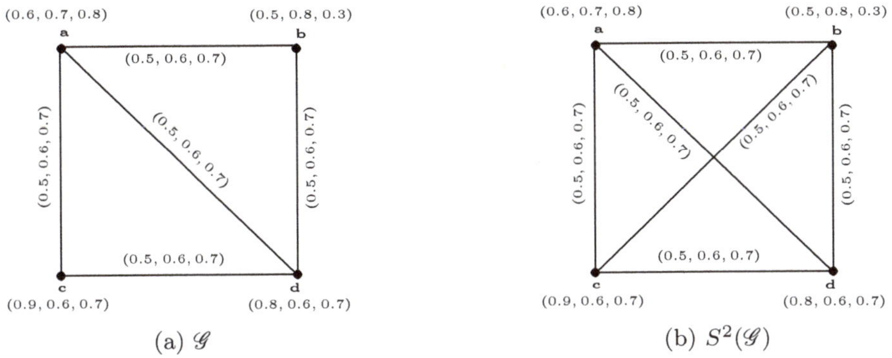

Figure 17. A 5-RPFG $\mathscr{G}$ and its square graph $S^2(\mathscr{G})$.

We see that $S^2(\mathscr{G})$ is (2.0, 2.4, 2.8)-edge regular 5-RPFG. However, since $d_{\mathscr{G}}(ab) = (1.5, 1.8, 2.1) \neq (2.0, 2.4, 2.8) = d_{\mathscr{G}}(ad)$. Hence, $\mathscr{G}$ is not edge regular.

Lemma 1. *The square q-RPFG $S^2(\mathscr{G})$ of a complete q-RPFG $\mathscr{G}$ is the q-RPFG $\mathscr{G}$ itself.*

Proof. The proof is obvious. □

Theorem 21. *If $\mathscr{G} = (\mathscr{P}, \mathscr{Q})$ be complete q-RPFG. Then, $\mathscr{G}$ is an edge regular q-RPFG if and only if $S^2(\mathscr{G}) = (\mathscr{P}, \mathscr{Q}_{sq})$ is an edge regular q-RPFG.*

Proof. Let $\mathscr{G} = (\mathscr{P}, \mathscr{Q})$ be complete q-RPFG. By Lemma 1, $\mathscr{G}$, and $S^2(\mathscr{G})$ are same. It follows that $\mathscr{G}$ is an edge regular q-RPFG if and only if $S^2(\mathscr{G}) = (\mathscr{P}, \mathscr{Q}_{sq})$ is an edge regular q-RPFG. □

Lemma 2. *The square q-RPFG $S^2(\mathscr{G})$ of a complete bipartite q-RPFG $\mathscr{G}$ is complete.*

Proof. Let $\mathscr{G} = (\mathscr{P}, \mathscr{Q})$ be a q-RPFG defined on a complete bipartite graph $G = (P, Q)$. Then, $\mu_{\mathscr{Q}}(u_iv_j) = \mu_{\mathscr{P}}(u_i) \wedge \mu_{\mathscr{P}}(v_j)$, $\eta_{\mathscr{Q}}(u_iv_j) = \eta_{\mathscr{P}}(u_i) \wedge \eta_{\mathscr{P}}(v_j)$, and $\nu_{\mathscr{Q}}(u_iv_j) = \nu_{\mathscr{P}}(u_i) \wedge \nu_{\mathscr{P}}(v_j)$ for all $u_i \in P_1$, and $v_j \in P_2$, where $P = P_1 \cup P_2$. Let $S^2(\mathscr{G}) = (\mathscr{P}, \mathscr{Q}_{sq})$ be the square q-RPFG of $\mathscr{G}$. Then, we have the following:

1. If $u_iv_j \in Q$, then $\mu_{\mathscr{Q}_{sq}}(u_iv_j) = \mu_{\mathscr{Q}}(u_iv_j)$, $\eta_{\mathscr{Q}_{sq}}(u_iv_j) = \eta_{\mathscr{Q}}(u_iv_j)$, and $\nu_{\mathscr{Q}_{sq}}(u_iv_j) = \nu_{\mathscr{Q}}(u_iv_j)$.
2. If $u_iv_j \notin Q$, and u, v are joined by a path of length less then or equal to 2 in G, then $\mu_{\mathscr{Q}_{sq}}(u_iv_j) \leq \mu_{\mathscr{P}}(u_i) \wedge \mu_{\mathscr{P}}(v_j)$, $\eta_{\mathscr{Q}_{sq}}(u_iv_j) \leq \eta_{\mathscr{P}}(u_i) \wedge \eta_{\mathscr{P}}(v_j)$, and $\nu_{\mathscr{Q}_{sq}}(u_iv_j) \leq \nu_{\mathscr{P}}(u_i) \vee \nu_{\mathscr{P}}(v_j)$.

Since each pair of vertices in a complete bipartite graph must connected by a path of length less then or equal to 2 (see Figure 18), therefore, above axioms imply that for all $u_i, v_j \in P$

$$\begin{aligned} \mu_{\mathscr{Q}_{sq}}(u_iv_j) &\leq \mu_{\mathscr{P}}(u_i) \wedge \mu_{\mathscr{P}}(v_j), \\ \eta_{\mathscr{Q}_{sq}}(u_iv_j) &\leq \eta_{\mathscr{P}}(u_i) \wedge \eta_{\mathscr{P}}(v_j), \\ \nu_{\mathscr{Q}_{sq}}(u_iv_j) &\leq \nu_{\mathscr{P}}(u_i) \vee \nu_{\mathscr{P}}(v_j). \end{aligned}$$

Hence, the square q-RPFG $S^2(\mathscr{G})$ is complete. This completes the proof. □

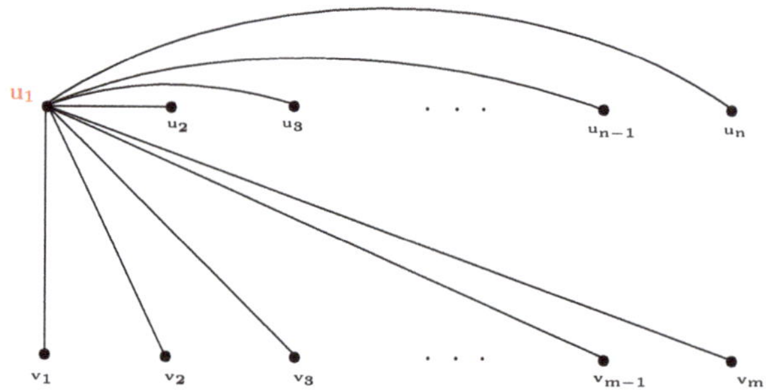

Figure 18. Role of a vertex u_1 in square graph of a complete bipartite graph.

Theorem 22. *If $\mathscr{G} = (\mathscr{P}, \mathscr{Q})$ is a complete bipartite q-RPFG such that $\mu_{\mathscr{P}}$, $\eta_{\mathscr{P}}$, and $\nu_{\mathscr{P}}$ are constant functions, then its square graph $S^2(\mathscr{G})$ is edge regular.*

Proof. Let $\mathscr{G} = (\mathscr{P}, \mathscr{Q})$ be a complete bipartite q-RPFG such that $\mu_{\mathscr{P}}$, $\eta_{\mathscr{P}}$, and $\nu_{\mathscr{P}}$ are constant functions. Assume that $S^2(\mathscr{G}) = (\mathscr{P}, \mathscr{Q}_{sq})$ is the square q-RPFG of $\mathscr{G}$. Then, by Lemma 2, $S^2(\mathscr{G})$ is a complete q-RPFG.

Let $\mu_{\mathscr{P}}(u) = c_1$, $\eta_{\mathscr{P}}(u) = c_2$, and $\nu_{\mathscr{P}}(u) = c_3$ for all $u \in P$. By definition of square q-RPFG, we have

$$\begin{aligned} \mu_{\mathscr{Q}_{sq}}(uv) &\leq \mu_{\mathscr{P}}(u) \wedge \mu_{\mathscr{P}}(v) = c_1 \wedge c_1 = c_1, \\ \eta_{\mathscr{Q}_{sq}}(uv) &\leq \eta_{\mathscr{P}}(u) \wedge \eta_{\mathscr{P}}(v) = c_2 \wedge c_2 = c_2, \\ \nu_{\mathscr{Q}_{sq}}(uv) &\leq \nu_{\mathscr{P}}(u) \vee \nu_{\mathscr{P}}(v) = c_3 \vee c_3 = c_3. \end{aligned}$$

Hence, $\mu_{\mathscr{Q}_{sq}}$, $\eta_{\mathscr{Q}_{sq}}$, and $\nu_{\mathscr{Q}_{sq}}$ are constant functions. Thus, by Theorem 8, $S^2(\mathscr{G})$ is edge regular. This completes the proof. □

Theorem 23. *If $\mathscr{G}$ is a q-RPFG on a cycle of length n such that $\mu_{\mathscr{Q}}$, $\eta_{\mathscr{Q}}$, and $\nu_{\mathscr{Q}}$ are constant functions, then its square q-RPFG $S^2(\mathscr{G})$ is edge regular.*

Proof. Let $\mathscr{G} = (\mathscr{P}, \mathscr{Q})$ be a q-RPFG defined on a cycle $u_1u_2u_3\ldots u_{n-1}u_nu_1$ (see Figure 19). Assume that $\mu_{\mathscr{Q}}$, $\eta_{\mathscr{Q}}$, and $\nu_{\mathscr{Q}}$ are constant functions, and $\mu_{\mathscr{Q}}(u_iu_{i+1}) = c_1$, $\eta_{\mathscr{Q}}(u_iu_{i+1}) = c_2$, and $\nu_{\mathscr{Q}}(u_iu_{i+1}) = c_3$ for all $i = 1, 2, \ldots, n$, where $n + 1 = 1$. The degree of edge u_iu_{i+1} can be computed as:

$$\begin{aligned} d_{\mathscr{G}}(u_iu_{i+1}) &= d_{\mathscr{G}}(u_i) + d_{\mathscr{G}}(u_{i+1}) - 2(\mu_{\mathscr{Q}}(u_iu_{i+1}), \eta_{\mathscr{Q}}(u_iu_{i+1}), \nu_{\mathscr{Q}}(u_iu_{i+1})) \\ &= (\mu_{\mathscr{Q}}(u_{i-1}u_i), \eta_{\mathscr{Q}}(u_{i-1}u_i), \nu_{\mathscr{Q}}(u_{i-1}u_i)) + 2(\mu_{\mathscr{Q}}(u_iu_{i+1}), \eta_{\mathscr{Q}}(u_iu_{i+1}), \nu_{\mathscr{Q}}(u_iu_{i+1})) + \\ &\quad (\mu_{\mathscr{Q}}(u_{i+1}u_{i+2}), \eta_{\mathscr{Q}}(u_{i+1}u_{i+2}), \nu_{\mathscr{Q}}(u_{i+1}u_{i+2})) - 2(\mu_{\mathscr{Q}}(u_iu_{i+1}), \eta_{\mathscr{Q}}(u_iu_{i+1}), \nu_{\mathscr{Q}}(u_iu_{i+1})) \\ &= (\mu_{\mathscr{Q}}(u_{i-1}u_i), \eta_{\mathscr{Q}}(u_{i-1}u_i), \nu_{\mathscr{Q}}(u_{i-1}u_i)) + (\mu_{\mathscr{Q}}(u_{i+1}u_{i+2}), \eta_{\mathscr{Q}}(u_{i+1}u_{i+2}), \nu_{\mathscr{Q}}(u_{i+1}u_{i+2})) \\ &= (c_1, c_2, c_3) + (c_1, c_2, c_3) = 2(c_1, c_2, c_3). \end{aligned}$$

Hence, $\mathscr{G}$ is edge regular. Let $S^2(\mathscr{G})$ be the square q-RPFG of $\mathscr{G}$. To prove that $S^2(\mathscr{G})$ is edge regular, there arise three cases:

Case-I When $n = 3$, we have a complete q-RPFG $\mathscr{G}$. By Lemma 1, $S^2(\mathscr{G})$ is $\mathscr{G}$ itself.
Case-II When $n = 4$, each pair of non-adjacent vertices is connected by two distinct paths of length 2. Thus, its square graph $S^2(\mathscr{G})$ will be a complete q-RPFG with $\mu_{\mathscr{Q}}$, $\eta_{\mathscr{Q}}$, and $\nu_{\mathscr{Q}}$ are constant functions. In this case, $S^2(\mathscr{G})$ is edge regular by Theorem 8.
Case-III Let $n \geq 5$. Consider a vertex u in $S^2(\mathscr{G})$. There are exactly two vertices which are at distance 2

from u. Therefore, u is adjacent to exactly two more vertices in $S^2(\mathscr{G})$. Hence, $S^2(\mathscr{G})$ is 4-regular. Then, for each uv in $S^2(\mathscr{G})$, we have

$$\begin{aligned} d_{S^2(\mathscr{G})}(uv) &= d_{S^2(\mathscr{G})}(u) + d_{S^2(\mathscr{G})}(v) - 2(\mu_{\mathscr{Q}_{sq}}(uv), \eta_{\mathscr{Q}_{sq}}(uv), \nu_{\mathscr{Q}_{sq}}(uv)) \\ &= 4(c_1, c_2, c_3) + 4(c_1, c_2, c_3) - 2(c_1, c_2, c_3) = 6(c_1, c_2, c_3). \end{aligned}$$

Hence, $S^2(\mathscr{G})$ is $6(c_1, c_2, c_3)$-edge regular. This completes the proof. □

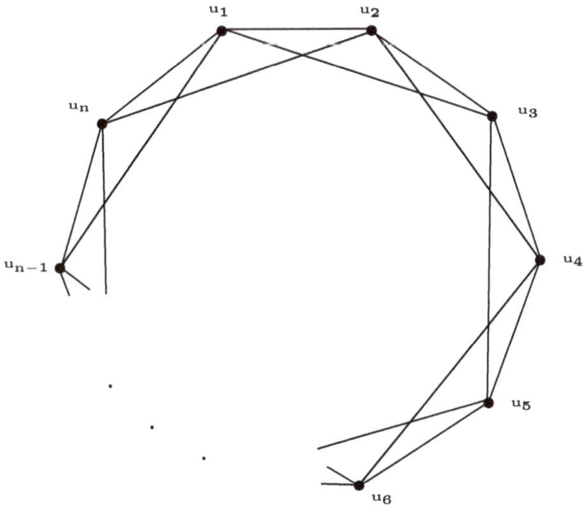

Figure 19. Square graph $S^2(C_n)$ of cycle C_n.

2.3. Edge Regular q-RPF Line Graphs

Following [45], this section gives a brief discussion on edge regular q-rung picture fuzzy line graphs (q-RPFLGs). Akram and Habib [35] defined the q-RPFLGs as follows:

Definition 24. *Let $\mathscr{G} = (\mathscr{P}, \mathscr{Q})$ be a q-RPFG defined on $G = (P, Q)$. The q-rung picture fuzzy line graph $L(\mathscr{G}) = (\mathscr{A}, \mathscr{B})$ with underlying graph $L(G) = (A, B)$ where $A = \{S_x = \{x\} \cup \{u_x v_x\} | x \in Q,\ x = u_x v_x,\ u_x, v_x \in P\}$ and $B = \{S_x S_y | S_x \cap S_y \neq \phi,\ x, y \in Q,\ x \neq y\}$ can be defined as*

(i) *For each vertex S_x in A,*

$$\mu_{\mathscr{A}}(S_x) = \mu_{\mathscr{Q}}(x),\ \eta_{\mathscr{A}}(S_x) = \eta_{\mathscr{Q}}(x),\ \nu_{\mathscr{A}}(S_x) = \nu_{\mathscr{Q}}(x).$$

(ii) *For each edge $S_x S_y$ in B,*

$$\begin{aligned} \mu_{\mathscr{B}}(S_x S_y) &= \min\{\mu_{\mathscr{A}}(S_x), \mu_{\mathscr{A}}(S_y)\}, \\ \eta_{\mathscr{B}}(S_x S_y) &= \min\{\eta_{\mathscr{A}}(S_x), \eta_{\mathscr{A}}(S_y)\}, \\ \nu_{\mathscr{B}}(S_x S_y) &= \max\{\nu_{\mathscr{A}}(S_x), \nu_{\mathscr{A}}(S_y)\}. \end{aligned}$$

Theorem 24. *Let $\mathscr{G} = (\mathscr{P}, \mathscr{Q})$ be a q-RPFG such that $\mu_{\mathscr{Q}}$, $\eta_{\mathscr{Q}}$, and $\nu_{\mathscr{Q}}$ are constant functions. Then, $d_{L(\mathscr{G})}(S_x S_y) = (c_1, c_2, c_3)[d_{L(G)}(S_x) + d_{L(G)}(S_y) - 2]$ for all $S_x S_y \in B$.*

Proof. Let $L(\mathscr{G}) = (\mathscr{A}, \mathscr{B})$ be a q-rung picture fuzzy line graph of a q-RPFG $\mathscr{G} = (\mathscr{P}, \mathscr{Q})$. For any $S_x S_y \in B$,

$$\begin{aligned} d_{L(\mathscr{G})}(S_x S_y) &= d_{L(\mathscr{G})}(S_x) + d_{L(\mathscr{G})}(S_y) - 2(\mu_{\mathscr{B}}(S_x S_y), \eta_{\mathscr{B}}(S_x S_y), \nu_{\mathscr{B}}(S_x S_y)) \\ &= \sum_{S_x S_z \in B} (\mu_{\mathscr{Q}}(S_x S_z), \eta_{\mathscr{Q}}(S_x S_z), \nu_{\mathscr{Q}}(S_x S_z)) + \sum_{S_y S_z \in B} (\mu_{\mathscr{Q}}(S_y S_z), \eta_{\mathscr{Q}}(S_y S_z), \nu_{\mathscr{Q}}(S_y S_z)) \\ &\quad - 2(\mu_{\mathscr{Q}}(S_x S_y), \eta_{\mathscr{Q}}(S_x S_y), \nu_{\mathscr{Q}}(S_x S_y)) \\ &= (c_1, c_2, c_3) d_{L(G)}(S_x) + (c_1, c_2, c_3) d_{L(G)}(S_y) - 2(c_1, c_2, c_3) \\ &= (c_1, c_2, c_3)[d_{L(G)}(S_x) + d_{L(G)}(S_y) - 2]. \end{aligned}$$

This completes the proof. □

Remark 10. *Consider a 6-RPFG $\mathscr{G} = (\mathscr{P}, \mathscr{Q})$ as displayed in Figure 20a. Its corresponding line graph $L(\mathscr{G}) = (\mathscr{A}, \mathscr{B})$ is displayed in Figure 20b.*

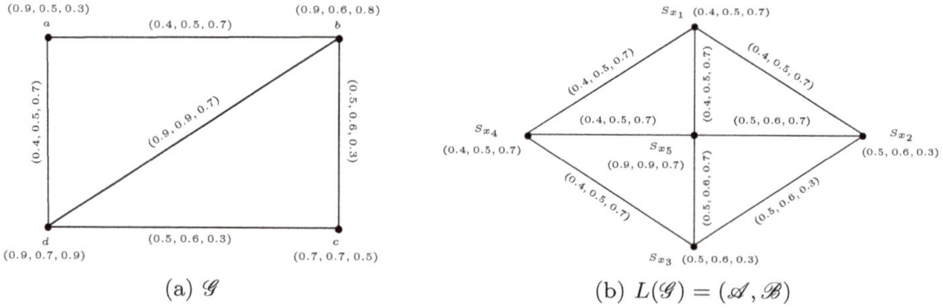

Figure 20. A 6-RPFLG of $\mathscr{G}$

We see that $\mathscr{G}$ is $(1.8, 1.8, 1.4)$-edge regular 6-RPFG. However, since $d_{L(\mathscr{G})}(S_{x_1} S_{x_5}) = (2.2, 2.7, 3.5) \neq (1.8, 2.2, 2.4) = d_{L(\mathscr{G})}(S_{x_1} S_{x_2})$; therefore, $L(\mathscr{G})$ is not an edge regular 6-RPFG.

We now develop a necessary condition for q-RPFLG to be perfect edge regular.

Theorem 25. *Every q-rung picture fuzzy line graph $L(\mathscr{G}) = (\mathscr{A}, \mathscr{B})$ of an edge regular q-RPFG $\mathscr{G} = (\mathscr{P}, \mathscr{Q})$ is perfect edge regular only if $\mu_{\mathscr{Q}}$, $\eta_{\mathscr{Q}}$, and $\nu_{\mathscr{Q}}$ are constant functions in $\mathscr{G}$.*

Proof. Let $\mathscr{G} = (\mathscr{P}, \mathscr{Q})$ be an edge regular q-RPFG defined on $G = (P, Q)$. Then, for each edge uv in $\mathscr{Q}$, $d(uv) = (p_1, p_2, p_3)$. Let $\mu_{\mathscr{Q}}$, $\eta_{\mathscr{Q}}$, and $\nu_{\mathscr{Q}}$ be constant functions. Then, for each edge uv in $\mathscr{Q}$, $\mu_{\mathscr{Q}}(uv) = c_1$, $\eta_{\mathscr{Q}}(uv) = c_2$ and $\nu_{\mathscr{Q}}(uv) = c_3$. Consider $L(\mathscr{G}) = (\mathscr{A}, \mathscr{B})$ as a q-RPFG as a q-rung picture fuzzy line graph of $\mathscr{G}$. Then, by definition of q-rung picture fuzzy line graph, each vertex S_x of $L(\mathscr{G})$ corresponding to an edge $uv = x$ of $\mathscr{G}$ has membership value $(\mu_{\mathscr{A}}(S_x), \eta_{\mathscr{A}}(S_x), \nu_{\mathscr{A}}(S_x)) = (\mu_{\mathscr{Q}}(uv), \eta_{\mathscr{Q}}(uv), \nu_{\mathscr{Q}}(uv)) = (c_1, c_2, c_3)$, and each edge of $L(\mathscr{G})$ has membership value

$$\begin{aligned} (\mu_{\mathscr{B}}(S_x S_y), \eta_{\mathscr{B}}(S_x S_y), \nu_{\mathscr{B}}(S_x S_y)) &= (\mu_{\mathscr{A}}(S_x) \wedge \mu_{\mathscr{A}}(S_y), \eta_{\mathscr{A}}(S_x) \wedge \eta_{\mathscr{A}}(S_y), \eta_{\mathscr{A}}(S_x) \vee \eta_{\mathscr{A}}(S_y)) \\ &= (c_1 \wedge c_1, c_2 \wedge c_2, c_3 \vee c_3) \\ &= (c_1, c_2, c_3). \end{aligned}$$

Let $u(x)$ and $v(x)$ be the end points of an edge $uv = x$ in $\mathscr{G}$. Since $\mathscr{G}$ is edge regular with constant functions $\mu_{\mathscr{Q}}$, $\eta_{\mathscr{Q}}$, and $\nu_{\mathscr{Q}}$, therefore, each vertex u of $\mathscr{G}$ is common vertex of α (say) edges in $\mathscr{G}$ (see Figure 21).

Let S_x be the vertex of $L(\mathscr{G})$ corresponding to an edge x in $\mathscr{G}$. The edge x has a vertex v_x in common with $\alpha - 1$ edges $v_1, v_2, \ldots, v_{\alpha-1}$. Similarly, u_x is common with $\alpha - 1$ edges $u_1, u_2, \ldots, u_{\alpha-1}$

in $\mathscr{G}$. Thus, each vertex S_x in $L(\mathscr{G})$ is common vertex of $\alpha - 1 + \alpha - 1 = 2\alpha - 2$ edges. Hence, for all S_x in $L(G)$

$$d_{L(G)}(S_x) = 2\alpha - 2. \tag{1}$$

By Theorem 24, we have $d_{L(\mathscr{G})}(S_xS_y) = (c_1, c_2, c_3)[d_{L(G)}(S_x) + d_{L(G)}(S_y) - 2]$. Then, by Equation (1), $d_{L(\mathscr{G})}(S_xS_y) = (c_1, c_2, c_3)[2\alpha - 2 + 2\alpha - 2 - 2] = 2(2\alpha - 3)(c_1, c_2, c_3)$, which implies that $L(\mathscr{G})$ is an edge regular q-RPFG. Moreover, $td_{L(\mathscr{G})}(S_xS_y) = d_{L(\mathscr{G})}(S_xS_y) + (\mu_{\mathscr{B}}(S_xS_y), \eta_{\mathscr{B}}(S_xS_y), \nu_{\mathscr{B}}(S_xS_y)) = 2(2\alpha - 3)(c_1, c_2, c_3) + (c_1, c_2, c_3) = (4\alpha - 5)(c_1, c_2, c_3)$, which shows that $L(\mathscr{G})$ is total edge regular q-RPFG. Consequently, $L(\mathscr{G})$ is perfect edge regular q-RPFG. This completes the proof. □

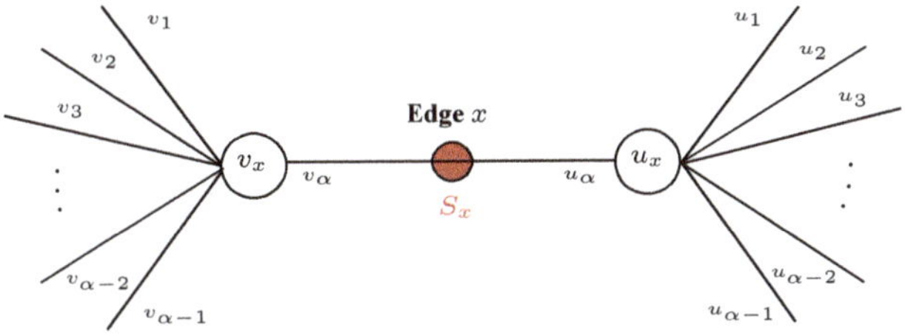

Figure 21. View of a graph towards its line graph.

Remark 11. *Consider a 4-RPFG $\mathscr{G} = (\mathscr{P}, \mathscr{Q})$ as displayed in Figure 22a, and its line graph $L(\mathscr{G}) = (\mathscr{A}, \mathscr{B})$ Figure 22b.*

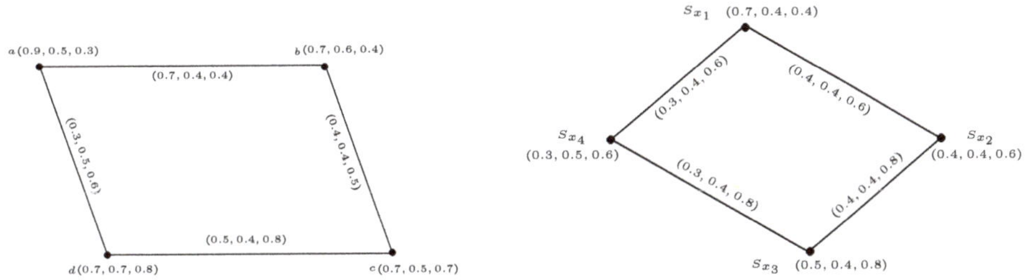

(a) $\mathscr{G} = (\mathscr{P}, \mathscr{Q})$ (b) $L(\mathscr{G}) = (\mathscr{A}, \mathscr{B})$

Figure 22. A 4-RPFLG of $\mathscr{G}$

We see that $L(\mathscr{G})$ is $(0.7, 0.8, 1.4)$-edge regular 4-RPFLG. However, since $d_{\mathscr{G}}(ab) = (0.7, 0.7, 1.1) \neq (1.2, 0.8, 1.2) = d_{\mathscr{G}}(bc)$; therefore, $\mathscr{G}$ is not an edge regular 4-RPFG.

Theorem 26. *If $L(\mathscr{G}) = (\mathscr{A}, \mathscr{B})$ is an edge regular q-rung picture fuzzy line graph of $\mathscr{G} = (\mathscr{P}, \mathscr{Q})$, and $\mu_{\mathscr{Q}}$, $\eta_{\mathscr{Q}}$, and $\nu_{\mathscr{Q}}$ are constant functions, then $\mathscr{G}$ is edge regular q-RPFG.*

Proof. The proof is similar to the proof of Theorem 25. □

Theorem 27. *Let $\mathscr{G} = (\mathscr{P}, \mathscr{Q})$ be a strong q-rung picture fuzzy graph such that $\mu_{\mathscr{P}}$, $\eta_{\mathscr{P}}$, and $\nu_{\mathscr{P}}$ are constant functions. If $\mathscr{G}$ is an edge regular q-RPFG, then its line graph $L(\mathscr{G}) = (\mathscr{A}, \mathscr{B})$ is an edge regular q-rung picture fuzzy graph.*

Proof. Let $\mathscr{G} = (\mathscr{P}, \mathscr{Q})$ be a strong edge regular q-rung picture fuzzy graph such that $\mu_{\mathscr{P}}$, $\eta_{\mathscr{P}}$, and $\nu_{\mathscr{P}}$ are constant functions. Then, the functions $\mu_{\mathscr{Q}}$, $\eta_{\mathscr{Q}}$, and $\nu_{\mathscr{Q}}$ must be constants. Consequently, the result follows from Theorem 25. □

3. Relations between Regular and Edge-Regular q-RPFGs

The fact that edge regularity property is a strong analog of regularity in fuzzy graphs is shown by many researchers. This section provides some additional properties relating regularity and edge regularity of q-RPFGs.

Remark 12. *Every regular q-RPFG need not be edge regular.*
For example, consider a 5*-RPFG* $\mathscr{G} = (\mathscr{P}, \mathscr{Q})$ *displayed in Figure 23.*

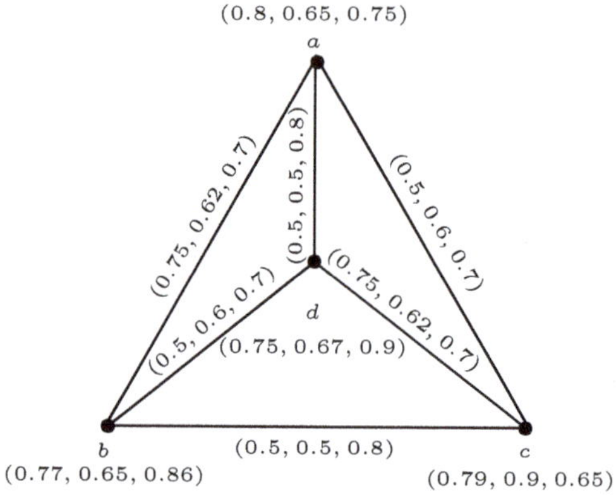

Figure 23. A regular but not edge regular 5-RPFG $\mathscr{G}$.

It is clear that $\mathscr{G}$ *is* $(1.75, 1.72, 2.2)$*-regular* 5*-RPFG, but* $\mathscr{G}$ *is not edge regular as* $d_{\mathscr{G}}(ab) = (2.0, 2.2, 3.0) \neq (2.5, 2.44, 2.8) = d_{\mathscr{G}}(bc)$.

Remark 13. *Every edge regular q-RPFG need not be regular.*
For example, consider a 6*-RPFG* $\mathscr{G}$ *as shown in Figure 10. It is clear that* $\mathscr{G}$ *is* $(1.4, 1.8, 1.6)$*-edge regular* 6*-RPFG but* $\mathscr{G}$ *is not regular as* $d_{\mathscr{G}}(u) = (0.8, 1.0, 0.6) \neq (1.4, 1.8, 1.6) = d_{\mathscr{G}}(v)$.

The above remarks show that one form of regularity can not imply another form. We now develop some results between regularity and edge regularity.

Theorem 28. *Let* $\mathscr{G} = (\mathscr{P}, \mathscr{Q})$ *be a regular q-RPFG on* $G = (P, Q)$*. Then,* $\mathscr{G}$ *is edge regular if and only if* $\mu_{\mathscr{Q}}, \eta_{\mathscr{Q}}$*, and* $\nu_{\mathscr{Q}}$ *are constant functions.*

Proof. Let $\mathscr{G} = (\mathscr{P}, \mathscr{Q})$ be a (k_1, k_2, k_3)-regular q-RPFG. Then, $d_{\mathscr{G}}(u) = (k_1, k_2, k_3)$ for all $u \in P$. Assume that $\mu_{\mathscr{Q}}, \eta_{\mathscr{Q}}$, and $\nu_{\mathscr{Q}}$ are constant functions, that is, $\mu_{\mathscr{Q}}(uv) = c_1$, $\eta_{\mathscr{Q}}(uv) = c_2$, and $\nu_{\mathscr{Q}}(uv) = c_3$. By definition of edge degree $d_{\mathscr{G}}(uv) = d_{\mathscr{G}}(u) + d_{\mathscr{G}}(v) - 2(\mu_{\mathscr{Q}}(uv), \eta_{\mathscr{Q}}(uv), \nu_{\mathscr{Q}}(uv)) = 2(k_1, k_2, k_3) - 2(c_1, c_2, c_3) = 2(p_1 - c_1, p_2 - c_2, p_3 - c_3)$ for all $uv \in Q$. Hence, $\mathscr{G}$ is edge regular.
Conversely, assume that $\mathscr{G}$ is (p_1, p_2, p_3)-edge regular. Then, $d_{\mathscr{G}}(uv) = (p_1, p_2, p_3)$, for all $uv \in Q$. By definition of edge degree $(p_1, p_2, p_3) = 2(k_1, k_2, k_3) - 2(\mu_{\mathscr{Q}}(uv), \eta_{\mathscr{Q}}(uv), \nu_{\mathscr{Q}}(uv))$. Thus, $(\mu_{\mathscr{Q}}(uv), \eta_{\mathscr{Q}}(uv), \nu_{\mathscr{Q}}(uv)) = \frac{(2k_1 - p_1, 2k_2 - p_2, 2k_3 - p_3)}{2}$, for all $uv \in Q$. Hence, $\mu_{\mathscr{Q}}, \eta_{\mathscr{Q}}$, and $\nu_{\mathscr{Q}}$ are constant functions. This completes the proof. □

Theorem 29. *Let* $\mathscr{G} = (\mathscr{P}, \mathscr{Q})$ *be a q-RPFG on* $G = (P, Q)$ *such that* $\mu_{\mathscr{Q}}, \eta_{\mathscr{Q}}$*, and* $\nu_{\mathscr{Q}}$ *are constant functions. If* $\mathscr{G}$ *is full regular q-RPFG, then* $\mathscr{G}$ *is full edge regular q-RPFG.*

Proof. Let $\mathscr{G} = (\mathscr{P}, \mathscr{Q})$ be a q-RPFG on $G = (P, Q)$, and $\mu_{\mathscr{Q}}(uv) = c_1$, $\eta_{\mathscr{Q}}(uv) = c_2$, and $\nu_{\mathscr{Q}}(uv) = c_3$, for all $uv \in Q$. Assume that $\mathscr{G}$ is full regular q-RPFG, that is, $d_{\mathscr{G}}(u) = (k_1, k_2, k_3)$, and $d_G(u) = k$ for all $u \in P$. Then, $d_G(uv) = d_G(u) + d_G(v) - 2 = 2r - 2$. Hence, G is edge regular graph. Now, $d_{\mathscr{G}}(uv) =$

$d_{\mathscr{G}}(u) + d_{\mathscr{G}}(v) - 2(\mu_{\mathscr{Q}}(uv), \eta_{\mathscr{Q}}(uv), \nu_{\mathscr{Q}}(uv)) = 2(k_1, k_2, k_3) - 2(c_1, c_2, c_3) = 2(k_1 - c_1, k_2 - c_2, k_3 - c_3)$. Hence, $\mathscr{G}$ is edge regular q-RPFG. Consequently, $\mathscr{G}$ is full edge regular q-RPFG. □

Remark 14. *The converse of the above theorem need not be true.*
For example, consider a 4*-RPFG* $\mathscr{G}$ *as displayed in Figure 24.*

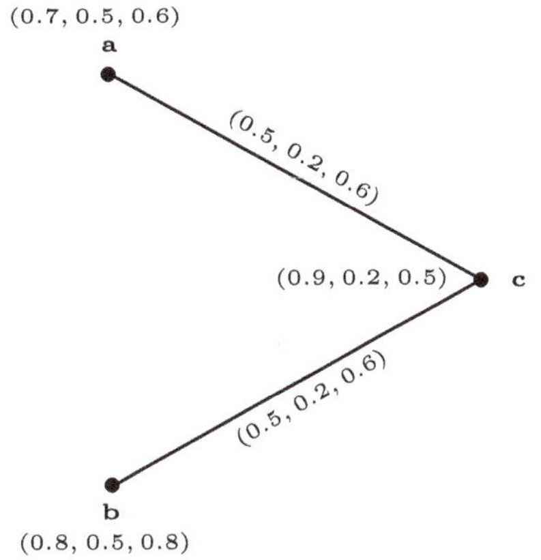

Figure 24. A 4-RPFG $\mathscr{G}$.

We see that $\mathscr{G}$ *is* $(0.5, 0.2, 0.6)$*-edge regular q-RPFG, and G is* 1*-edge regular graph. Thus,* $\mathscr{G}$ *is full edge regular q-RPFG. However, since* $d_{\mathscr{G}}(a) = (0.5, 0.2, 0.6) \neq (1.0, 0.4, 1.2) = d_{\mathscr{G}}(c)$*, therefore,* $\mathscr{G}$ *is not regular. Moreover, G is not regular. Thus,* $\mathscr{G}$ *is not full regular.*

Theorem 30. *Let* $\mu_{\mathscr{Q}}, \eta_{\mathscr{Q}}$*, and* $\nu_{\mathscr{Q}}$ *are constant functions in a q-RPFG* $\mathscr{G} = (\mathscr{P}, \mathscr{Q})$*. If* $\mathscr{G}$ *is regular, then* $\mathscr{G}$ *is perfect edge regular.*

Proof. Let $\mathscr{G} = (\mathscr{P}, \mathscr{Q})$ be a q-RPFG on $G = (P, Q)$ with $\mu_{\mathscr{Q}}(uv) = c_1$, $\eta_{\mathscr{Q}}(uv) = c_2$, and $\nu_{\mathscr{Q}}(uv) = c_3$, for all $uv \in Q$. Assume that $\mathscr{G}$ is (k_1, k_2, k_3)-regular. Then, $d_{\mathscr{G}}(u) = (k_1, k_2, k_3)$, for all $u \in P$. By definition of edge degree, $d_{\mathscr{G}}(uv) = d_{\mathscr{G}}(u) + d_{\mathscr{G}}(u) - 2(\mu_{\mathscr{Q}}(uv), \eta_{\mathscr{Q}}(uv), \nu_{\mathscr{Q}}(uv)) = 2(k_1 - c_1, k_2 - c_2, k_3 - c_3)$, for all $uv \in Q$. Moreover, by definition of total edge degree, $td_{\mathscr{G}}(uv) = d_{\mathscr{G}}(u) + d_{\mathscr{G}}(u) - (\mu_{\mathscr{Q}}(uv), \eta_{\mathscr{Q}}(uv), \nu_{\mathscr{Q}}(uv)) = (2k_1 - c_1, 2k_2 - c_2, 2k_3 - c_3)$, for all $uv \in Q$. Hence, $\mathscr{G}$ is perfect edge regular. □

Remark 15. *The converse of above theorem need not be true.*
For example, consider 11*-RPFG* $\mathscr{G}$ *as shown in Figure 25.*

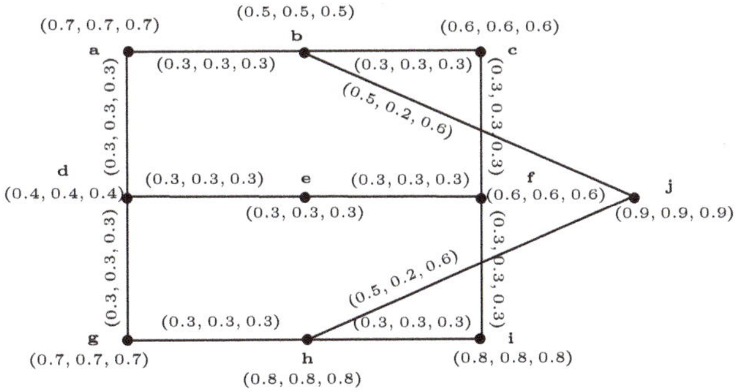

Figure 25. An 11-RPFG $\mathscr{G}$.

We see that $\mathscr{G}$ is $(0.9, 0.9, 0.9)$-edge regular, and $(1.2, 1.2, 1.2)$-total edge regular 11-RPFG, but $\mathscr{G}$ is not regular 11-RPFG as $d_{\mathscr{G}}(a) = (0.6, 0.6, 0.6) \neq (0.9, 0.9, 0.9) = d_{\mathscr{G}}(b)$.

Theorem 31. [35] *Let $\mathscr{G} = (\mathscr{P}, \mathscr{Q})$ be a q-RPFG such that $\mu_{\mathscr{Q}}, \eta_{\mathscr{Q}}$, and $\nu_{\mathscr{Q}}$ are constant functions. Then, $\mathscr{G}$ is regular q-RPFG if and only if $\mathscr{G}$ is partially regular q-RPFG.*

Theorem 32. *If a q-RPFG $\mathscr{G} = (\mathscr{P}, \mathscr{Q})$ is perfect edge regular, then $\mathscr{G}$ is regular if and only if $\mathscr{G}$ is partially regular.*

Proof. Let $\mathscr{G} = (\mathscr{P}, \mathscr{Q})$ be perfect edge regular. Then, by Theorem 15, $\mu_{\mathscr{Q}}, \eta_{\mathscr{Q}}$, and $\nu_{\mathscr{Q}}$ are constant functions. Thus, Theorem 31 implies that $\mathscr{G}$ is regular if and only if $\mathscr{G}$ is partially regular. □

Remark 16. *The converse of above theorem need not be true.*
For example, consider a 5-RPFG $\mathscr{G}$ as shown in Figure 26.

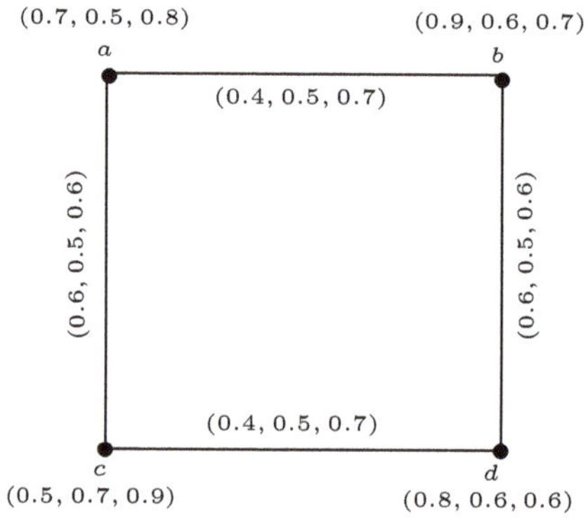

Figure 26. A 5-RPFG $\mathscr{G}$.

We see that $\mathscr{G}$ is $(1.0, 1.0, 1.3)$-regular 5-RPFG. Moreover, $\mathscr{G}$ is partially regular. However, it is easy to observe that $\mathscr{G}$ is neither edge regular nor total edge regular.

Theorem 33. *Let $\mathscr{G} = (\mathscr{P}, \mathscr{Q})$ be the k-partially regular q-RPFG. Then, $\mu_{\mathscr{Q}}, \eta_{\mathscr{Q}}$, and $\nu_{\mathscr{Q}}$ are constant functions if and only if $\mathscr{G}$ both regular, and perfect edge regular.*

Proof. Consider a q-RPFG $\mathscr{G} = (\mathscr{P}, \mathscr{Q})$ defined on $G = (P, Q)$ such that G is k-regular. Assume that $\mu_{\mathscr{Q}}(uv) = c_1$, $\eta_{\mathscr{Q}}(uv) = c_2$, and $\nu_{\mathscr{Q}}(uv) = c_3$, for all $uv \in Q$. Then, $d_{\mathscr{G}}(u) = \sum_{uv \in Q}(\mu_{\mathscr{Q}}(uv), \eta_{\mathscr{Q}}(uv), \nu_{\mathscr{Q}}(uv)) = \sum_{uv \in Q}(c_1, c_2, c_3) = (c_1, c_2, c_3)d_G(u) = k(c_1, c_2, c_3)$ for all $u \in P$. Hence, $\mathscr{G}$ is regular q-RPFG. Thus, by Theorem 30, $\mathscr{G}$ is perfect edge regular.

Conversely, let $\mathscr{G}$ be both regular, and perfect edge regular. Then, suppose that $d_{\mathscr{G}}(u) = (k_1, k_2, k_3)$ for all $u \in P$, $d_{\mathscr{G}}(uv) = (p_1, p_2, p_3)$, and $td_{\mathscr{G}}(uv) = (q_1, q_2, q_3)$ for all $uv \in Q$. By definition of edge degree,

$$\begin{aligned} d_{\mathscr{G}}(uv) &= d_{\mathscr{G}}(u) + d_{\mathscr{G}}(u) - 2(\mu_{\mathscr{Q}}(uv), \eta_{\mathscr{Q}}(uv), \nu_{\mathscr{Q}}(uv)), \\ (p_1, p_2, p_3) &= 2(k_1, k_2, k_3) - 2(\mu_{\mathscr{Q}}(uv), \eta_{\mathscr{Q}}(uv), \nu_{\mathscr{Q}}(uv)), \\ (\mu_{\mathscr{Q}}(uv), \eta_{\mathscr{Q}}(uv), \nu_{\mathscr{Q}}(uv)) &= \frac{(2k_1 - p_1, 2k_2 - p_2, 2k_3 - p_3)}{2}. \end{aligned}$$

Hence, $\mu_{\mathscr{Q}}, \eta_{\mathscr{Q}}$, and $\nu_{\mathscr{Q}}$ are constant functions. This completes the proof. □

Lemma 3. *If $\mathscr{G} = (\mathscr{P}, \mathscr{Q})$ is perfectly regular (or regular) q-RPFG, and $\mu_{\mathscr{Q}}, \eta_{\mathscr{Q}}$, and $\nu_{\mathscr{Q}}$ are constant functions, then $\mathscr{G}$ is perfect edge-regular.*

Proof. Let $\mathscr{G} = (\mathscr{P}, \mathscr{Q})$ be a perfectly regular q-RPFG defined on $G = (V, E)$, and $\mu_{\mathscr{Q}}, \eta_{\mathscr{Q}}$, and $\nu_{\mathscr{Q}}$ be constant functions. Then, $(\mu_{\mathscr{Q}}(uv), \eta_{\mathscr{Q}}(uv), \nu_{\mathscr{Q}}(uv)) = (c_1, c_2, c_3)$ for all $uv \in Q$. Since $\mathscr{G}$ is regular q-RPFG, therefore, $d_{\mathscr{G}}(u) = (k_1, k_2, k_3)$ for all $u \in V$. Then, for any edge uv in $\mathscr{G}$, $d_{\mathscr{G}}(uv) = d_{\mathscr{G}}(u) + d_{\mathscr{G}}(v) - 2(\mu_{\mathscr{Q}}(uv), \eta_{\mathscr{Q}}(uv), \nu_{\mathscr{Q}}(uv)) = 2(k_1, k_2, k_3) - 2(c_1, c_2, c_3) = 2(k_1 - c_1, k_2 - c_2, k_3 - c_3)$. Hence, $\mathscr{G}$ is edge regular q-RPFG. Now, the total degree of an edge in q-RPFG is $td_{\mathscr{G}}(uv) = d_{\mathscr{G}}(uv) + (\mu_{\mathscr{Q}}(uv), \eta_{\mathscr{Q}}(uv), \nu_{\mathscr{Q}}(uv)) = 2(k_1 - c_1, k_2 - c_2, k_3 - c_3) + (c_1, c_2, c_3) = (2k_1 - c_1, 2k_2 - c_2, 2k_3 - c_3)$ for all $uv \in Q$. Thus, $\mathscr{G}$ is total edge-regular q-RPFG. Consequently, $\mathscr{G}$ is perfect edge-regular q-RPFG. This completes the proof. □

Remark 17. *If $\mathscr{G} = (\mathscr{P}, \mathscr{Q})$ is totally regular q-RPFG, and $\mu_{\mathscr{Q}}, \eta_{\mathscr{Q}}$, and $\nu_{\mathscr{Q}}$ are constant functions, then $\mathscr{G}$ may not perfect edge-regular.*
For example, consider a 3-RPFG $\mathscr{G}$ as shown in Figure 27.

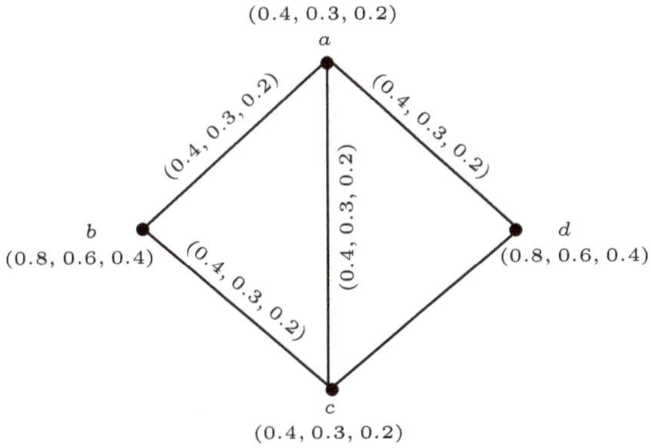

Figure 27. A 3-RPFG $\mathscr{G}$.

We see that $\mathscr{G}$ is totally regular since $td_{\mathscr{G}}(a) = td_{\mathscr{G}}(b) = td_{\mathscr{G}}(c) = td_{\mathscr{G}}(d) = (1.6, 1.2, 0.8)$, and $\mu_{\mathscr{Q}}(uv), \eta_{\mathscr{Q}}(uv)$, and $\nu_{\mathscr{Q}}(uv)$ are constant functions as $\mu_{\mathscr{Q}}(uv) = 0.4, \eta_{\mathscr{Q}}(uv) = 0.3$, and $\nu_{\mathscr{Q}}(uv) = 0.2$ for all $uv \in Q$. However, $d_{\mathscr{G}}(ab) = (1.2, 0.9, 0.6) \neq (1.6, 1.2, 0.8) = d_{\mathscr{G}}(ac)$ leads $\mathscr{G}$ to be not edge-regular 3-RPFG. Thus, $\mathscr{G}$ is not perfect edge-regular.

Theorem 34. *If $\mathscr{G} = (\mathscr{P}, \mathscr{Q})$ is complete perfectly regular q-RPFG, then $\mathscr{G}$ is perfect edge-regular.*

Proof. Let $\mathscr{G} = (\mathscr{P}, \mathscr{Q})$ be a complete perfectly regular q-RPFG defined on $G = (P, Q)$. Then, for each vertex u of $\mathscr{G}$, $(\mu_{\mathscr{P}}(u), \eta_{\mathscr{P}}(u), \nu_{\mathscr{P}}(u)) = (k_1, k_2, k_3)$. In addition, completeness of $\mathscr{G}$ implies that, for every edge uv of $\mathscr{G}$, $(\mu_{\mathscr{Q}}(uv), \eta_{\mathscr{Q}}(uv), \nu_{\mathscr{Q}}(uv)) = (\mu_{\mathscr{P}}(u) \wedge \mu_{\mathscr{P}}(v), \eta_{\mathscr{P}}(u) \wedge \eta_{\mathscr{P}}(v), \nu_{\mathscr{P}}(u) \wedge \nu_{\mathscr{P}}(v))$. Combining the two facts above, we obtain $(\mu_{\mathscr{Q}}(uv), \eta_{\mathscr{Q}}(uv), \nu_{\mathscr{Q}}(uv)) = (k_1, k_2, k_3)$ for all $uv \in Q$. It shows that $\mu_{\mathscr{Q}}, \eta_{\mathscr{Q}}$, and $\nu_{\mathscr{Q}}$ are constant functions. Hence, by Lemma 3, $\mathscr{G}$ is perfect edge-regular q-RPFG. This completes the proof. □

Next, we relate the concepts of regularity and edge regularity in q-RPF line graphs.

Observation 1. *The total degree of an edge x in $\mathscr{G}$ is equal to the sum of the membership values of corresponding vertex S_x, and its adjacent vertices in $L(\mathscr{G})$.*

Theorem 35. *Let $\mathscr{G} = (\mathscr{P}, \mathscr{Q})$ be a q-RPFG such that $\mu_{\mathscr{Q}}$, $\eta_{\mathscr{Q}}$, and $\nu_{\mathscr{Q}}$ are constant functions. If $\mathscr{G}$ is a regular q-RPFG, then its line graph $L(\mathscr{G}) = (\mathscr{A}, \mathscr{B})$ is an edge regular q-RPFG.*

Proof. Let $\mathscr{G} = (\mathscr{P}, \mathscr{Q})$ be a regular q-RPFG such that $\mu_{\mathscr{Q}}$, $\eta_{\mathscr{Q}}$, and $\nu_{\mathscr{Q}}$ are constant functions. Then, by Theorem 28, $\mathscr{G}$ is an edge regular q-RPFG. Consequently, by Theorem 25, the q-RPF line graph $L(\mathscr{G}) = (\mathscr{A}, \mathscr{B})$ is edge regular. □

Remark 18. *The converse of the above theorem need not be true.*
Consider a 4-rung picture fuzzy graph $\mathscr{G} = (\mathscr{P}, \mathscr{Q})$, *and its line graph* $L(\mathscr{G}) = (\mathscr{A}, \mathscr{B})$ *as shown in Figure 28.*

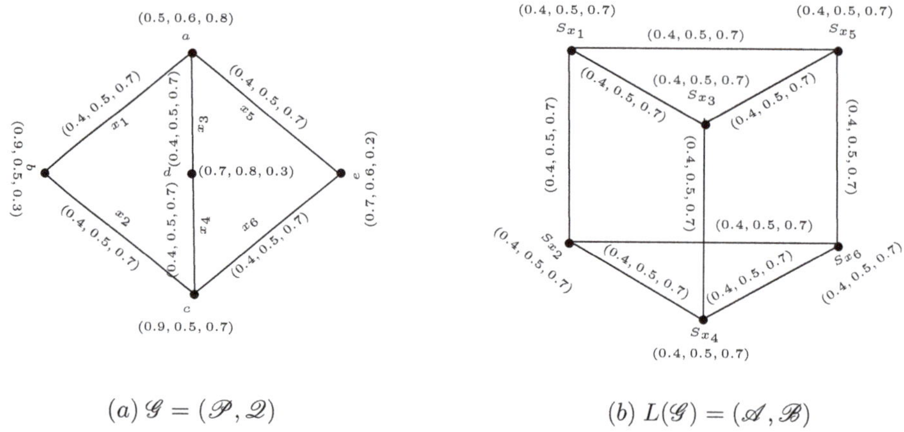

Figure 28. A not regular 4-RPFG $\mathscr{G}$ and its edge regular line graph $L(\mathscr{G})$.

We see that $\mu_{\mathscr{Q}}$, $\eta_{\mathscr{Q}}$, *and* $\nu_{\mathscr{Q}}$ *are constant functions, that is, for each uv in Q,* $(\mu_{\mathscr{Q}}(uv), \eta_{\mathscr{Q}}(uv), \nu_{\mathscr{Q}}(uv)) = (0.4, 0.5, 0.7)$. *Since each edge of* $L(\mathscr{G})$ *has degree* $(1.6, 2.0, 2.8)$, *therefore,* $L(\mathscr{G})$ *is* $(1.6, 2.0, 2.8)$*-edge regular 4-rung picture fuzzy line graph of* $\mathscr{G}$. *However,* $d_{\mathscr{G}}(b) = (0.8, 1.0, 1.4) \neq (1.2, 1.5, 2.1) = d_{\mathscr{G}}(a)$ *leads to the fact that the 4-RPFG* $\mathscr{G}$ *is not regular.*

4. Applications

Graph theory, graph-partitioning, and graph-based computing are characterized by detecting different network structures. The unstoppable growth of social networks, and the huge number of connected users, has made these networks some of the most popular, and successful domains for a large number of research areas. The different possibilities, volume, and variety that these social networks offer has made them essential tools for everyday working, and social relationships. The number of users registered in different social networks, and the volume of information generated by them in a more explanatory manner are increasing day by day. The analysis over the whole network becomes extremely difficult due to this fact. For instance, if every people has 5 close friends, then in a town of 10,000 people, there will be 50,000 close friendship ties to study. In order to extract the knowledge from such a large network, some relevant works have focused the attention on 'ego-networks'. On the other hand, if someone does not want to understand a whole community, but just what individual people do, that is, one only finds some people in a community interesting (leaders, teenagers, artists, etc.), also lead to study ego-networks, which tells us about social structure of entire population, and its sub-populations.

An *Ego-network* is a social network composed of one user centering the graph (called *ego*), all the users connected to this ego (called *alters*), and all the relations between these alters (called *ties*). Ego can be considered as an individual 'focal' node. It can be a person, a group, an organization, or a whole society. A network has as many egos as it has nodes.

q-Rung Picture Fuzzy Social Network

An ordinary social network cannot represent the power of employees, and the degree of relations among employees within an organization. As the powers, and relationships have no defined boundaries, it is desired to represent them in the form of fuzzy set. The fuzzy social networks are used to express interactions between different nodes. It is obvious that the fuzzy graph model is not enough to fully illustrate any phenomenon represented by networks/graphs. Several extensions of fuzzy set have been introduced in this context. Thus, to deal with the situations where opinions are not only yes or no, but there are some abstinence, and refusal too; recently, the q-rung picture fuzzy graph model was introduced, providing a vast depiction space of triplets. We now discuss the concept of ego-networks under q-rung picture fuzzy environment.

Algorithm 1 illustrates the extraction of q-rung picture fuzzy ego-networks from a q-rung picture fuzzy social network $\mathscr{G} = (\mathscr{P}, \mathscr{Q})$. The complexity of algorithm is $O(n^2)$, where $n = |P|$.

Algorithm 1 Extraction of q-rung picture fuzzy ego-networks

INPUT: A q-RPF social network $\mathscr{G} = (\mathscr{P}, \mathscr{Q})$.
OUTPUT: A q-RPF ego-network $\mathscr{G}_i = (\mathscr{P}_i, \mathscr{Q}_i)$.
function $\mathscr{G}_i = (\mathscr{P}_i, \mathscr{Q}_i)$
for $i = 1$ *to* n **do**
  for $j = 1$ *to* n **do**
    if $\mathbf{u_i}(\mu_{\mathscr{P}}(u_i), \eta_{\mathscr{P}}(u_i), \nu_{\mathscr{P}}(u_i))$ is q-RPF ego **then**
      choose all $u_j(\mu_{\mathscr{P}}(u_j), \eta_{\mathscr{P}}(u_j), \nu_{\mathscr{P}}(u_j))$ such that $i \neq j$ and
$(\mu_{\mathscr{Q}}(u_i u_j), \eta_{\mathscr{Q}}(u_i u_j), \nu_{\mathscr{Q}}(u_i u_j)) \neq (0, 0, 0)$
Define a q-RPF vertex set
$\mathscr{P}_i = \{\mathbf{u_i}(\mu_{\mathscr{P}}(u_i), \eta_{\mathscr{P}}(u_i), \nu_{\mathscr{P}}(u_i)), \mathbf{u_j}(\mu_{\mathscr{P}}(u_j), \eta_{\mathscr{P}}(u_j), \nu_{\mathscr{P}}(u_j))|\ for\ all\ j\}$
and a q-RPF edge set
$\mathscr{Q}_i = \{\mathbf{u_i u_j}(\mu_{\mathscr{Q}}(u_i u_j), \eta_{\mathscr{Q}}(u_i u_j), \nu_{\mathscr{Q}}(u_i u_j)), \mathbf{u_j u_{j'}}(\mu_{\mathscr{Q}}(u_j u_{j'}), \eta_{\mathscr{Q}}(u_j u_{j'}), \nu_{\mathscr{Q}}(u_j u_{j'}))|j < j',\ for\ all\ j\}$
    end if
  end for
end for
end function

We can use q-RPFG to examine the social relationships among different groups of people. By the concept of q-RPF ego-networks, we can focus on a single entity to investigate its potential with other members of a social group. We can also examine the percentage of relationships under a q-rung picture fuzzy environment.

Consider a large company in Japan. The typical structure of executive titles is illustrated as a set of employees P = {*Chairman (C), CEO, President (C), Deputy president (DP)/Senior executive (SE), Executive vice president (EVP), Senior vice president (SVP), Vice president (VP)/general manager (GM)/Department head (DH), Deputy general manager (DGM), Manager (M)/Section head (SH), Assistant manager (AM)/Team leader (TL), Staff (S)*} for that company. Let $\mathscr{P}$ be the 4-RPFS on P given by Table 1.

Table 1. A 4-RPFS of employees.

	C	CEO	P	DP/SE	EVP	SVP	VP/GM/DH	DGM	M/SH	AM/TL	S
$\mu_{\mathscr{P}}$	0.9	0.8	0.7	0.6	0.6	0.45	0.55	0.65	0.5	0.4	0.3
$\eta_{\mathscr{P}}$	0.6	0.5	0.6	0.55	0.56	0.8	0.5	0.6	0.8	0.7	0.7
$\nu_{\mathscr{P}}$	0.5	0.7	0.7	0.85	0.88	0.6	0.8	0.9	0.7	0.9	0.6

Figure 29 displays a 4-RPFG representing the social network composed of 11 different employees.

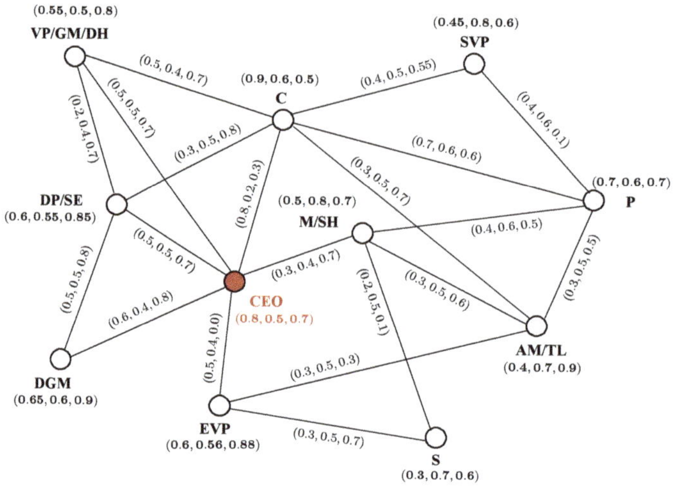

Figure 29. A 4-rung picture fuzzy social network.

There are 11 different q-RPF ego networks/graphs associated with the given q-RPF network/graph, where the vertices/nodes represent the powers of employees in the company, and edges/links represent their relationships. The membership functions $\mu_{\mathscr{P}}$, $\eta_{\mathscr{P}}$, and $\nu_{\mathscr{P}}$ in triplet $(\mu_{\mathscr{P}}, \eta_{\mathscr{P}}, \nu_{\mathscr{P}})$, assigned to each node, indicate their positive impact, their neutral behavior, and their negative impact within the company, respectively. For example, $(0.8, 0.5, 0.7)$ tells the status level of the CEO as: the CEO possesses 40.96% positive impact, and 24.01% negative impact within the company. A total of 6.25% of his behavior is neutral. While the refusal degree $\pi = \sqrt[4]{1 - 0.8^4 - 0.5^4 - 0.7^4}$ of CEO indicates that only 28.78000001%, he has the ability to refuse to participate in the company's matters. However, the positive, neutral, and negative membership degrees $\mu_{\mathscr{Q}}$, $\eta_{\mathscr{Q}}$, and $\nu_{\mathscr{Q}}$ of edges depict the positive associations, abstinence to associate, and negative associations among employees. For example, the edge between president and assistant manager is assigned by triplet $(0.3, 0.5, 0.3)$. The relationships between them can be translated by a means of q-rung picture fuzzy analysis as follows: they have only 0.81% positive attitude. The neutral and negative behavior have same the weight, i.e., 6.25% and 86.69% of interactions translate the extent of the clash between them. Similar considerations can be extracted from other edges. The corresponding adjacency matrix is shown in Table 2, where the degrees for each edge uv in $\mathscr{G} = (\mathscr{P}, \mathscr{Q})$ are evaluated using the following relations:

$$\begin{aligned}
\mu_{\mathscr{Q}}(uv) &\leq \mu_{\mathscr{P}}(u) \wedge \mu_{\mathscr{P}}(v), \\
\eta_{\mathscr{Q}}(uv) &\leq \eta_{\mathscr{P}}(u) \wedge \eta_{\mathscr{P}}(v), \\
\nu_{\mathscr{Q}}(uv) &\leq \nu_{\mathscr{P}}(u) \vee \nu_{\mathscr{P}}(v).
\end{aligned}$$

Table 2. Adjacency matrix corresponding to Figure 29.

	C	CEO	P	DP/SE	EVP	SVP	VP/GM/DH	DGM	M/SH	AM/TL	S
C	(0.0, 0.0, 0.0)	(0.8, 0.2, 0.3)	(0.7, 0.6, 0.6)	(0.3, 0.5, 0.8)	(0.0, 0.0, 0.0)	(0.4, 0.5, 0.55)	(0.5, 0.4, 0.7)	(0.0, 0.0, 0.0)	(0.0, 0.0, 0.0)	(0.3, 0.5, 0.7)	(0.0, 0.0, 0.0)
CEO	(0.8, 0.2, 0.3)	(0.0, 0.0, 0.0)	(0.0, 0.0, 0.0)	(0.5, 0.5, 0.7)	(0.5, 0.4, 0.0)	(0.0, 0.0, 0.0)	(0.5, 0.5, 0.7)	(0.6, 0.4, 0.8)	(0.3, 0.4, 0.7)	(0.0, 0.0, 0.0)	(0.0, 0.0, 0.0)
P	(0.7, 0.6, 0.6)	(0.0, 0.0, 0.0)	(0.0, 0.0, 0.0)	(0.0, 0.0, 0.0)	(0.0, 0.0, 0.0)	(0.4, 0.6, 0.1)	(0.0, 0.0, 0.0)	(0.0, 0.0, 0.0)	(0.4, 0.6, 0.5)	(0.3, 0.5, 0.5)	(0.0, 0.0, 0.0)
DP/SE	(0.3, 0.5, 0.8)	(0.5, 0.5, 0.7)	(0.0, 0.0, 0.0)	(0.0, 0.0, 0.0)	(0.0, 0.0, 0.0)	(0.0, 0.0, 0.0)	(0.2, 0.4, 0.7)	(0.5, 0.5, 0.8)	(0.0, 0.0, 0.0)	(0.0, 0.0, 0.0)	(0.0, 0.0, 0.0)
EVP	(0.0, 0.0, 0.0)	(0.5, 0.4, 0.0)	(0.0, 0.0, 0.0)	(0.0, 0.0, 0.0)	(0.0, 0.0, 0.0)	(0.0, 0.0, 0.0)	(0.0, 0.0, 0.0)	(0.0, 0.0, 0.0)	(0.0, 0.0, 0.0)	(0.3, 0.5, 0.3)	(0.3, 0.5, 0.7)
SVP	(0.4, 0.5, 0.55)	(0.0, 0.0, 0.0)	(0.4, 0.6, 0.1)	(0.0, 0.0, 0.0)	(0.0, 0.0, 0.0)	(0.0, 0.0, 0.0)	(0.0, 0.0, 0.0)	(0.0, 0.0, 0.0)	(0.0, 0.0, 0.0)	(0.0, 0.0, 0.0)	(0.0, 0.0, 0.0)
VP/GM/DH	(0.5, 0.4, 0.7)	(0.5, 0.5, 0.7)	(0.0, 0.0, 0.0)	(0.2, 0.4, 0.7)	(0.0, 0.0, 0.0)	(0.0, 0.0, 0.0)	(0.0, 0.0, 0.0)	(0.0, 0.0, 0.0)	(0.0, 0.0, 0.0)	(0.0, 0.0, 0.0)	(0.0, 0.0, 0.0)
DGM	(0.0, 0.0, 0.0)	(0.6, 0.4, 0.8)	(0.0, 0.0, 0.0)	(0.5, 0.5, 0.8)	(0.0, 0.0, 0.0)	(0.0, 0.0, 0.0)	(0.0, 0.0, 0.0)	(0.0, 0.0, 0.0)	(0.0, 0.0, 0.0)	(0.0, 0.0, 0.0)	(0.0, 0.0, 0.0)
M/SH	(0.0, 0.0, 0.0)	(0.3, 0.4, 0.7)	(0.4, 0.6, 0.5)	(0.0, 0.0, 0.0)	(0.0, 0.0, 0.0)	(0.0, 0.0, 0.0)	(0.0, 0.0, 0.0)	(0.0, 0.0, 0.0)	(0.0, 0.0, 0.0)	(0.3, 0.5, 0.6)	(0.2, 0.5, 0.1)
AM/TL	(0.3, 0.5, 0.7)	(0.0, 0.0, 0.0)	(0.3, 0.5, 0.5)	(0.0, 0.0, 0.0)	(0.3, 0.5, 0.3)	(0.0, 0.0, 0.0)	(0.0, 0.0, 0.0)	(0.0, 0.0, 0.0)	(0.3, 0.5, 0.6)	(0.0, 0.0, 0.0)	(0.0, 0.0, 0.0)
S	(0.0, 0.0, 0.0)	(0.0, 0.0, 0.0)	(0.0, 0.0, 0.0)	(0.0, 0.0, 0.0)	(0.3, 0.5, 0.7)	(0.0, 0.0, 0.0)	(0.0, 0.0, 0.0)	(0.0, 0.0, 0.0)	(0.2, 0.5, 0.1)	(0.0, 0.0, 0.0)	(0.0, 0.0, 0.0)

The CEO is a company's top decision-maker, and all other executives answer to him/her. Next, we want to study the powers of the CEO in this company, and his associations with other executives. For this, we consider the CEO as ego, and extract a q-rung picture fuzzy ego network from a q-rung picture fuzzy social network (see Figure 29) according to Algorithm 1.

Figure 30 displays a q-rung picture fuzzy ego network $\mathscr{H} = (\mathscr{A}, \mathscr{B})$ for selected ego, i.e., CEO (colored node).

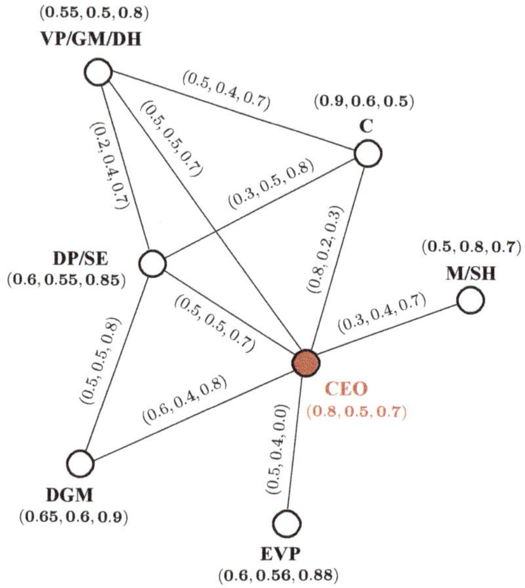

Figure 30. A 4-rung picture fuzzy ego-network.

The CEO influences the executives who are board members, and jointly supervise the activities of company. The titles designate an individual as an officer of the company with specific responsibilities that make them legally accountable in their position. The following investigations provide a q-RPF social network analysis for the CEO:

- The impact of CEO is equally distributed with the vice president, and deputy president as the positive, neutral, and negative associations of both are 6.25%, 6.25%, and 24.01%, respectively.
- The CEO greatly refuses to value the inputs of executive vice present for strategic initiatives as there is 91.19% refusal part while no negative associations between them, where the positive associations are 6.25%, and 2.56% is the neutrality in their behavior.
- By means of the positive and negative associations of the CEO with the Chairman, the founder of the company, we can this interpret as: the Chairman works 40.96% on the CEO's opinion, for 0.81%, he works opposite of his opinion, for 0.16%, his behavior is neutral with the CEO, and, for 58.07%, the chairman refuses to trust his part.
- The reporting relationships between general manager, and CEO is 12.96% positive, 2.56% neutral, and 40.96% negative, while, for 56.48% of his own part, the general manager refuses to report to him.
- Between the vice president and deputy president, there is less positive (0.16%), and high negative (24.01%) associations, which indicate their history of conflict.
- The manager being a section head has to resolve all cases in his section to answer to the CEO. The absence of edge with the executive vice president show that he has no concern with the executive vice president in company's matters.

The similar investigations can be seen between executives (alters) in the q-RPF ego-network (Figure 30).

5. Conclusions

Real life situations are often very uncertain and vague in nature. Due to a lack of information, the future state of system might not be completely known. The notion of fuzzy set offers a suitable departure point for the creation of such a framework which has potentially wider scope of applicability. The picture and spherical fuzzy models are a good way to tackle the uncertain information, when people think about not only the reaction is positive or negative, but there are some abstinence and refusal. The q-RPF model is more efficient than picture and spherical fuzzy models because it provides a wide space of permissible triplets. In the present study, we have explored some graph-theoretic ideas under q-RPF circumstances. Regularity is one of the attributes that permits many of the challenges connected with graph analysis to be addressed. The present study has provided a novel description on edge regularity of q-RPFGs and developed several related results. In particular, the edge regularity of square q-RPFGs and q-RPF line graphs has been illustrated. These graphs have eventually helped to link certain q-RPF systems and crisp systems allowing for greater ease in computing properties of these q-RPF systems for modeling purposes or optimizing q-RPF networks. Moreover, the q-RPFLGs can be applied to work out extremal problems. We have also established the relationships between regular and edge regular q-RPFGs. Finally, we have presented the concept of q-RPF ego-networks to extract knowledge from large social networks as an application. Our future work will expand on these terms: (1) Interval-valued q-RPFGs; (2) Bipolar-valued q-RPFGs; and (3) Hesitant q-RPFGs.

Author Contributions: M.A., A.H. and A.N.A.K. conceived of the presented concept. M.A. and A.H. developed the theory and performed the computations. A.N.A.K. verified the analytical methods.

Funding: This research received no external funding.

Conflicts of Interest: The authors declare that they have no conflict of interest regarding the publication of the research article.

References

1. Zadeh, L.A. Fuzzy sets. *Inf. Control* **1965**, *8*, 338–353. [CrossRef]
2. Cuong, B.C. Picture fuzzy sets—First results, Part 1. In *Seminar Neuro-Fuzzy Systems with Applications*; Preprint 03/2013; Institute of Mathematics, Vietnam Academy of Science and Technology: Hanoi, Vietnam, 2013.
3. Cuong, B.C. Picture fuzzy sets—First results, Part 2. In *Seminar Neuro-Fuzzy Systems with Applications*; Preprint 04/2013; Institute of Mathematics, Vietnam Academy of Science and Technology: Hanoi, Vietnam, 2013.
4. Atanassov, K. Intuitionistic fuzzy sets: Theory and Applications. *Fuzzy Sets Syst.* **1986**, *20*, 87–96. [CrossRef]
5. Son, L.H. DPFCM: A novel distributed picture fuzzy clustering method on picture fuzzy sets. *Exp. Syst. Appl.* **2015**, *2*, 51–66. [CrossRef]
6. Thong, N.T. HIFCF: An effective hybrid model between picture fuzzy clustering and intuitionistic fuzzy recommender systems for medical diagnosis. *Exp. Syst. Appl.* **2015**, *42*, 3682–3701. [CrossRef]
7. Cuong, B.C.; Kreinovich, V. Picture fuzzy sets—A new concept for computational intelligence problems. In Proceedings of the 3rd World Congress on Information and Communication Technologies (WICT 2013), Hanoi, Vietnam, 15–18 December 2013; ISBN 918-1-4799-3230-6.
8. Cuong, B.C. Picture fuzzy sets. *J. Comput. Sci. Cybern.* **2014**, *30*, 409–420.
9. Cuong, B.C.; Hai, P.V. Some fuzzy logic operators for picture fuzzy sets. In Proceedings of the Seventh International Conference on Knowledge and Systems Engineering, Ho Chi Minh City, Vietnam, 8–10 October 2015. [CrossRef]
10. Garg, H. Some picture fuzzy aggregation operators and their applications to multicriteria decision-making. *Arab. J. Sci. Eng.* **2017**, *42*, 5275–5290. [CrossRef]
11. Son, L.H.; Viet, P.V.; Hai, P.V. Picture inference system: A new fuzzy inference system on picture fuzzy set. *Appl. Intell.* **2017**, *46*, 652–669. [CrossRef]

12. Wang, C.Y.; Zhou, X.Q.; Tu, H.N.; Tao, S.D. Some geometric aggregation operators based on picture fuzzy sets and their application in multiple attribute decision making. *Ital. J. Pure Appl. Math.* **2017**, *37*, 477–492.
13. Zhang, H.; Zhang, R.; Huang, H.; Wang, J. Some picture fuzzy Dombi Heronian mean operators with their application to multi-attribute decision-making. *Symmetry* **2018**, *10*, 593. [CrossRef]
14. Yager, R.R.; Abbasov, A.M. Pythagorean membership grades, complex numbers, and decision making. *Int. J. Intell. Syst.* **2013**, *28*, 436–452. [CrossRef]
15. Yager, R.R. Generalized orthopair fuzzy sets. *IEEE Trans. Fuzzy Syst.* **2017**, *25*, 1222–1230. [CrossRef]
16. Gündoğdu, F.K.; Kahraman, C. Spherical fuzzy sets and spherical fuzzy TOPSIS method. *J. Intell. Fuzzy Syst.* **2018**. [CrossRef]
17. Li, L.; Zhang, R.; Wang, J.; Shang, X.; Bai, K. A novel approach to muti-attribut group decision-making with q-rung rung picture linguistic information. *Symmetry* **2018**, *10*, 172. [CrossRef]
18. Ashraf, S.; Abdulla, S.; Mahmood, T.; Ghani, F.; Mahmood, T. Spherical fuzzy sets and their applications in multi-attribute decision making problems. *J. Intell. Fuzzy Syst.* **2018**. [CrossRef]
19. Liu, P.; Wang, P. Some q-rung orthopair fuzzy aggregation operators and their applications to multiple-attribute decision making. *Int. J. Intell. Syst.* **2017**, *33*, 259–280. [CrossRef]
20. Mahmood, T.; Ullah, K.; Khan, Q.; Jan, N. An approach toward decision-making and medical diagnosis problems using the concept of spherical fuzzy sets. *J. Neutral Comput. Appl.* **2018**. [CrossRef]
21. Xu, Y.; Shang, X.; Wang, J.; Wu, W.; Huang, H. Some q-rung dual hesitant fuzzy Heronian mean operators with their application to multiple attribute group decision-making. *Symmetry* **2018**, *10*, 472. [CrossRef]
22. Zadeh, L.A. Similarity relations and fuzzy ordering. *Inf. Sci.* **1971**, *3*, 177–200. [CrossRef]
23. Kaufmann, A. *Introduction a la Theorie des Sousensembles Flous*; Massonet Cie Paris: Paris, France, 1973.
24. Rosenfeld, A. Fuzzy graphs. In *Fuzzy Sets and Their Applications to Cognitive and Decision Processes*; Academic Press: New York, NY, USA, 1975; pp. 77–95.
25. Karunambigai, M.G.; Parvathi, R. Intuitionistic fuzzy graphs. In *Advances in Soft Computing: Computational Intelligence, Theory and Applications, Proceedings of the 9th Fuzzy Days International Conference on Computational Intelligence*; Springer: Berlin/Heidelberg, Germany, 2006; Volume 20, pp. 139–150.
26. Akram, M.; Davvaz, B. Strong intuitionistic fuzzy graphs. *Filomat* **2012**, *26*, 177–196. [CrossRef]
27. Naz, S.; Ashraf, S.; Akram, M. A novel approach to decision-making with Pythagorean fuzzy information. *Mathematics* **2018**, *6*, 95. [CrossRef]
28. Habib, A.; Akram, M.; Farooq, A. q-Rung orthopair fuzzy competition graphs with application in soil ecosystem. *Mathematics* **2019**, *7*, 91. [CrossRef]
29. Akram, M. *m-Polar Fuzzy Graphs: Theory, Methods and Applications*; Studies in Fuzziness and Soft Computing; Springer: Cham, Switzerland, 2019; Volume 371, pp. 1–284.
30. Akram, M.; Ashraf, A.; Sarwar, M. Novel applications of intuitionistic fuzzy digraphs in decision support systems. *Sci. World J.* **2014**, *2014*, 904606. [CrossRef] [PubMed]
31. Akram, M.; Naz, S. Energy of Pythagorean fuzzy graphs with applications. *Mathematics* **2018**, *6*, 136. [CrossRef]
32. Akram, M.; Habib, A.; Ilyas, F.; Dar, J.M. Specific types of Pythagorean fuzzy graphs and application to decision-making. *Math. Comput. Appl.* **2018**, *23*, 42. [CrossRef]
33. Akram, M.; Dar, J.M.; Farooq, A. Planar graphs under Pythagorean fuzzy environment. *Mathematics* **2018**, *6*, 278. [CrossRef]
34. Sokolov, S.; Zhilenkov, A.; Chernyi, S.; Nyrkov, A.; Mamunts, D. Dynamics models of synchronized piecewise linear discrete chaotic systems of high order. *Symmetry* **2019**, *11*, 236. [CrossRef]
35. Akram, M.; Habib, A. q-Rung picture fuzzy graphs: A creative view on regularity with applications. *J. Appl. Math. Comput.* **2019**. [CrossRef]
36. Gani, A.N.; Ahamed, M.B. Order and size in fuzzy graphs. *Bull. Pure Appl. Sci.* **2003**, *22*, 145–148.
37. Radha, K.; Kumaravel, N. On Edge regular fuzzy graphs. *Int. J. Math. Arch.* **2004**, *5*, 100–112.
38. Cary, M. Perfectly regular and perfect edge regular fuzzy graphs. *Ann. Pure Appl. Math.* **2018**, *16*, 461–469.
39. Akram, M.; Dudek, W.A.; Yousaf, M.M. Regularity in vague intersection graphs and vague line graphs. In *Abstract and Applied Analysis*; Hindawi: Cairo, Egypt, 2014. [CrossRef]
40. Akram, M.; Dudek, W.A. Regular bipolar fuzzy graphs. *Neutral Comput. Appl.* **2012**, *21*, 197–205. [CrossRef]
41. Ashraf, S.; Naz, S.; Rashmanlou, H.; Malik, M.A. Regularity of graphs in single valued neutrosophic environemnt. *J. Intell. Fuzzy Syst.* **2017**. [CrossRef]

42. Gani, A.N.; Radha, K. Regular property of fuzzy graphs. *Bull. Pure Appl. Sci.* **2008**, *27*, 425–423.
43. Gani, A.N.; Radha, K. On regular fuzzy graphs. *J. Phys. Sci.* **2008**, *12*, 33–40.
44. Karunambigai, M.G.; Palanivel, K.; Sivasankar, S. Edge regular intuitionistic fuzzy graph. *Adv. Fuzzy Sets Syst.* **2015**, *20*, 25–46. [CrossRef]
45. Radha, K.; Kumaravel, N. On Edge regular fuzzy line graphs. *Int. J. Comput. Appl. Math.* **2016**, *11*, 105–118.
46. Radha, K.; Kumaravel, N. The degree of an edge in cartesian product and composition of two fuzzy graphs. *Int. J. Appl. Math. Stat. Sci.* **2013**, *2*, 65–78.
47. Sanjeevi, G. Some results on square fuzzy graphs. *Int. J. Math. Arch.* **2017**, *8*, 124–128.

MDPI
St. Alban-Anlage 66
4052 Basel
Switzerland
Tel. +41 61 683 77 34
Fax +41 61 302 89 18
www.mdpi.com

Symmetry Editorial Office
E-mail: symmetry@mdpi.com
www.mdpi.com/journal/symmetry

www.ingramcontent.com/pod-product-compliance
Lightning Source LLC
LaVergne TN
LVHW070735100526
838202LV00013B/1242
* 9 7 8 3 0 3 9 2 8 7 0 8 6 *